Reinventing Los Ang

Urban and Industrial Environments

Series editor: Robert Gottlieb, Henry R. Luce Professor of Urban and Environmental Policy, Occidental College

For a complete list of books published in this series, please see the back of the book.

Reinventing Los Angeles

Nature and Community in the Global City

Robert Gottlieb

The MIT Press
Cambridge, Massachusetts
London, England

For information about special quantity discounts, please e-mail <special_sales@mitpress.mit.edu>.

This book was set in Bembo by SNP Best-set Typesetter Ltd., Hong Kong.
Printed on recycled paper and bound in the United States of America.

Library of Congress Cataloging-in-Publication Data

Gottlieb, Robert, 1944–
Reinventing Los Angeles : nature and community in the global city / Robert Gottlieb.
p. cm.—(Urban and industrial environments)
Includes bibliographical references and index.
ISBN 978-0-262-07287-8 (hardcover : alk. paper)—ISBN 978-0-262-57243-9 (paperback : alk. paper)
1. City planning—California—Los Angeles. 2. Sociology, Urban—California—Los Angeles. 3. Urban ecology—California—Los Angeles.
I. Title.
HT168.L6G68 2007
307.1′2160979494—dc22

2007001996

10 9 8 7 6 5 4 3 2 1

Contents

Acknowledgments

Reinventing Los Angeles: Nature and Community in the Global City was inspired by a number of collaborations, including with the many partner organizations, friends, and colleagues who helped shape the ideas, gave feedback, and participated in the actions and events described in the book. I especially want to thank my fellow colleagues and former colleagues at the Urban and Environmental Policy Institute who have been engaged with me in this idea of reinventing this global city. They include Andrea Azuma, Moira Beery, Corey Bowers, Andrea Brown, Sylvia Chico, Francesca de la Rosa, Peter Dreier, Mahnaz Ghaznavi, Cyrus Grout, Jessica Gudmundson, Margaret Haase, Anupama Joshi, Kelly Lamkin, Sandra Martinez, Michelle Mascarenhas, Martha Matsuoka, Elizabeth Medrano, Angela Namkoong, Sandy Ramirez, Marcus Renner, Billi Romain, Lucia Sanchez, Amanda Shaffer, Peter Sinsheimer, and Mark Vallianatos. A number of colleagues at Occidental College and other researchers and activists have helped frame these issues and identify the key arenas for change. They include Maria Avila, Beth Braker, Tim Brick, Marianne Brown, Rick Brown, Bill Deverell, Regina Freer, Dorothy Green, Andrea Hricko, Angelo Logan, Anastasia Loukaitou-Sideris, Joe Linton, Lewis MacAdams, Alexis Moreno, Penny Newman, Gretchen North, Sam Pedroza, Jenny Price, and James Rojas. Material for the chapters on the Los Angeles River and the Arroyo Seco Parkway were derived in part from collaborative work with Andrea Azuma and Anastasia Loukaitou-Sideris that has been previously published. Marcus Renner was crucial to the ArroyoFest freeway walk and bike ride and shared with me in retrospect his observations of this improbable yet compelling event. Clay Morgan of MIT Press has been a longstanding and marvelous collaborator. And most central to my life have been Marge, Andie, and Casey, who continually remind me that our connections and engagements with those around us represent both deeply personal and social parts of our lives.

Reinventing Los Angeles

The Reinvention of Los Angeles: An Introduction

Hollywood Forever

About five years ago, while strolling through the Hollywood Forever Cemetery on the way to the cemetery's annual summer classic film festival, I first conceived of a book about nature and community in a global city. The Hollywood Forever Cemetery, founded in 1899, has become a Los Angeles icon. Its large and eclectic assortment of gravestones mark "the resting places of hundreds of Hollywood's stars," as its Web page proclaims. Each Saturday during the summer, as the sun sets, those Hollywood stars come to life on a screen that is projected onto the white wall of a large mausoleum, with palm trees creating the backdrop to this unusual open-air theater. These screenings in the cemetery have evolved from modest beginnings as a word-of-mouth event featuring such films as Alfred Hitchcock's *Strangers on a Train* and *Breakfast at Tiffany's* with Audrey Hepburn into an L.A. happening that draws thousands through the surface streets along Santa Monica Boulevard to transform an open field into a compact, grassy knoll theater.[1]

Aside from enjoying the experience of such an L.A. moment, what also resonated for me that evening was a comment made in passing by my friends, Marianne and Rick Brown, long-time activists and researchers in the health field. As we made our way through the cemetery, we talked about L.A., its multiple identities, and the possibility of finding a common thread for a city and region that for so long had been characterized as a place without a sense of place. "What about the L.A. River?" Rick asked. "Couldn't its renewal provide that thread?" While attitudes about the river had begun to change, it was still primarily derided as a concrete flood channel that was filled with debris and was fenced off, often with barbed wire, from its surrounding neighborhoods as it flowed north to south through the heart of Los Angeles. Rick's remark, coming from someone who was not a traditional environmentalist or nature advocate, suggested to me that the growing struggle to transform the L.A. River was capable of becoming "the very symbol of L.A.'s own renewal," as historian Bill Deverell had put it during a forum on the history of the L.A. River that I had earlier moderated.[2]

In the weeks and months that followed my evening at the Hollywood Forever Cemetery, I became heavily engaged in an effort to close a freeway to host a bike ride and a walk on the freeway itself, an event that was designed to culminate in a community festival in a nearby park. Like the L.A. River, freeways in Los Angeles have become dividing lines separating rather than connecting communities. Their construction requires the bulldozing of greenspace and livable places, and their operation contributes to the most polluted air basin in the United States. Like the history and contentious politics of water, the history and politics surrounding L.A.'s freeways have powerfully influenced the ways in which nature and community in Los Angeles are conceived. To close the freeway for a walk and bike ride seemed more than just a symbolic act. It also cut to the heart of the issues of nature and community in a city and region like Los Angeles.

Nature in the city. Community in the city. These concepts seem contradictory or at least counterintuitive, particularly for a city or region like Los Angeles.[3] How can people find nature and community on a freeway, where destination substitutes for place, or presume that the L.A. River, that straightjacketed water freeway, is a form of nature that can help build community? And can anyone argue that this other iconic representation of Los Angeles, the Hollywood Forever Cemetery, might also be considered representative of nature in the city or at least of what some urban environmentalists might characterize as urban "open space"? And can a cemetery where movies are screened also build community?

A Question from Bill Cronon

In the 1920s, my wife's grandfather and his two hunting buddies built a cabin on undeveloped land along the Peshtigo River about seventy miles north of Green Bay, Wisconsin, and seventeen miles from the small town of Crivitz. For many summers, my wife, our children, and I traveled from L.A. to the Peshtigo. We fished and swam in the river, lazed in the cabin, read in a hammock in the cabin porch, and enjoyed our time away from the city. The cabin was located close to the area of the massive Peshtigo fire of 1871, which occurred the same day as the great Chicago fire that redefined that city.

During the summer I was completing my book *Forcing the Spring*, which describes the history of the environmental movement and its diverse urban and industrial roots, I read *Nature's Metropolis: Chicago and The Great West*, environmental historian William Cronon's signature book about the development of Chicago in the mid-nineteenth century and its relation to its resource hinterland, including the white pine forest lands by the Peshtigo River in northern Wisconsin.[4] I was especially interested in how *Nature's Metropolis* argued that the urban forces that transformed Chicago also transformed the countryside, challenging the idea that nature could be set apart from its connection to human activity.

The summer after *Forcing the Spring* was published, I had the opportunity to meet with Bill Cronon at the University of Wisconsin campus dining area. Cronon had argued in his writings that the concept of nature was as much "a human idea as a non-human thing" and that environmentalism was "as much a cultural prospect as a 'natural' one." A number of environmentalists had argued that Cronon's reinventing nature concept had provided ammunition for the environmental movement's opponents since policies designed to protect nature, particularly wilderness areas and their special status as untouched environments, might now be effectively undermined. I knew that Cronon was deeply attached to many of the environmentalist goals, but in our discussion it also became clear that he was worried about the implications for a wilderness-focused and protection-oriented environmental movement once the argument about nature as a "human idea" was translated into practice. "What then would be left?" he wondered out loud about this environmentalist dilemma.[5]

From 1993, when this encounter took place, and 2007, as I write this introduction, environmental advocacy has not been diminished or constrained when the concept of nature set apart from human activity is challenged. Even where the arguments about protecting nature have been highly public, as with the fight to protect the Arctic National Wildlife Refuge from oil drilling, they have often been associated with human concerns as well, such as the need to protect the livelihood of indigenous people. More significantly, the language of environmental justice that references concerns about quality of life, pollution of the air and water, or negative land uses has largely been embraced by much of the environmental movement. Nevertheless, environmentalists, including environmental justice advocates, have had some difficulty responding to the core of Bill Cronon's concern—namely, how to establish a language and a politics that addresses what the late Alexander Wilson called "the culture of nature" regarding "the ways we think, teach, talk about and construct the natural world." The same could be said of "community," that other overused and uncertain and historically evolving concept that has been so central to the language and politics of environmental and social justice movements. And the concepts of nature and community both appear particularly opaque or unclear when placed in an urban context: constructing and promoting nature in the city or community in the city are not readily apparent as reference points for environmental action.[6]

By the time I made my excursion to the Hollywood Forever Cemetery, I knew that the arguments I developed in *Forcing the Spring* and my subsequent book *Environmentalism Unbound* only partially addressed this question of nature and community in the city. In *Forcing the Spring*, I highlighted those radicals, social reformers, and other historical environmental justice advocates for whom environment was embedded in the daily life issues of the city and the workplace. I also discussed other environmentalist roots, including the classic divide between the wilderness preservationist and the resource-management tendencies symbolized by John Muir and Gifford Pinchot. But I sought to situate the concepts of nature and community within this broader definition of environmentalism as a response to urban and industrial change. In *Environmentalism Unbound*, I extended the argument by discussing how contemporary environmental discourse—the use of language itself—needed to become an essential part of the process of reconstructing nature and community within the context of everyday life, what environmental justice advocates characterize as "where we live, work, and play."[7]

Still missing, however, has been a more direct elaboration of the position that nature and community do indeed have an urban face—that there is nothing "unnatural about New York City," as David Harvey has put it. By addressing the question of nature and community in the city, environmental and social justice advocates as well as mainstream environmentalists may be better able to demonstrate, through their language as well as through their actions, an effective response to the question posed by Bill Cronon—namely, what would be left if nature or community were to be defined as socially constructed and historically situated?[8]

A Land of Magical Improvisation

If the struggle to remake nature and community in the city can be incorporated into the practice of environmental and social justice movements, what better place to evaluate that opportunity and the extraordinary challenges involved in doing so than Los Angeles? Los Angeles has long been critics' prime illustration of urban sprawl, urban environmental degradation, and the loss of community, The history of Los Angeles during the twentieth century is seen by such critics as representative of the process of urbanization where undeveloped land (open space) is forever being subdivided, where rivers and streams are turned into concrete thoroughfares, where community is overwhelmed by the omnipresent automobile or reshaped by the continuing migrations into the city, and where the word *sprawl* has become synonymous with what passes for planning. For these critics, Los Angeles represents the absence of nature and community in the city. In this sense, Los Angeles has become representative of the contemporary urban form in the United States.

Yet is the picture this bleak? In 2005, three of my colleagues from Occidental College and I published *The Next Los Angeles: The Struggle for a Livable City*, a book that describes the history of progressive social movements in Los Angeles and their various efforts to reinvent the city. As its most insightful observer, Carey McWilliams, wrote fifty years ago, Los Angeles is a "land of magical improvisation," a characteristic that more recently can be extended to such extraordinary initiatives as the effort to reenvision the Los Angeles River as the centerpiece of building community and reimagining nature in the city. Transforming Los Angeles and reinventing it as a place that demonstrates how nature and community in the city can be redefined and flourish also might provide insights into how other cities and regions can be reinvented along similar lines.[9]

Global Los Angeles

When I settled in Los Angeles in 1969, I was aware that the city and the region had begun to change and that change was occurring rapidly. The city had recently experienced a highly charged and divisive mayoral election that pitted a popular African American candidate, City Council member Tom Bradley, against Sam Yorty, the incumbent mayor since 1961. Yorty had once been a populist Democrat but shifted sharply to the right after the 1965 Watts riots, and his sharp tongue, provocative racial statements, and irascible manner created an explosive environment, particularly regarding racial politics. Although Bradley lost that election, he came back to win a rematch in 1973 at a point when political, economic, cultural, and demographic changes in Los Angeles had become even more apparent. Los Angeles, it appeared, was prepared to undergo a major redefinition, perhaps as significant as its earlier shift from pueblo to the boom-and-bust Anglo town of the late nineteenth century.

At the heart of this redefinition were the ways in which Los Angeles had begun to be redefined as a global city. As these changes became increasingly pronounced during the 1970s and 1980s, the global-city concept emerged as a critical framework for analysis, interpretation, and debate among academics, policy makers, and various multinational corporate interests. The global-city concept had two significant reference points. One involved the expansion of trade, investment, capital flows, and the shift of urban economies like Los Angeles from manufacturing to finance, tourism, and a wide variety of specialized service firms. The second reference point had to do with enormous changes in population and migration. In this regard, Los Angeles was becoming not so much a melting pot but a city of multiple languages, cultures, and attitudes about daily life, including nature and community in the city.

Like anyone whose children attended public schools during this period, I witnessed some of those changes firsthand. When my son first began to attend elementary school in a mixed-race neighborhood in Santa Monica in the late 1980s, the demographics of the school were already noticeable. Three years later, when my daughter entered kindergarten, this demographic shift of diverse languages and nationalities had come to dominate the school environment. My daughter's kindergarten class had sixty-two students, thirty-eight of whom spoke a language other than English at home. Fifteen of the children spoke Spanish, and among the remaining twenty-three children, twelve other languages were spoken. When my daughter graduated from Santa Monica College fifteen years later, sixty-two different countries of origin were represented among the students in her graduating class. As her experience indicated, it became clear to me that in doing research for a book about nature and community in Los Angeles, such a book would need to explore what issues are associated with a global city and how Los Angeles, similar to other global cities around the world, had changed.[10]

Research and Action

In 1980, I was appointed by the City of Santa Monica to the board of directors of the Metropolitan Water District (MWD) of Southern California, the massive water wholesaler that played a crucial role in defining the urban landscape of Los Angeles and the six-county region of Southern California. Two years later, I began to teach at UCLA in the Graduate School of Architecture and Urban Planning (later the urban planning department) after having served the previous year as a client for an urban planning student group project analyzing the institutional issues related to the MWD. Once situated at UCLA, I sought to carve out a role as researcher, educator, and activist rather than a more narrowly defined traditional academic who would be less focused on community engagement. Toward that end, I cofounded in 1991, with Professors John Froines and David Allen, an interdisciplinary environmental center called the Pollution Prevention Education and Research Center (PPERC) that linked the departments of chemical engineering, public health, and urban planning. PPERC had two distinctive components: (1) a research and teaching aspect that drew on the technical and research capacity of the different disciplines to address urban and industrial environmental issues and (2) an "action research" program designed to help develop new public policies and establish linkages with key stakeholders, including community-based organizations. With a research focus that addressed the work environment and industry issues and an environmental-justice focus that addressed urban daily life concerns, PPERC operated as an interdisciplinary and action-oriented center at the edges of the University of California's complex system of research centers and departmental silos. Because UCLA is located on the west side of Los Angeles in a wealthy neighborhood, most UCLA researchers tended to focus less on neighborhood and regional issues and more on national and global questions, a research orientation that began to change when the 1992 civil disorders in Los Angeles sent shock waves through the region.[11]

In 1997, I had the opportunity to move to Occidental College, attracted in part by the nature of the appointment and the college's mission, location, and focus on Los Angeles. Occidental is a small, diverse liberal arts college located in the Eagle Rock neighborhood of northeast Los Angeles, not far from the area where the L.A. River begins to enter downtown L.A. With my new position defined as part teaching, part research, and part community engagement regarding environmental issues in Los Angeles, I was able to reestablish several of the projects of the Pollution Prevention Center at Occidental. With colleagues from Occidental and managers from several of our UCLA projects who had also joined me at Occidental, we decided to expand the center's focus and rename it the Urban and Environmental Policy Institute (UEPI). UEPI became a visible example of Occidental's own commitment to community engagement and "learning by doing." Thus, while UEPI was located directly within an academic program (urban and environmental policy), it was able to strengthen and significantly expand its community emphasis. It also sought to define itself as a multifaceted, social-change-oriented institute that provided a place where faculty, students, organizers, community partners, researchers, and policy analysts could collaborate. We thus defined UEPI as part academic center with strong community ties and part community-based organization with a strong research and policy-development capacity.[12]

This book is partly derived from the research, organizing, actions, and events I've been engaged in during the past decade through my work with UEPI. It reflects that "scholarship of engagement" that includes the "action research" and "community-based participatory research" approaches or orientations to research and scholarship. Most definitions of community-based participatory research place the participant as an outsider seeking to establish a coequal relationship with community partners. Similarly, many action or policy-oriented entities like UEPI that are based at higher-education institutions or that function as nonprofit think tanks, are often associated with an action research model, albeit one that provides guidance on the development of policies rather than participates directly in action. While UEPI shares many of those same attributes, it has also defined itself as a hybrid that is situated between researcher and community group and that functions simultaneously, or at least in tandem, as outside participant and inside actor. Thus, to do research or to write a book derived from the UEPI experience, I sought to represent those experiences by observation and analysis and also by a narrative of events and actions that includes firsthand description. This would need to be a book of research and action.

Framing the Book

Several events and actions were drawn from the UEPI experience for this book. Three areas of engagement—water politics, cars and freeways, and immigration and globalization—form the book's core. The first involves the debates around the L.A. River and the critical battles regarding how this river (and other urban rivers and streams) should be characterized and perceived as part of the urban landscape. These discourse battles, as we came to identify them at UEPI, provide one important illustration of the culture-of-nature concept that helps situate some of the ways we think about and ultimately act regarding urban nature. Along these lines, from the summer of 1999 through the fall of 2000, UEPI, along with the key L.A. River advocacy group, the Friends of the Los Angeles River, engaged in a multifaceted series of events called "Re-Envisioning the L.A. River: A Program of Ecological and Community Revitalization." These events, programs, and actions came at a point in time when a transition had begun to take place that led to a shift in the ways in which L.A. River issues came to be addressed. The Re-Envisioning program, in turn, furthered that shift—not just with respect to the river but also the communities bordering the river that had long perceived the river as off limits, if not hostile territory. Beyond the L.A. River, the issues of how water has been and continues to be secured and land is developed has also loomed large, significantly defining the geography of cities and regions and influencing the ways in which urban nature and community in the city are defined.

If water—where it flows, how it flows, and how it is used—helps define the geography and political economy of Los Angeles and other urban regions, the car and the freeway (affecting how people and goods flow through the region) are two more of a city and region's defining features. For forty years, from the late 1930s up through the 1970s, transportation policy in Los Angeles and Southern California had been focused almost exclusively on where and how to build and expand the freeway system. Similar to the large water-supply projects and extended flood-control programs that were built in this same period and that put in place the water infrastructure for Los Angeles, highway construction molded and shaped the land-use patterns, commercial and industrial activities, and spatial identities of the region. By the 1980s and 1990s, however, massive construction projects no longer seemed feasible, due to cost, political and legal barriers, and enormous environmental impacts. The freeway system's focus shifted from system expansion to system management, as congestion—one- to two-hour and longer freeway commutes from house to work—became the issue of the moment. Though more concerted efforts have been developed in the past decade to expand bus and rail alternatives to the freeway, the overwhelming concern among those who live and work in the region remains how to get there by car and how to avoid a massive freeway tie-up.

The first freeway of the West, California 110 or the Pasadena Freeway (stretching from near the Rose Bowl in Pasadena to the northern edge of downtown Los Angeles near the current site of Chinatown), was built in the late 1930s. It was designed partly as a parkway (the Arroyo Seco Parkway) to establish a visual connection to the surrounding landscape and communities and partly as a modern freeway that would significantly improve the speed in which drivers could arrive at their destination. This hybrid model quickly gave way to the massive freeway system that eliminated any relationship to parkway design or visual connection. L.A.'s freeways, like the channelized L.A. River or the massive aqueduct systems built to import water into the region, served a singular purpose—the flow—while simultaneously redefining and ultimately undermining the experience of both nature and community in the city. Still, that first hybrid, the Pasadena Freeway/Arroyo Seco Parkway, found itself a prisoner of its original design and subsequent use. As congestion increased, the graceful ride providing visual connection became a dangerous exercise in navigating curves or entering or exiting the freeway. During the 1990s, community groups mobilized to recreate some of the parkway features of this first freeway of the West and, by so doing, also addressed related community, water, and landscape issues. As those issues were debated, UEPI, allied with a number of community, environmental, and alternative transportation groups, developed plans for an unprecedented event for the region: a Sunday stroll and bike ride *on* the freeway. This event, called "ArroyoFest," by its organizers, serves as the second set of activities profiled in the book. Beyond ArroyoFest, the freeway system—how it rose and how it influences daily life in the region—provides the second important backdrop for the discussion of nature and community in the city.

In 2003, the same year that ArroyoFest took place, a bus mechanics' strike in Los Angeles paralyzed the bus transit system. Unlike the subway strike in New York City that took place shortly before Christmas in 2005, paralyzing that city's overall transportation system and causing major economic dislocation, the L.A. bus strike had little impact on most of Southern California's freeway-oriented commuters. It also revealed how extensively the L.A. transit system primarily (and, in many areas, exclusively) served low-income residents, including large numbers of immigrants who were its primary customers. Moreover, unlike New York, press coverage was less intense, the political response by elected officials was tepid, and the common perception about the strike was that it underlined the region's dependence on the car and the freeway and that the bus riders were a powerless and marginalized community of users.

Shortly before the bus strike, a major debate was unfolding about immigrants and the environment within the largest of the mainstream or traditional environmental groups, the Sierra Club. The population-control advocates within the Sierra Club, through an initiative they had developed requiring a referendum vote among club members, argued that rapid population increases create significant negative environmental impacts, including sprawl, an unsustainable use of resources, and reduced quality of living for all. Immigrant issues had long been a volatile topic in Los Angeles, and as the city and the region became increasingly Latino and Asian, they continued to spill over into both political and environmental battlegrounds.[13]

UEPI also became engaged in the immigration battles around this time, primarily through one of its centers, the Migration Policy and Resource Center, which had been established just prior to the Sierra Club immigration referendum. Through connections it had established with a wide range of immigrant groups, such as day-laborer organizations, immigrant gardeners, immigrant-rights groups, and immigrant-based community-development organizations, the Migration Center had begun to explore through research and action the issues concerning immigrants, the environment, and the global city. Other UEPI projects, such as the student and parent organizing about food-justice issues, necessarily engaged immigrants, documented and undocumented, whose presence had changed the face of the city and the region in less than three decades.

The UEPI approach sought to counter the environmental-impact assertions of the conservationists and to identify the crucial ways in which immigrants contributed to building community and establishing a different kind of culture of nature. In this context, immigrants—in Los Angeles and elsewhere—could be seen as a core constituency for helping redefine and reinvent nature and community in the global city. The examples in this book of the immigrants' place in the global city illustrate that opportunity for reinvention. At the same time, the massive demographic changes, fueled by the influx of immigrants from Latin America and Asia beginning in the 1960s and 1970s, became a key component of the transformation of Los Angeles into a global city. This aspect of globalization—the growing movement of people crossing borders—represented the other side of globalization's enormous economic and environmental effects related to the flow of capital and the movement of goods across borders through such exit and entry points as the Port of Los Angeles and its sister Port of Long Beach. While a backlash against immigrants influenced the political climate in California and the United States as a whole, globalization impacts associated with global capital, outsourcing of goods and production, and the movement of goods and commodities from places like China to outlets like WalMart were often lauded by politicians and policy analysts as benefiting communities, regions, and the country as a whole. This juxtaposition between a charged political environment regarding immigrants and a much touted global economy also serves as a backdrop to the discussion of Los Angeles and the ways in which the changes and conflicts in this global city anticipate changes and conflicts in the country as a whole.

These three sets of issues and themes—water and the river; cars and freeways and their alternatives; and immigrants, globalization, and the global city—constitute the basis for this book's exploration into possibilities for reinvention. The barriers for reinvention seem obvious, while the difficulty of constructing a new language and new or reinvented forms of nature and community in a global city may well appear insurmountable. But change can appear suddenly, often after a long, undetected process of innovation and empowerment, to make possible what had previously seemed so unlikely. "Be realistic. Demand the impossible!" the students and workers who took to the streets in Paris in May 1968 chanted. This book explores how the seemingly impossible yet critically realistic scenario of reinventing Los Angeles might happen—by discovering nature and community in this global city, by valuing the complexities and differences among its residents, and by helping construct social movements across borders.

A long-time friend and colleague told me after reading a draft of *Reinventing Los Angeles* that the book feels more like a journey of discovery than a traditional academic thesis, exposition, and conclusion. That insight has been particularly helpful in situating for me the book's structure of narrative, analysis, and contemporary history. The narrative describes the events, campaigns, activities, and related battles that constitute the action dimension of the book's research-action framework. The analysis draws on theoretical discussions about nature and community and on the book's central issues of how water and land use, transportation, and migration and globalization have shaped Los Angeles. And both the narratives and more theoretical discussions are elaborated in the context of the contemporary history of Los Angeles and other urban regions in transition and in the context of a new politics of social change within and across borders that has begun to challenge the very forces that have shaped the global city.

1

Nature in the City

Nature is perhaps the most complex word in the language.

—*Raymond Williams*[1]

Walking through a Cornfield

On a warm day in October 2005, with the brown haze of the polluted Los Angeles skyline above us, I arrived at the thirty-two-acre site that was just to the east of Chinatown and north of downtown Los Angeles and that had become emblematic of the quest for reinventing nature in Los Angeles. The Cornfield, as it was known, was the proximate site of the original *zanja madre*, the mother ditch that had diverted water from the Los Angeles River as it flowed past the site to serve the fields near the location of the original mission. Alternatively known as the Chinatown Yards and recently named the Los Angeles State Historic Park, the Cornfield had been prized for its choice location and its availability as undeveloped land—the single largest plot of land in Los Angeles that was capable of being developed. As a result, an intense battle had been waged between, on the one side, community and environmental activists, including L.A. River renewal advocates who dreamed of parks and open space adjoining the river, and, on the other side, a powerful and well-connected developer with political allies, such as then Mayor Richard Riordan, who wanted to see warehouses and light industry on the site.[2]

This conflict was resolved in 2001 when the developer sold the land to the state of California. The transfer of ownership to the state, acting largely on behalf of the river-renewal advocates, was designed to turn the land into a state park commemorating its varied history, including its recent struggles. But planning such a park was complicated by an insistent and sometimes contending group of advocates, a state parks bureaucracy not used to the idea of developing a park in the heart of the city, and an unprecedented type of planning process required to establish the various uses for the park that were still to be determined. With expectations high and the slow process of planning the park beginning to create tensions, park officials were intrigued by the offer to temporarily house a public art installation. The funding for the project would be provided by the Annenberg Foundation, made possible in part by the fact that the artist, Lauren Bon, was also a member of the family and a trustee of the foundation.[3]

Though there was uncertainty about whether the site had actually been a cornfield in its earlier configuration, the name for the site had stuck. As a teenager growing up in Los Angeles, Bon had biked along the rail yards and had felt a connection to the site. After living abroad for a number of years, she had recently returned to Los Angeles and become fascinated by the myriad of players, including the artists, poets, and cultural activists who had embraced the cause of renewing the Los Angeles River and transforming the Cornfield into a public space. Bon's idea was to actually create a field of corn as a living sculpture and entitle it *Not a Cornfield*, to signify that her project would be temporary and not the ultimate signature of the park. A cornfield, for Bon, served as a "potent metaphor for those of us living in this unique megalopolis," as Bon described it on her *Not a Cornfield* Web site. The artist portrayed the project as a living landscape, and although the corn would not be harvested for consumption, given the concern about prior contamination of the land when it had been a railroad yard, plans were made to use the corn, after harvesting, for other uses.[4]

With her Annenberg resources and grand vision, Bon saw her art project as both active and reflective—building on a legacy of radical art while contributing to the quest to reinvent nature and community in the city as part of the perennial search for defining Los Angeles and locating its center. Located at the northern edge of downtown itself, the view from the Cornfield extended south to City Hall and the skyscrapers of downtown; west to the Chinatown stop of the new Gold Line rail line and the Chinatown commercial center of shops, restaurants, and dense housing; east to where the handful of warehouses gave way to the low-flung structures of the William Mead Homes low-income public housing project that extended into the Latino immigrant neighborhood of Lincoln Heights; and north to the majestic Broadway Avenue bridge and the channelized Los Angeles River as it made its way south through the eastern edge of downtown towards Boyle Heights and out to the ocean at Long Beach.[5]

Bon's concept was an ambitious one. To accomplish her goal, she imported 1,500 truckloads of earth to plant a million seeds and hired more than two dozen youth from the William Mead Homes to serve as community docents, while fending off criticism that she was simply a wealthy matron dropping in on a site that had such a long and protracted history of struggle and expectation. The corn grew throughout the summer, and small pathways between the rows of stalks provided routes within the site for joggers, bikers, and walkers through the field.

When I arrived that October, the corn was nine feet high, and the experience of walking through the cornfield was both surreal and invigorating. I knew the status of the cornfield was temporary and that the space would not ultimately become a form of urban agriculture, though the opportunity to create gardens on or near the site was available and a small plot of land had, separately, been dedicated for a native garden. The symbolism seemed uncertain, as with much of the goals of the project, both of which continued to change as new circumstances and demands came to the fore. But the site was also reminiscent of Carey McWilliams's characterization of Los Angeles as continually reinventing itself. For people who walked through the rows of corn and looked back at the City Hall skyline through the high stalks, the idea of reinvention seemed tangible.[6]

Six years earlier, when it had been vacant land, dusty and unappealing, our Re-Envisioning the L.A. River program had sponsored a walking tour through the Cornfield site to the Broadway Bridge and the view of the L.A. River. The walking tour had included elementary students from the nearby Chinatown school, Castelar Elementary. The site was then bleak and forbidding, but the walk had been a hopeful event, with the children later discussing and imagining and drawing pictures of what a park—for an area that had no parks—could be like. Now six years later, with the corn stalks providing a landscape of improvisation, visitors could look at the City Hall skyline through the rows of corn and envision a type of reconstructed nature in the city, a scene both fleeting yet imbued with possibility.

The Language of Nature

If, as Raymond Williams has argued, *nature* is the most complex word in the English language, that's partly due to how its meanings and reference points are continually changing. Nature is not just in the eye of the beholder but also in the language used to describe what one sees. The English poet laureate of nature, William Wordsworth, spoke of the beauty of the Alps mountains in 1844 while opposing the construction of a railway that could despoil this "region of the Heavens," but earlier travelers voyaging through the same area made no mention of the Alps' beauty. They instead spoke of "precipitous rocks and mountains [that were] objects of dislike and fear," as Phil Macnaghten and John Urry noted. Definitions of nature change in relation to place. The English, Macnaghten and Urry point out, have defined nature in relation to the land, while Americans have focused on scenery. "To the European eye," they comment, "there is no countryside, only land. American farms [in contrast] are not sights of visual enticement; rather it is to the 'natural' scenery of the deserts and canyons, mountains and ravines, that the eye has been mainly drawn."[7]

During the 1950s and 1960s, environmental groups like the Sierra Club and the Wilderness Society often utilized a language of nature as a scenic resource and spoke of the need to bring city residents to those scenic resources, given the absence of nature in the city. Subsequently, the rise of the predominantly urban-based environmental-justice movement began to have its influence on environmental discourse. The language regarding nature shifted to the extent that groups like the Sierra Club now spoke of "[saving] those smaller green places close to home from pollution and overdevelopment" as well as seeing both "big majestic places" as well as "everyday places" as places in nature.[8]

But the divide between nature and human activity still seemed to prevail in both language and action. The environmental-justice groups, for example, sought to disassociate themselves from what they perceived to be the common reference point for nature ("birds and bees") as distinct and ultimately detracting from their core concerns regarding polluted places and inequitable impacts. The nature-oriented big environmental groups, on the other hand, still focused on a language and policy framework to protect nature from unwanted human activity. Both types of environmental groups were not able to get beyond the distinction between nature and human activity that established a divide in relation to advocacy, policy, and resources as well as language. By maintaining that separation, environmentalists were caught in a bind, often forced to choose between being either a nature movement or a people-centered social movement rather than both.

This type of separation of the social and the natural could be traced historically with the rise of industrial society and the growth of cities, particularly in Western society. "Non-Western cultures have never been interested in nature; they have never adopted it as a category," argued French philosopher Bruno Latour in his discussion of Western science. "Most earlier ideas of Nature had included, in an integral way, ideas of human nature," the great English social critic, Raymond Williams, wrote in the same vein. But the rise of a market-dominated urban and industrial order transferred nature to a place "out there," without "[considering] very deeply what this reshaping might do to men," as Williams put it. Separating humans from nature had become "incorporated into our dichotomous way of thinking," wrote the landscape historian and cultural critic J. B. Jackson. "It is a 19th century aberration," Jackson argued, "and in time it will pass."[9]

But the consequences of such dichotomous thinking have continued to be felt in multiple ways. The dichotomy, for example, between the natural sciences and the social sciences, according to Latour, produced a world of "social facts" and "natural facts." "Nature became separated from society," Latour argued, "in order to be scientifically studied, and ultimately tamed, and the world was separated into things natural (the objects of study of natural sciences) and things social (the objects of study of social sciences)." Moreover, nature, as a distinctive category, became itself a shifting target, dependent on the strategies designed to control it and the different perceptions of what it represented as a world apart. Different frameworks regarding how to define, relate to, and manage nature were also developed. The preservationist concept (save nature from human activity since human intervention or interaction inevitably despoils it) was first contrasted with the idea of managing nature as a type of resource to be more effectively and efficiently utilized for human activity, even as it continued to be defined separately.[10]

Both the preservationist and management or conservationist discourses, however, were not able to contend with continuing impacts related to changes in urban and industrial society. The preservationist concept was effectively a losing proposition, conceived as a defensive strategy to save what was left in the face of the inexorable expansion of urban and industrial activity that encroached on an untouched wilderness. During the 1950s and 1960s, key environmental figures like David Brower constructed their arguments around the ideas that little of the earth's surface remained as wilderness and that the goal of environmental policy was to protect as much of what remained as possible through legislation like the 1964 Wilderness Act. But even those efforts remained problematic as such legislation remained incomplete in its intended outcomes and contending forces such as oil and mining interests or land developers continued to covet the areas to be protected. Already by 1989,

New Yorker writer Bill McKibben would proclaim that nature, as a category set apart, had ended and that even in his Maine woods retreat he could witness human impact. This concern about an end to nature was further reinforced when startling impacts began to show up in what had been considered the most untouched areas. This included, for example, the discovery that polar bears in the Arctic region, particularly in their remote dens near the North Pole, had high concentrations of a number of toxic compounds such as polybrominated diphenyl ethers (PBDEs), toxic flame retardants produced in the United States that were among the hundreds of other pesticides and industrial chemicals that had been carried to the Arctic by northbound winds and ocean currents. This "end of nature" fear was even more pronounced in relation to some of the anticipated impacts from global warming, such as the melting of the hunting grounds of Alaskan polar bears and the discovery that some of them had begun to drown.[11]

The conservationist/management approaches have not fared much better from an environmentalist vantage point. Already by the 1950s, industry groups such as chemical manufacturers and government agencies such as the Bureau of Reclamation intent on building large dam projects were using the language of "resource management" and "conservation" to justify projects that had significant environmental impacts. By the 1970s and 1980s, organizations advocating under the banner of the Sagebrush Rebellion and acting on behalf of mining, grazing, and other development interests used the resource-management language of "wise use" as part of their argument for opening up federal government lands previously protected from development. This further shifted the language of "conservation" and "management" toward the realities of development and extraction. Even the more recent language of "sustainability" or sustainable management, with its explicit environmental intent to reduce environmental harms, faced the challenge of figuring out how to reduce the "ecological footprint" on nature, given the continuing acceptance of the nature/social divide; that is, the more human activity, the larger the footprint. Much of the sustainability discourse relied on a combination of technology and market forces in the pursuit of "ecological modernization" strategies to achieve a state of sustainable development. But despite their faith in the capacity of technology and good corporate practices to achieve environmental goals and notwithstanding the rhetoric about protecting future generations, many of the sustainability advocates shared the pessimistic assumptions that this alternative development scenario would, at best, staunch the bleeding. Nature, once set apart, was destined, even with the most effective techniques, to remain wounded or seriously impaired.

Faced with this sense of impending loss, a revised language of ecological *restoration* emerged during the twentieth century, stimulated by the writings and activities of such diverse figures as Jens Jensen, Frederick Law Olmsted, and Aldo Leopold. Their approaches to the nature/human-activity divide, as well as the new types of environmental-management strategies that began to be devised in places like the national parks and the national forests, influenced the development of a discrete new field of the natural sciences called "restoration ecology" or "ecosystem restoration." Instead of a focus on slowing habitat loss or preventing the collapse of ecosystems, the restorationists sought to try to "reclothe the earth" by repairing ruptures in the landscape and reconnecting its parts. Landscapes are themselves hybrids since they are "worked, lived on, meddled with, [and] developed, [and thus require] human intervention and care," Alexander Wilson said of the logic of a restorationist approach.[12]

The hybrid argument was a key to the notion of restoration. "Partly artificial and partly natural, the restored landscape is not exactly either," argued William R. Jordan III, the editor of the journal *Ecological Restoration*. "It is rather, a landscape of ambiguity—the very place, we might suppose, where established identities are challenged and where relationship and community begin—yet a place where it seems we have been ill equipped either to recognize these opportunities or to take advantage of them." In this context, restoration ecology has sought to make a partial step towards overcoming the nature/human-activity divide, recognizing that its efforts have attempted not so much to save but to recover ecosystems that had otherwise been damaged or destroyed—a process that, in its most ambitious forms, could reproduce nature by taking into account the necessity of human intervention. But for critics of the restoration approach who felt that ecological change was part of "an evolutionary process in normal time" but *without human intervention*, the ecological restoration movement was simply another example of "humanity's attempt to control the natural world," as Eric Katz put it. The result was less a landscape of ambiguity than a form of "fake nature," the critics argued, more similar to an art forgery than a new form of art.[13]

Both critics and advocates agreed that the language of ecological restoration forced us to reconsider the notion of what was "natural" and how and why it should be valued, which might well be the movement's most significant accomplishment. But the limits to ecological restoration have also become immediately apparent, both in relation to the opportunities available for restoration in urban areas (a kind of "ecological junk-picking," according to Jordan), as well as in the scale and the social context in which restoration efforts have taken place. The restoration approach—with its requirements for sufficient resources, available sites, and acceptable forms of intervention—has often needed to operate at a much smaller scale than the types of repairs that might otherwise have been required, given the broader-scale impacts from polluting industrial practices, global warming–related changes, major land-use changes, or the significant consequences from globalization policies and practices. Moreover, as Cindi Katz has argued, given the resources required and the circumstances often associated with a restoration approach, "[restoration ecology] has a tendency to privilege certain landscapes and land use practices," failing to account for the environmental-justice argument of the powerful changes in the urban and industrial environment that weigh most heavily in relation to class, race, and ethnicity.[14]

Given the limits of these different approaches and the continuing difficulty of overcoming this residual and deeply embedded notion of a nature/human-activity divide, the environmental movement has created a dilemma for itself. The persistence of this divide potentially marginalizes what Latour calls the "non-humans" from the political sphere that has otherwise been the domain of human activity. Although trees might have standing in certain limited legal contexts and strategies around sustainable development or ecological restoration provide some partial relief to the notion that human activity inevitably impacts negatively the separate domain of nature, the divide ultimately shortchanges both sides: polluted environments, loss of green space, the exploitation and alienation of labor, traffic congestion, the urban heat-island effect, and global warming, to name just a few, are all visible by-products, inadequately addressed. To overcome the divide, a new type of politics, what Latour and others have defined as "political ecology," needs to emerge that would allow for an integrated approach. By doing so, nature can then be construed as multidimensional and occurring in multiple spheres and places, whether city or countryside, urban or suburban, active or passive park uses, undeveloped lands or built environments, or other places where people live, work, or play. The land, Aldo Leopold famously commented, itself constitutes "a community to which we belong." To define nature, then, is to define the social, cultural, and natural spheres as one—to recognize, as Raymond Williams put it, that the idea of nature "contains, though often unnoticed, an extraordinary amount of human history . . . both complicated and changing, as other ideas and experiences change."[15]

Urban and Suburban Natures

During the twentieth century, the idea of nature was significantly influenced first by dramatic population shifts from the countryside to the city and then, in the later half of the century in the United States, by shifts from city to suburb. Urban places had long been considered the antithesis of the natural. The rise of the industrial city in the late eighteenth and nineteenth centuries was continually subject to withering commentary, with descriptions of how these urban "lowlands," as John Muir called urban places, represented unimaginable assaults on nature and community. Such urban places also represented assaults on the senses: nineteenth-century London, John Ruskin wrote, was a "great foul city . . . stinking—a ghastly heap of fermenting brickwork, pouring out poison at every pore." Lewis Mumford, in his biting commentary on the nineteenth-century industrial city, was equally emphatic about the kinds of foul discharges that polluted every aspect of the urban environment and turned urban rivers into a "flood of liquid manure." What passed for nature in the city—the air, the water, and the land—had become fully subordinated to the factory, Mumford argued, and "a pitch of foulness and filth was reached [in the working-class neighborhoods] that the lowest serf's cottage scarcely achieved in medieval Europe." Yet nature outside the city in the countryside was also, in nineteenth-century England, "rich with odours, of farm animals, raw sewage, rotting vegetables, smoke, stagnant water, and so on," Macnaghten and Urry point out, suggesting that even this conception of nature in the industrial city as a set of smells and other sensory perceptions was socially constructed. In the United States, the urban/nonurban contrast was not so much a city/countryside distinction as the depiction of nature as located in those areas not settled.[16]

The divide between nature and human activity was often viewed, through much of the nineteenth and twentieth centuries, as an urban/nature split. Even urban reform advocates during the Progressive Era such as the settlement house reformers like Jane Addams and the "garbage lady" Mary McDowell who promoted playgrounds, municipal bathhouses, sanitation reform, and other urban environmental initiatives, saw the industrial city of the early twentieth century as "unnatural" and obviously and desperately in need of reform. For the settlement house movement, "fresh air" and any form of green space was a welcome relief from the crowded and polluted cities like Chicago that still shared many of the characteristics of Ruskin's "foul city" of nineteenth-century London. But the Progressive Era reformers also took the existence of the city as a given, and their focus was less on escape from the city than on changing the conditions in which people lived, whether in relation to housing, the work environment, or the need for recreational places like parks and playgrounds.[17]

During the first half of the twentieth century, many of the efforts to reform the city, led by the regional planning movement, appeared essentially to be exercises in rescaling its size and composition (what the garden city advocates called "the cosmopolitan city of scale") and in remaking the relationship between city and countryside or the urban/nature interface. Part of their argument was dependent on the new technologies of transportation such as the automobile that they hoped would allow a more seamless connection between the city and nature (the "townless highway," Benton MacKaye called it, promoting the idea that car travel to and from the city would encourage an appreciation of and connection to nature). The scale argument also identified the need for a more self-contained city where parks and green space, people, and industry could coexist.[18]

But already by the 1940s, the garden city was quickly becoming the garden suburb, while a place like Radford, New Jersey (the original showcase of the regional planners), was turning into the mass-produced Levittown. As opposed to the streetcar suburbs of the nineteenth and early twentieth centuries, the most visible and influential of the new suburbs of the post–World War II era were the automobile suburbs—new developments that were located at the urban edge rather than within the urban core and in relation to their proximity to the highway rather than the streetcar. At the same time, the growth of the automobile suburbs extended and in some case revised the definition of an urban place as encompassing a new kind of continually expanding and changing urban area. This "supercity," as the publisher of the *Los Angeles Times* boasted of the rapidly expanding Los Angeles region in 1960, would eventually stretch from Santa Barbara to the Mexican border. Such metropolitan regions included both inner-city areas and outer suburbs continually stretching outward, creating what one critic called the "100-mile city." The contrast between the urban core (dense, polluted, and lacking green space) and the automobile suburbs (whose landscapes and street designs sought to create the impression of a reconnection to nature) also provided a contrast between what came to be called the "brownfields" of the inner cities and the "greenfields" of the suburbs.[19]

The nature motif of the automobile suburbs was a constant theme that was heavily promoted by developers, in news stories, or in films or television shows. While whole new planned developments were seen as surrounded by nature and open space, they were often placed in areas that reduced farmland or the open spaces at the edges of the expanding urban borders. This juxtaposition of the suburban search for nature coming at the expense of the loss of open space and farmland was particularly prominent in the case of Los Angeles. While Los Angeles was not the only metropolitan area that featured the development of the automobile suburb, the city's boosters celebrated how its new roadways could break "the wild virgin areas of Southern California . . . to the uses of progress, yielding up their beauties to the motoring public," as the *Los Angeles Times* described it in 1933. A decade later, the L.A. region would fully symbolize the association of the post–World War II suburbs with the dynamics of automobile-induced sprawl and the search for a reconnection with nature as a suburban amenity. In the process, Los Angeles also experienced its rapid loss of farmland (declining from one of the largest to one of the smallest agricultural counties in California between 1930 and 1960), major changes to what biologists and forest managers characterized as the wildland/urban interface and a range of new "natural" hazards such as debris flows stimulated by the erosion of the chaparral owing to fires in forest lands abutting the new developments.[20]

The development of the new automobile suburb of Westlake at the Los Angeles and Ventura County line provides a good illustration of the search for nature and community in the automobile suburbs. Westlake was designed in the mid-1960s as a self-contained city, situated off the recently competed section of the Ventura freeway on land that had a long and storied history as open space, cattle range, and an open-air location for such movies and television shows as *Robin Hood*, *King Rat*, *Laredo*, *Gunsmoke*, and *Bonanza*. In 1963, a developer, the American-Hawaiian Steamship Company, working with the Prudential Insurance Company, bought the land from the Albertson Ranch and immediately commissioned the development of a master plan to create what the developers called "a city in the country."[21]

The key claim underlying the development of this automobile suburb, initially called "The City with No Limits," was its core goal of not "[despoiling] the land." The homes to be built would be situated on "curving cul-de-sac streets clustered along broad greenbelts and natural open areas," as one favorable *Los Angeles Times* article proclaimed. The developers asserted that Westlake would maintain "all the rolling hillocks, the ancient oaks, [and] the glittering streams," while adding "thousands more trees and bushes and flowers." The initial neighborhood was called "The Park." A core attraction for the new suburban homesteaders was its newly constructed lake, especially designed for an area that experienced limited rainfall and long dry seasons that extended late into the fall. This would be situated in the heart of what had been the 12,000-acre Albertson Ranch in order to create eight miles of shoreline by importing water through the new city's connection to the large regional water agency, the Metropolitan Water District, and by also diverting water from a stream that ran through the property. Designed by the Bechtel Corporation, the lake would be made possible by removing 50,000 cubic yards of alluvial soil and by constructing a dam site consisting of 30,000 cubic yards of concrete that would be 700 feet in length and 40 feet high. The 2,000 boat slips that would be made available for residents would allow them, with their newly created access to lake water, to fish, go boating, and enjoy the lengthy dry season.[22]

By building what became the eighty-second city to be incorporated in Los Angeles County, along with the residential homes in their cul-de-sacs, the lake, the eleven golf courses, the ten tennis courts, and the hiking trails and access to nature, Westlake Village was also representative of the dilemma of the automobile suburb. In its search for a connection to nature and community, this newly constituted suburb/city was also associated, from its very origins, with stretching the urban boundary and thereby lengthening the automobile commute, intensifying the continuing search for an increased imported water supply, finding itself subject to sudden and massive fires at the end of the dry season, and ultimately reinventing the notion of what constituted open space and "natural areas." The development of Westlake Village, as with so many other automobile suburbs, underlined the notion that nature was continually being reinvented through, and as an extension of, human activity. This activity was itself a product of a market-driven system that turned the "natural area" into increased property value and created a broad array of environmental hazards, including "natural" occurrences that market economists called externalities but that were part of the very fabric of the new automobile suburbs.

The automobile suburb at the outer edge of the urban boundary, however, was just part of Los Angeles' urban-suburban continuum of places that maintained an uncertain relationship to nature. Suburban development that emphasized its "natural surroundings" occurred early in the twentieth century in Los Angeles and elsewhere in places like Bel Air and Palos Verdes Estates. These "restricted suburbs," characterized by their various rules imposed on each residence, were designed to attract a wealthier constituency. Restrictions included prohibitions on certain land uses (such as farm animals and other agricultural activities) and limits on potential buyers designed to also keep certain people out (such as nonwhites and lower-class Angelenos).[23]

By the 1920s, Los Angeles had also established what its planners called "suburban-industrial clusters." Associated with the growth of new manufacturing centers like the oil and aircraft industries, these blue-collar or working-class suburbs were also spatially dispersed according to streetcar service and industry location decisions. The difference was that these working-class suburbs were primarily designed as manufacturing hubs that provided residences for its workers—working-class bedroom communities defined by their connection to industry but without schools, libraries, churches, or parks.[24]

These suburban-industrial clusters were made possible in part because of the scattered development patterns of Los Angeles that predated the construction of the freeways that would subsequently become prominent in the development of the automobile suburbs. As late as 1940, just as the great manufacturing boom in Los Angeles began to take off, more than a third of the subdivided property still remained vacant, providing ample opportunities for the siting of large manufacturing plants and the development of the new working-class suburbs. The key to these developments was the association with the workplace and the centrality of the home. "To workers in 1920s Los

Angeles," Becky Nicolaides wrote of these working-class suburbs, "a home represented independence, a goal valued in both American and immigrant traditions." Even the first of the suburban-industrial clusters, like the cities southeast of downtown Los Angeles such as Huntington Park and South Gate that had been developed as part of the industrial districts designated along the L.A. River, reinforced the connection between work and home that was absent in the automobile suburbs. At the same time, the connection to nature, given the more dense and compact development, was limited to the home environment, primarily through backyard gardens often designed as functional and edible self-provisioning landscapes.[25]

As these inner- and outer-ring suburbs sought to establish their different kinds of connections to nature, urban-core or inner-city areas also witnessed important though uneven efforts on behalf of reclaiming or reinventing an urban nature. This included a handful of planned communities in the Los Angeles area, such as Baldwin Hills Village (currently called "The Village Green") that was established in 1942 and modeled as a type of Garden City seeking to establish a community environment based on social and political equality. The Village's original 630 row houses and apartments were arranged in an S-like formation around garden courts, tree-shaded malls, and three central greens, with automobiles confined to the periphery of the development. Even after the 1963 flood that significantly impacted the Village and much of the surrounding area, the homes were largely restored to correspond to some of their original design features. Though a modest shift toward more parking and garage spaces has since occurred, the Village Green has maintained much of the original intent of the design to establish an integration of green space, public space, and private homes. The core of the Village constitutes "an oasis of pedestrian calm," as Mike Davis puts it, celebrating this design of a "democratic community" that contrasted with Los Angeles' ever expanding land mass dedicated to the automobile. In his otherwise noir assessment of Los Angeles in *Ecology of Fear: Los Angeles and the Imagination of Disaster*, Davis argued that at every level of its organization, Baldwin Hills Village attained a design that sustained "a superb dialectic between private and communal space" and remains one of "Los Angeles's most vibrant neighborhoods."[26]

But Baldwin Hills Village was clearly more the exception than the dominant form of development and land use in Los Angeles, especially with the region's enormous expansion in population and subdivided land after the 1940s. Los Angeles' loss of green space, farmland, and open space was staggering and contributed to its post–World War II reputation as the capital of sprawl and the leading edge of environmental degradation. It was the site of the first massive urban air pollution episode in 1943, it saw the rise of the most intricate and extensive urban freeway system in the world by the 1960s, it channeled its rivers and streams and polluted its bays in a manner that was both aesthetically unappealing and indicative of the prevalence of the flood-control and sewage-discharge functions over any other purpose such as the planting of trees along the channel, and its officials and business leaders focused on paving and subdividing and traffic flows and congestion, rather than parklands or public places, greenbelts or open spaces. *Development* became the buzzword; polluted landscapes the outcome. Urban Los Angeles was assumed to be bereft, with a few exceptions, of urban nature. It came to represent by the 1970s the worst fears of environmentalists—urban places that symbolize the loss of nature.

The primary areas of concern for L.A.'s mainstream environmentalists led by middle-class homeowners were the remaining large open spaces and undeveloped coastal areas still accessible within the urban boundaries of Los Angeles, such as the Santa Monica Mountains and Santa Monica Bay. During the 1960s and 1970s, developers and their political allies waged fierce battles against open-space advocates and their political allies who sought to place strict limits on development and maintain wildlife corridors, nature trails, and other environmental amenities in the Santa Monica Mountains. The battleground over the coastal areas was even fiercer, resulting in a sharply contested but ultimately successful statewide ballot initiative in 1972 to protect the coast from development and provide public access to the beaches. Pollution from sewage outfalls and industrial waste discharges also remained a constant threat to the coast. Partly in response to those issues, working-class communities bordering or adjacent to the coast, including the working-class suburbs that had been established during the post–World War II manufacturing and housing booms, began to mobilize around the emissions and discharges from the nearby oil refineries, pesticide manufacturing plants, and aircraft plants. Two sets of issues came to be linked to two types of movements: (1) discharges and emissions that led to the development of local, ad hoc neighborhood groups and (2) environmental groups focused on the discharges that had contributed to the ocean pollution and marred beaches that had undermined Los Angeles's reputation as the "land of sunshine" and "outdoor living" so celebrated by its boosters.[27]

Both of these types of movements expanded significantly during the 1980s and 1990s as the pollution and discharge problems affecting the coast and nearby communities became increasingly visible and contested. Heal the Bay, a key middle-class-based local environmental group, first took shape after a high school teacher who was also a surfer began to challenge the failure of the United States Environmental Protection Agency to enforce the Clean Water Act in relation to sewage and industrial discharges into the bay. Along with other ocean-protection groups like the Surfrider Foundation and Santa Monica Baykeeper, Heal the Bay successfully forced a change in the operation of the major sewage-treatment plant along the coast, organized coastal cleanup days, established environmental report cards for the beach areas along the coast, and focused on water conservation and other development-related environmental "best practices" to reduce the amount of polluted runoff that made its way to the bay.[28]

In contrast to the middle-class-based groups that focused on protection of important environmental amenities such as the ocean and mountain areas ringing the urban core, neighborhood-based working-class and low-income groups also emerged in Los Angeles to focus on the pollution that was impacting their homes and communities. These polluted environments were impacted by the discharges from industrial plants, the diesel-belching truck traffic from the Ports of Los Angeles and Long Beach, and the noise and air pollution associated with the industrial corridors that dotted parts of the coast as well as much of the southern, central, and eastern parts of inner-city Los Angeles. This new type of environmental-justice movement that took root during the 1980s and 1990s in Los Angeles and other areas around the country consisted primarily of place-based movements whose members wanted protection for their own places—homes, families, and communities. Although groups like Heal the Bay, the Natural Resources Defense Council, and Environmental Defense embraced some of the language and the focus of the environmental justice groups, the distinction nevertheless remained, in some subtle and not so subtle ways, between a primarily nature-focused (the ocean and mountains) and a people-focused (home and community) urban environmental politics.[29]

The main area where the different environmental groups seemed to converge was around the question of open space *in the heart of the city* (rather than just at its edge or along the coast). In the late 1990s, both sets of groups began to talk about the need for public places—parks, community gardens, green spaces, a revitalized L.A. River—that could address the need for more livable places in the city. New coalitions were established, and a stronger urban environmental politics that was at once place-based and focused on green and open spaces seemed possible. And Los Angeles, the symbol of environmental degradation and the absence of nature in the city, became one of those urban places where this new politics began to take root.[30]

GREEN SPACES

If Los Angeles is to become a green city, as its current mayor continually predicts, it needs to establish green spaces where its residents live, work, and play. Among those spaces, lawns, gardens, trees, and parks are most visible. How such spaces are designed, constructed, accessed, and integrated into everyday life can tell us how such an urban greening agenda is able to reinvent or fail to prevent further impairment of what passes as nature in the city.

Lawns

For many of L.A.'s residents, the idea of urban nature, particularly nature in the automobile suburbs, begins with the front lawn—what one lawn-promotion group calls "your own little piece of the earth."[31] The well-kept, highly manicured, pesticide-laden, weed-free, grassy lawn has become a quintessential American phenomenon. "Nowhere in the world are lawns as prized as in America," writes Michael Pollan in his discussion of how lawns and gardens have come to represent a form of "second nature." "Like the interstate highway system, like fast-food chains, like television," Pollan argues, "the lawn has served to unify the American landscape." This "American obsession," as the lawn's historian, Virginia Scott Jenkins, also characterized the lawn phenomenon, has especially become a manufactured form of suburban nature. It is also a target of opportunity for those industry and development interests that see the industrial lawn, as this manufactured piece of the earth is called by some, as both a profit center and an extension of the suburban connection to nature as green space.[32]

Until the twentieth century, lawns were associated with opulent European creations signifying wealth or royalty or both. The first major effort to overcome the problem of introducing and maintaining a durable, weed-free grassy lawn came about through a collaboration, in the period just prior to World War I, among the U.S. Department of Agriculture, the U.S. Golf Association, and the Garden Club of America to try to identify which type of grass or combinations of grass could best provide the type of lawn required for golf courses. Around this time, Frederick Winslow Taylor, the promoter of scientific management and the standardization of work and also an avid golfer, sought to apply his shop techniques to create a uniform, lawn-based putting surface with "a longer and more velvety turf." Standardization became the key for both golfer and home owner. After a hiatus in lawn development from World War I through the Depression and World War II, the connection between golf course and lawn combined with the widespread development of golf courses as part of the automobile suburbs helped jump start a new phenomenon that would soon sweep suburban America: the closely cropped manicured lawns for golfers and the neatly kept, standardized front lawn of the suburban home.[33]

For many of the suburban developers, the lawn became central to the character and identity of the suburban home. Part private space, representing the individual's connection to a well kept nature, and part public space that helped define the suburban community in relation to its aesthetic and "natural" look, the lawn served as a symbol of nature and community in the suburb. The "supergreen lawn look" that emerged during the 1950s paralleled the growing introduction of brightly colored consumer products such as hot pink cars, both of which, as Ted Steinberg points out, became status symbols. At the same time, the lawn became a kind of quasi-regulatory instrument to maintain that look and that uniformity. Almost from the outset of the development of the post–World War II suburbs, residents came to serve as gatekeepers who complained about any lawns that were not well maintained by their neighbors.[34]

This powerful association of lawn with suburb greatly stimulated the growth of the lawn-care market. Increasingly in urban and suburban settings, land cover became lawn cover, with as much as 23 percent of urban land cover represented by lawns. To feed and maintain these lawns, the market in residential lawn care alone amounted to more than $4 billion within a $35 billion lawn and garden market. It also extended to such products as chemical fertilizers and pesticides, with estimates ranged from an average of three to twenty pounds of fertilizer and between five to ten pounds of pesticides applied each year per lawn. More than twenty million acres of residential lawns came to be planted, served by an increasingly concentrated turf and lawn-care business.[35]

The environmental consequences of the development of these industrial lawns have been staggering. One environmental assessment evaluated the industrial lawn's different inputs (fossil-fuel energy, irrigation water, fertilizer, pesticides, grass seed, sod, and so forth) and outputs (carbon dioxide emissions, surface-water runoff that included pesticide and fertilizer residues, solid waste from grass clippings, and so on) and called it the dark side of the lawn's evolution. As one example, the assessment cited a California Air Resources Board (CARB) estimate from the early 1990s of the amount of fossil-fuel energy used to mow a lawn during a one-hour period as representing the equivalent of the amount of pollutants emitted by driving a car 350 miles. According to CARB, the annual emissions of all lawn machines in California were equivalent to the emissions of 3.5 million 1991 model automobiles, each driven 16,000 miles during the year.[36]

Parallel to the concerns that had developed regarding suburban sprawl, the growing awareness of the lawn's negative environmental impacts created concerns within the lawn-care industry that led to the search for new marketing and product-innovation strategies for their product. This was particularly the case of the industry leader, the Scotts Company, which had been the first to patent Kentucky Bluegrass, had taken over a number of lawn-care products and brand names (such as Miracle-Gro, Ortho, and Smith and Hawken) and by 2005 had more than $2.3 billion in sales. Along with other industry players, the Scotts Company sought to deflect environmental concerns by claiming that lawns provided major environmental benefits and that "for most of suburban America, it's the foundation of the neighborhood's ecosystem," as the company Web site put it. New and aggressive marketing strategies sought to manipulate consumer concerns by associating the aesthetics of intensive lawn care with "community, family, and environmental health," as Paul Robbins and Julie T. Sharp argued, while industry groups established new public relations initiatives, such as the development of the Project EverGreen alliance, to fend off environmental attacks.[37]

More recently, the development of genetically engineered lawns, including a pesticide-resistant strain of grass that would enable lawn owners to massively spread pesticides on their lawns to kill any persistent weeds, has emerged as the latest version of the industrial lawn. The Scotts Company has pursued this development in collaboration with the world's leading biotech company, the Monsanto Corporation, which had been eagerly marketing its genetically engineered "Roundup Ready" seed to farmers. This product was designed to make plants, like corn and soybeans, resistant to Roundup, Monsanto's bestselling pesticide on the market. Scotts, as part of its deal with Monsanto to acquire Ortho from Monsanto, had also purchased the exclusive U.S. distribution rights for Roundup Ready. The Scotts/Monsanto collaboration then resulted in the development of a grass seed product designed to allow golf course managers to "control their weed problems with a single herbicide," as a Scotts spokesman put it.[38]

The idea of a genetically engineered lawn immediately generated opposition from critics of the biotechnology products and from those who were seeking to develop an alternative approach to the industrial lawn. Scotts, aware of the increasingly vocal opposition to any products that used the term *biotechnology* or *genetically engineered*, instead sought to identify their new products as "superior plants" but without much success. The genetically engineered lawn label, or "frankenlawn" as its critics also called it, stuck, with research indicating that genetically engineered grass could cross-pollinate with wild grasses, potentially creating herbicide-resistant superweeds. As opposition grew, the genetically engineered lawn promoters sought to make a strategic retreat by temporarily withdrawing their product, though they still hoped that the industrial lawn would continue to be accepted as the leading example of a manufactured suburban nature.[39]

Beyond the issue of the genetically engineered lawn, the industrial lawn has continued to be challenged. In Canada, more than eight municipalities passed laws that either banned or severely restricted the use of pesticides for lawn care. "The consequences to the professional lawn care industry in Canada have been huge," one trade publication said of these municipal initiatives that also caused concern among lawn-care industry groups in the United States. Within the United States, in one highly symbolic confrontation in 2003, a homeowner in one of L.A.'s well known working-class suburbs, the appropriately named Lawndale, planted cactuses, roses, sage scrub, and fruit trees, refused to use pesticides, and created a jungle of native plants and shrubs. The city decided to issue a citation for "excessive overgrown vegetation in [her] front yard" and told the owner to "cut all overgrown grass and weeds." After the threat of legal action and support for the owner from a number of city residents, the city backed down and allowed the plants to be left intact, while the owner agreed to trim when appropriate. Advocates for water conservation, antitoxic garden chemicals, and native plants began to speak of a "landscape ethic" that questioned the water- and pesticide-intensive lawn and promoted the use of native plants. The giant Metropolitan Water District, not previously known for encouraging alternative strategies regarding water use, introduced in 2002 a "native plant" incentive program for nurseries that also extended to individual homeowners. But perhaps in the biggest blow yet to the industrial lawn was an important shift among its core constituency. As a November 2005 *New York Times* article noted, some homeowners began to pave over their lawns to make room for the ubiquitous—and more demanding (of space)—automobile. Indeed, as the article described it, "paving over has become so commonplace that it is spurring differences between neighbors and debates within households about whether to dispense with the lawn." The battle between car and turf had been joined.[40]

GARDENS

If the industrial lawn could be considered environmentally problematic, the garden in contrast evokes strong associations regarding the human connection to the land. For Michael Pollan, the garden best conveyed his idea of "second nature," arguing that gardens can provide a road map for understanding how people "fit into nature." Cultivating small plots of land to grow plants or food, whether as part of the back yard in a home or on a vacant lot transformed into a community garden, has continued to be one of the few opportunities for urban residents to establish a bit of green space in the city.[41]

Gardens have often reflected the changing class and ethnic dynamics of urban areas like Los Angeles. Dolores Hayden, in her discussion of urban landscapes, points out that the early working-class suburbs in the late nineteenth century and early twentieth century often consisted of small front gardens and rear yards that were intensively cultivated. Ethnic and working-class neighborhoods "could be identified by their plantings and the varied delights of ethnic kitchen gardens," Hayden wrote, contrasting those gardens with the "exotic landscapes and flower gardens of the elite in the borderlands and picturesque enclaves." The homes of the poor in the eastside districts bordering the L.A. River, including the "foreign districts" of immigrants, for example, often consisted of "single cottages with dividing fences and flowers in the front yard, and oftentimes with vegetables in the backyard" according to one 1907 observation, by City Beautiful advocate Dana Bartlett. These were "homes for the people; pure hearts for pure hearth stones," Bartlett argued. Many of the homes occupied by Mexicans in the eastside districts were seen as "garden spots, a wealth of flowers and vegetables providing an inspiring contrast to the hideous, jammed, foul-smelling courts of New High, Alameda, Olivera [sic], North Broadway, and other streets near the heart of the city," another commentator remarked. Gardens as a food source in this period thus represented a key dimension of the working-class home. Becky Nicolaides, in her history of South Gate, pointed out that the "food produced in backyard gardens was consumed at home, sold and bartered [since] home consumption made good economic sense for working class residents."[42]

While working-class gardens were cultivated primarily as a private source of food, Progressive Era reformers during the late nineteenth and early twentieth centuries promoted the notion of gardens as community building as well as food-provision enterprises. Unlike the private home gardens, these involved community activities designed to meet particular social needs. Detroit's potato patch gardens in the 1890s provided a source of food in the wake of a major economic depression; the United States garden army during World War I and the victory gardens during World War II were designed as civic enterprises contributing to the expansion of food sources during war time (including home gardens redefined for their social purpose); and school gardens during the first two decades of the twentieth century were established in part to build character through discipline and good work habits while also enabling a new generation of urban school children to understand and appreciate what was otherwise considered a rural experience.[43]

Similar to the expansion of the lawn-care industry after World War II, the home garden, especially the suburban garden, emerged as a new industry segment. While lawn herbicides guarded against the unwanted weed, chemical pesticides were introduced into the suburban garden to fight the intrusion of pests, while nitrogen fertilizers aimed to increase the capacity of the soil to grow the various exotics and other plants that became the garden's standard items. Public gardens tended to be limited to ornamental displays next to places like office buildings, commercial centers, or amusement parks that were less community spaces than private landscapes for public display.

In contrast to these trends, a new urban-based focus on gardens as community places began to reappear during the 1960s and 1970s as part of a broader social movement concerned about the loss of green space and the desire for environmental renewal in the cities. The community-garden movement took root in nearly all major urban areas over the next three decades, stimulated by a growing interest in urban food and urban greening issues. In Los Angeles, an advisory body, the Los Angeles Food Security and Hunger Partnership (LAFSHP), was established in 1996 to provide food-related policy and programmatic advice to the mayor and the City Council. As part of its focus, the LAFSHP, which I served on during its brief tenure, developed a community-garden policy statement that was designed to address barriers to gardens (such as water hook-up and service fees) as well as possible strategies to ensure that existing community gardens would not be eliminated on the basis of competing claims on the land and to identify opportunities to expand the land available for gardens.[44]

The LAFSHP proposals, however, were not acted on at the time, and many of the community-garden sites remained vulnerable to market forces that sought to capture the land, often made more valuable after community gardens had been introduced. Nevertheless, community gardens still maintained their reputation as an effective "urban greening" and community-building, community-stabilization strategy for remaking vacant or abandoned land, reintroducing green space, and creating a small number of public gathering places. The key barrier was that market-driven land-use policies, reflective of what Dolores Hayden called the "constant production of urban space," often trumped the causes of urban greening, community food security, or the social value of establishing community and public places. Most of the significant battles around community gardens were defensive in nature and sought to stave off new efforts by developers and other economic interests to secure or resecure the property. Those initiatives that were most successful were often established in abandoned and barren sites on existing public lands or even in public parks in places where there had been little or no upkeep.[45]

However, the future of urban community gardens, both as a community strategy to "fit with nature" and as a type of urban agriculture, does not appear entirely bleak. In the past decade, innovative types of urban gardens have begun to be established, particularly involving immigrant or ethnic groups such as the Hmong, Puerto Ricans, Dominicans, or Mexicans and Central Americans who have drawn on their own traditions and connections to the land and to farming. At the same time, the rise of an urban-based environmental-justice politics—and justice-oriented language regarding access to resources—has also begun to influence more mainstream environmental groups involved in land-use issues. These include groups like the Trust for Public Land (which established an active program in a number of urban areas) and similar community land trusts engaged in acquiring urban sites from private developers or third parties (and then sometimes selling the land back or creating some kind of trust arrangement with a public agency or local or state government). The urban land trusts have been able to save and maintain some of the most visible of the community garden sites, as in New York City when the Trust for Public Land, together with the actress Bette Midler, was able to stave off the dismantling of hundreds of gardens owing to the threat by the administration of Rudy Giuliani to sell off a number of the publically owned sites that had been turned into and maintained, sometimes for years, as community gardens in the city. In Los Angeles, community-garden and urban-park advocates have successfully lobbied the city to establish and provide modest funding for increasing the availability of land for community gardens and pocket parks and have advocated for and designed innovative public gathering places such as small benches dotted with planter gardens.[46]

These efforts might have appeared minimal or even insignificant given the enormous reach of the paved hardscape, much of it the result of the massive amounts of land dedicated to automobile uses, as well as the power of the land developers to shape and control land markets. Land sites available for gardens, particularly in low-income areas, also have had to contend with contamination of the soil from prior uses or from emissions drifting from adjacent freeways. But while the number and the size of community gardens in Los Angeles has remained small and the barriers have appeared enormous, the range and diversity and sometimes unusual places for gardens and their strategies for survival nevertheless have reinforced the notion of "improvisation" and "reinvention" of land use and place.

Trees

Similar to the lawn and the garden, trees have been powerful landscape symbols. English aristocrats in the eighteenth century, as Henry Lawrence has noted, invested trees "with anthropomorphic attributes" and planted them "to commemorate the births and deaths of loved ones [while] their cutting was protested vigorously by those who valued them as much as their favorite pets." In the process, "trees became symbols of the possession of landed property; their management was a statement of its economic improvement."[47] In the United States, trees associated with forest lands in wilderness areas as well as in the rapidly growing urban centers were assumed to have a strong economic value. When Gifford Pinchot made his transition back to the United States in the late nineteenth century, fresh from managing wooded estates in Germany, he also evoked a language of economic value from the careful management of forest lands. During that same period, trees were also assumed to provide important economic benefits to urban areas as well. Tree plantings, including those seeded as part of the development of urban parks, "were sold on the grounds they enhanced everyone's property," Daniel Rodgers argued in his comparative study of European and U.S. approaches towards urban development. But trees also provided a powerful link to what its varied advocates considered to be a more sacred form of nature with its life-enhancing spiritual and aesthetic associations, whether in forest lands or in urban streetscapes. John Muir's evocation of the forest lands of Yosemite as a "fountain of life" and "holy temple" paralleled the promotion of trees and other forms of green space by progressive reformers and City Beautiful advocates as an aesthetic and live-giving force that provided relief to the otherwise dirty, noisy, sewage-laden, and polluted urban environments where the poor were concentrated.[48]

Though coveted for their role in increasing property values and for beautifying neighborhoods, already by the 1950s the urban forest concept and even suburban trees had begun to lose some of their luster among developers, residents, and local governments. Maintenance in urban areas became a costly problem because of cracked sidewalks, pest infestations, falling branches, and irrigation needs. Meanwhile, the reduction of native species, the importing of exotics or nonnatives like eucalyptus, and the encroachment of developments adjacent to forest lands created fire hazards in suburban and exurban areas. Bureaucracies charged with building freeways, ensuring traffic flow, or establishing and maintaining flood channels were quick to eliminate trees from their landscape planning if such trees interfered with the dominant purpose of those single-purpose agencies, such as moving traffic or channelizing rivers and streams. By the late 1980s, a U.S. Forest Service study of eleven cities indicated a limited—and declining—tree cover in the cities. If a tree grown in Brooklyn appeared to be a rare but welcome sight, as a popular book from the 1940s put it, then a tree in South Central Los Angeles and many other inner-city communities in the early 1990s was just as unusual a sighting.[49]

While the lack of green space, including trees, was not a defining issue with the civil disorders in Los Angeles in 1992 (daily life issues such as food insecurity, police-community relations, and lack of jobs weighed most heavily in neighborhoods like South Los Angeles), they nevertheless stimulated a new interest in trees and parks. Tree advocates had already become, prior to 1992, an important environmental force in Los Angeles, led in part by Andy and Kate Lipkis and their organization, Tree People. Founded in the early 1970s as a response to the dying off of trees in the San Bernardino Mountains, which was caused by air pollution drift from Los Angeles, Tree People initially focused on tree planting as a strategy to fight air pollution and to provide a more aesthetically pleasing urban landscape and eventually began to influence policy initiatives around those objectives. The goal of planting one million trees, for example—one of Antonio Villaraigosa's "greening L.A." platform items in 2005—was initially put forth in 1981 as part of the South Coast Air Quality Management District's Air Quality Plan in response to the need to comply with federal Clean Air Act standards.[50]

Beyond the critical issue of air quality, a number of other social and environmental goals have been touted by urban tree advocates. Gary Moll, a vice president of the nonprofit group American Forests, calls trees the "ultimate urban multi-taskers." This includes their capacity to reduce carbon dioxide pollution, to generate energy savings by cooling buildings like schools, to reduce the urban heat-island effect, to limit the volume of stormwater runoff by providing shade and by reducing paved areas, to lower the decibel level of noise in the city, and to create a different kind of connection to place in areas that have had limited or no green space. Aside from parks, schools, and other easily identifiable and beneficial locations for shade trees, tree advocates have also explored the possibilities for trees to renew vacant lots, traffic islands and medians, and even outdoor parking lots, which remain almost invariably treeless in many urban areas while parked cars bake in the sun. These benefits, researchers note, outweigh the variety of costs associated with planting and maintenance.[51]

The connection to place became a key stimulus for other tree-advocacy organizations in L.A., such as Northeast Trees, which focused on tree planting, school garden development, and other urban greening strategies in the neighborhoods north and east of downtown Los Angeles, including the development of pocket parks and shade trees by the Los Angeles River. Both Tree People and Northeast Trees, as well as urban-park advocates who included tree planting as part of an overall strategy for reconstructing green space in the city, have helped strengthen the notion that urban environmental advocacy could and should appropriately focus on incremental yet still important greening strategies, including in inner-city communities where trees and green space are in short supply.

Parks

The urban-parks issue has been especially noteworthy in relation to this environmental-justice argument about the lack of green space in poor communities, and this argument has translated into a growing advocacy on behalf of more inner-city parks. This environmental justice approach, which focuses primarily on constructing small green spaces in dense urban-core areas, contrasts with the historical view dating back to the nineteenth and early twentieth centuries that conceived of urban parks as "large landscape parks" or "pleasure grounds," such as Central Park in New York and Griffith Park in Los Angeles. These lands were set aside to provide residents with their first or only exposure to nature by simulating a rural or countrylike setting and thereby offering a place to escape the hardscape of the city. Such parks, often several hundred acres in size, were developed as early as the 1850s in places like New York, New Orleans, Cincinnati, and Hartford, Connecticut. Designed to serve as a kind of nature oasis, such large urban parks were initially situated at the periphery of the built-up area of the city, though over time new developments rose up around the parks, reinforcing their strong urban association. Furthermore, many of the sites, such as Chicago's South Park or San Francisco's Golden Gate Park, were selected because they were deemed "unusable for other purposes."[52]

During the Progressive Era and into the New Deal period, a second type of urban park emerged—the "reform park," as park historian Galen Cranz characterized it—to provide public places for recreation and outdoor use, particularly for the low-income immigrant neighborhoods in cities like Chicago, Milwaukee, and New York. This included a "park as playground" concept where the parks were far smaller in size (often just one to ten acres), had some paved areas for sports activities, and tended to deemphasize the connection to "nature" while providing for physical activity opportunities, places for seniors, and a "children's right to play."[53]

The reform parks became less prominent by World War II as urban parks increasingly needed to justify their existence as contributing to the economic well-being of the neighborhood and the city and to the welfare of its residents. Innovations in park use, such as the pageants and communitywide events during the Depression or the vegetable gardens planted in park grounds during the war, kept the reform park alive during the 1930s and 1940s. But from the post–World War II period through the 1950s and 1960s, planners and park designers primarily emphasized the recreational uses of parks, while establishing a less costly, more standardized design—what Cranz called "parkway picturesque"—that included "lawns and spotting of trees and shrubs."[54]

The 1960s witnessed a revival of urban parks as public spaces that included largely spontaneous events such as the be-ins in New York's Central Park. Community struggles to create green spaces out of vacant land or land to be paved over also erupted, symbolized by the 1969 People's Park fight in Berkeley. By the 1980s and 1990s, the reinvention of the urban park had become part of the environmental-justice agenda that included a demand for green space in areas where trash-strewn streets, contaminated land, and freeways that crisscrossed neighborhoods prevailed. Los Angeles, in particular, became an important environmental-justice battleground regarding parks since the city compared poorly with other cities in relation to park acreage per thousand residents, park space as a percentage of land in the city, and park expenditure per resident. Further, resources in the low-income districts compared unfavorably with less wealthy districts within the city, a situation that had been compounded by the passage in 1978 of Proposition 13 in California. That initiative, which limited the amount of property tax available to local governments (which had also been a key source of funding for park development and maintenance), compounded the equity issue. While wealthier areas could access additional funds by establishing user fees or through linkage or impact fees that tied the supply of new parks to fees on development, these mechanisms were less available or not utilized in inner-city communities. Thus, areas that had the greatest need for open space, given the constraints on street life owing to crime concerns, the absence of backyard space to play, and schools that had eliminated recreational periods and had no green space in any case, found fewer resources to purchase land for parks or to maintain the few parks that were already established.[55]

This environmental-justice urban-parks movement extended the classic environmental-justice and civil rights–oriented arguments about unequal resources, risks, and burdens to call for the renewal of urban and community life through access to parks and green space. "The opportunity to establish some green space provided a way for us to feel connected again to our community," environmental-justice leader Penny Newman commented about her group's support of a new park and public space as part of a Superfund settlement regarding a nearby toxic-waste site. Like the Progressive Era social reformers, Depression-era pageant makers, World War II vegetable gardeners, and 1960s counterculture performers before them, urban-park advocates situated their claims around "quality-of-life" concerns, their approach seen as a quest to redefine and resituate nature as well as a sense of place in the city and the larger metropolitan region.[56]

Representing Nature

How nature reestablishes itself in the city can be influenced, as Alexander Wilson suggests, in how we talk about, imagine, conceptualize, and represent it. Nature might be described as a threat, felt as a loss, or seen as a life-giving and healing force, and each approach can influence certain kinds of outcomes. An urban nature agenda, then, needs to take into account how those representations are translated into action.

Urban Nature as Hazard

In the summer of 1995, a heat wave struck the Chicago area. Those who died were primarily the elderly and poor African Americans who lived in urban core neighborhoods in Chicago without air conditioning. The weather had been exceedingly hot, with temperatures as high as 105 degrees. Yet the Cook County chief medical examiner's conservative estimate of 465 deaths was challenged by Mayor Daley, who didn't want the episode to be "blown out of all proportion." The mayor argued that to blame every death as heat-related was not "really real," since many of the deaths could have happened anyway. Others argued that the deaths could also have been caused by such factors as cloudless skies and the urban heat-island effect since highways, streets, tall buildings, and other paved surfaces produced significant changes in temperature in urban areas. But as Eric Klinenberg argued in his book *Heat Wave: A Social Autopsy of Disaster in Chicago*, heat-wave-related social factors such as poverty, housing, governmental response, or cultural differences could not be separated from the heat wave as a natural hazard. This "environmentally stimulated but socially organized catastrophe" exposed social as well as natural fault lines.[57]

The 1995 Chicago heat wave raises the question of what constitutes a natural hazard and how or even whether the natural and the social can be distinguished. Ten years after the Chicago heat wave, the winds and the floods—and the breach of the levees—caused by Hurricane Katrina even more directly transformed the language and representation of what constituted a natural hazard. The physical geography of New Orleans has always defined the nature of this "unnatural metropolis," according to environmental historian Craig Colten, whose book by that name was published just months before Katrina struck. Colten makes the distinction between *resources* and *hazards* that often is made by geographers when analyzing the relationship between nature and human activity. When humans interact with nature to accomplish what would be considered positive outcomes (such as creating levees or damming a river to allow for development), then that relationship is assumed to produce resources, while negative outcomes (as when the levees break, and homes and lands are flooded) can define that same set of relationships as producing hazards. The literature on nature as hazard, in fact, assumes an a priori division between a "natural events system" ("the array of wind, water, and earth processes [that] functions largely independently of human activities and is an object of scientific inquiry in its own right by meteorologists, hydrologists, and geologists") and a "social system" that operates independently of natural events.[58]

But can the two systems be separated? When the 1994 L.A. earthquake (the city's most prominent natural hazard) occurred, the event's most direct effect was the collapse of a section of the heavily traveled Santa Monica Freeway. L.A.'s leadership immediately prioritized rebuilding this freeway segment, and its completion occurred rapidly and without any questions. The earthquake became a major hazard not simply owing to its magnitude but how it affected the city, while the response, given the central importance of freeways as part of L.A.'s built environment, reflected the desire to contain that effect and therefore the hazard. The representation and response to a hazard in a place like Los Angeles, with its propensity for fires, floods, debris flow, and extended dry periods, can be magnified when trying to separate nature from human activity. John McPhee, in his compelling discussion of the destruction to lives and property from the debris flow after heavy rains in the San Gabriel Mountains, argues in fact that the human manipulation of the environment (the "control of nature") is what establishes nature as hazard. Each of these events—earthquakes, rains, and debris flows in the Los Angeles region, hurricanes in the Gulf Coast, and Chicago heat waves—becomes dependent on the social context in which it takes place, while the actions taken in response serve to define the nature and extent of the hazard.[59]

Urban Nature as Nostalgia

When I first considered writing this book, I knew that I wanted to focus on the Los Angeles River and the recent initiatives to reenvision the river and the communities that surround it. I broached the idea with my editor at MIT Press, Clay Morgan, while we waited to be seated at Junior's Deli, a well-known Los Angeles gathering place. A man who was perhaps in his seventies was seated next to us in the waiting area, overheard the discussion, and began to tell us stories about catching tadpoles by the river's edge. His stories were a bit vague, but the nostalgia was striking and not unusual. Part of the power of the efforts to restore or at least modify the L.A. River's current state as concrete channel has been the power of nostalgia, the desire to undo the continuous loss of urban nature.

The nostalgia factor has been identified as one generation's evocation of a loss that came to be experienced as that generation grew older. Raymond Williams situated this sense of loss within generational terms. "The centuries-old appeal to nature has in fact reflected nostalgia for a visual and social world which existed thirty years previously," Williams wrote in *The Country and the City*. Fraser Harrison further connects the nostalgia about cultural loss with "the loss of each species or habitat [amounting] to a blow struck at our own identity." The term *nostalgia* is sometimes situated in negative terms as "a perversion of historical or biographical truth for the sake of cheap or gratuitous feelings," which could also be applied to this presumed loss of nature. But Harrison argues that the word's actual etymology (the Greek word *nostos* means "return home", and *algos* means "pain") provides a deeper and potentially liberating desire for reconnection. The nostalgia factor also suggests that the experience and loss of nature in the city is connected to a sense of place. How people see a loss of nature—or a connection to nature—depends on how they experience the material world itself, whether planting a garden in the city, hiking in nearby mountains, or imagining polar bears in Alaska drowning because of global warming. Although the nostalgia for nature may reflect a sense of loss that incapacitates, it can also, as has happened with the L.A. River, stimulate the desire for change and an agenda of renewal.[60]

Urban Nature as Health Promoting

Perhaps the most frequent association with nature is its healing powers. This is reflected, for example, in the arguments of the green psychologists who have documented how the presence of nature or green space can contribute to a sense of well-being in urban areas. Studies evaluating the impact of green space or its absence in such places as hospitals, prisons, apartment residences, or workplaces have shown that "providing a view, and especially one that includes vegetation, has positive implications for health and well-being," according to Rachel and Stephen Kaplan and Robert L. Ryan, leading figures in the field of green psychology. Conversely, these researchers argue, the absence of nature or the presence of an environment that undermines a connection to nature in the city (such as highways) can lead to symptoms like mental fatigue or road rage.[61]

Within this green psychology paradigm, one striking connection between health and well-being and access to green space has been the development of gardens as a form of *horticultural therapy* for veterans, the homeless, or domestic-violence victims. While these gardens have provided a fresh food source as well as an opportunity to be outdoors in a safe environment, they are seen as directly contributing to the participants' mental as well as physical health. Along these lines, UEPI undertook an initiative in conjunction with the California Department of Health Services to develop gardens and other food-provision programs at women's shelters. Working in the soil, one of the formerly abused participants said that her shelter's garden made her feel "as if the earth [was absorbing] negativity." "The garden gives me air," another shelter gardener commented. "I return to life. I stop now and notice the flowers." But urban greening initiatives are constrained by the limited availability of land for uses other than those associated with highest market value. Urban greening as a health-based strategy is often confined as solace and retreat from the demands of urban life instead of affirming opportunities to "fit into nature." As we discovered with the women's shelter program, nature as a healing force can be a powerful part of an affirmative urban nature agenda linking greening the city with a strategy for health and livability.[62]

LIVELIHOOD AND LIVABILITY

One of the great parodoxes of Los Angeles is that it has come to symbolize the absence and loss of nature in the city while at the same time enjoying a rich abundance of what would otherwise be considered a feast of nature: mountains, ocean, rivers and streams, oaks and cottonwoods, and at one time willows, duckweed, and watercress bordering the Los Angeles River. Nature in Los Angeles has always provided an opportunity for that "constant production of urbanized space." In 1891, a short commentary in the *Los Angeles Times* captured the region's celebration of nature and parallel desire to remake it. "Stretching away from the foot of the hill upon which you stand," the *Times* writer said of this L.A. panorama, "East Los Angeles looks like a vast forest or park, so thickly is it embowered in shade trees. To your left you get a fine view of Boyle Heights. On the north and east the scenery is striking in the extreme. Cutting its narrow passage through the high hills from the north flows the Los Angeles River. You can trace the valley as it opens out toward Burbank, above the mouth of the Arroyo Seco, down which ravine comes the mountain stream of that name from Pasadena, a portion of which city is visible. In the background are a succession of mountains, ending in the Sierra Madre, which from this point appear quite near. There rises in the mind of the beholder the thought: What a magnificent site for a big hotel!"[63]

"Nature has had a mixed career in Los Angeles," writes Roger Keil, a statement that could be extended to other metropolitan regions where nature in the city also appears problematic. Part of the difficulty for urban environmentalists has been the way that urban nature itself is characterized and valued—as places or lands that are set aside from the rest of urbanized space, whether workplaces, freeways, parking lots, houses, schools, or flood-control channels. One difficult and contentious conflict among urban social-justice and environmental-justice advocates, for example, has been the competition for scarce land that is potentially available to build schools, affordable housing, urban parks, or community gardens. Can the development of new green spaces be possible given the framework for urban land-use decisions and the enormous social needs (for schools and housing) not otherwise met? The answer may well depend on how these different needs are addressed by the social movements working on their behalf and in relation to each other.[64]

The questions of what constitutes urban nature and how to advocate for it force urban environmentalists to confront the dilemma of the nature/human-activity divide. Bruno Latour has argued that such a divide has led to the construction of three separate domains—facts (the domain of science), power (the domain of politics), and discourse (the domain of language and culture). Since nature is situated within the domain of facts but outside the human domain of politics and power, he warns environmentalists that the separation between the human and the nonhuman will continually exclude or at best limit how nature (seen as the nonhuman world) is addressed. Thus nature also becomes "victim," a perspective that permeates the thinking of the environmental movement, as Alexander Wilson has pointed out. Nowhere is this more apparent than in an urban context where saving nature had long been assumed by environmentalists to be a lost cause. Even with the environmental-justice movement's embrace of urban greening as an important new dimension of environmental advocacy, the acceptance of the divide undercuts the capacity to rethink how to advocate for—and reinvent and reimagine—nature in the city.[65]

Latour argues that overcoming the divide can be accomplished by creating what he calls "mixtures between two entirely new types of beings, hybrids of nature and culture," that are mobilized and assembled into networks that weave the natural and social worlds—humans and nonhumans—into a seamless fabric that may provide the basis for a different kind of advocacy. One way to define nature, Raymond Williams argues in his book *Keywords*, is to reference it as "the whole material world, and therefore to a multiplicity of things and creatures." "Any full history of the uses of nature," Williams further argues, "would be a history of a large part of human thought." Nature, then, is best understood not as a unitary concept (separate from human activity, outside of history or social life) but as multiple reference points—natures rather than nature, hybrids rather than something that is pure and untouched in its "natural" state. Overcoming the divide by establishing new networks and a common language means that to "respect nature," as David Harvey argues, "is to respect ourselves. To engage with and transform nature is to transform ourselves." A common agenda for transformation needs to be located, and as a number of contemporary social movements have begun to suggest, the concepts of livelihood and livability may provide a useful starting point.[66]

Livability references how we affect the material world around us and how it in turn affects our own lives. Livability can be distinguished from the concepts of sustainability and sustainable development, which often are used to suggest management strategies that help companies find profitable ways to reduce pollution but that fail to challenge the structures of power and the dynamic and logic of the existing production system (including the production of urbanized space). A livability agenda has the capacity to seek changes that can transform that production system from its most local level (creating a community garden linked to an affordable housing project) to its most global (creating a social change across borders model of development, as a number of the World Social Forum advocates have begun to explore).

Similarly, the concept of livelihood encompasses an idea of work that is vested with a sense of ownership, is connected to what is produced, and also embodies the idea of justice or "equitable livelihood." Livelihood in an urban context also refers spatially to where we work as well as where we live and play. As Peter Evans has defined it, "Livelihood means jobs close enough to decent housing with wages commensurate with rents and access to services that make for a healthful habitat."[67]

Livelihood and livability represent both an environmental agenda and a social-justice agenda, a politics of nature and an agenda for social transformation. It can be linked to Bruno Latour's notion of "political ecology" (welcoming nonhumans into politics) and Raymond Williams's concept of "the new politics of equitable livelihood." It is also an agenda for urban change, not for bringing nature back into the city but for reinventing nature in the city as a way to better understand and help transform that material world around us.

2

Community in the City

Community is where community happens.

—*Thomas Bender*[1]

Jogging around a Cemetery, Watching the Film *Crash*

Approaching Evergreen Cemetery, at the eastern edge of the Boyle Heights neighborhood of East Los Angeles, visitors are inevitably struck by the contrast between the extensive, open cemetery land and the dense, compact neighborhood that surrounds it. The cemetery is the largest open space in East L.A. A jogging path extends nearly a mile and a half around its perimeter. The path is well lit and has a rubberized cement pavement (made from recycled automobile tires) that covers the large cracks that had made jogging somewhat like running an obstacle course in the past. This cemetery's open space and exterior jogging path provide Los Angeles with a unique type of community-nature link that is steeped in the little-known histories of the neighborhood and the city.

Evergreen Cemetery, established in 1877, is the oldest cemetery in Los Angeles. Over the years, the cemetery has come to reflect the lives of those who lived in the many communities that have surrounded it. Its plots are segregated now by the gravestones of residents of the ethnic groups that have resided in Boyle Heights over the past one hundred thirty years, but it has also been the home of gang members, drug deals, and sporadic though persistent violence. It has been celebrated for its historic sites, such as the Chinese Cemetery shrine, the oldest Chinese shrine in the United States. It also houses the unmarked grave of Biddy Mason, the former slave who became the first African American woman to own land in Los Angeles and who cofounded and financed in 1872 the First African Methodist Episcopal Church of Los Angeles, the city's oldest African American church. Although the Boyle Heights neighborhood today is nearly all Latino, the cemetery provides "a connection with people who do not [or] no longer live in Boyle Heights but continue to return to the cemetery to commemorate loved ones on anniversaries," as an exhibit on Boyle Heights neighborhood sites by the Japanese American National Museum put it.[2]

Although it is a sacred place, a place of danger, and one infused with a distinctive history, Evergreen Cemetery has always represented an informal community and public space, even while it was in disrepair and dangerous because of how poorly lit and maintained it had become. A couple of joggers who had lived in Boyle Heights all their lives and had used the blocks around the cemetery to jog as best they could decided in 2002 that the sidewalks surrounding the cemetery needed to be fixed and that the area needed to become more attractive as a public space. The timing seemed right. The local councilman, Nick Pacheco, was being challenged by Antonio Villaraigosa, who was making a political comeback after losing a heated race for mayor in 2001. Villaraigosa also spoke of his desire to "build community," "green the city," and in particular reinvigorate the Boyle Heights area where he had grown up. Recognizing the opportunity, several of the residents joined with a group of Latino urbanists associated with the group, the Latino Urban Forum, to create the Evergreen Jogging Path Coalition. By talking to residents throughout the neighborhood and on the sidewalk of the cemetery or adjoining streets, the coalition was able to obtain several hundred signatures in support of their plan. They then created an imaginative logo with a smiling, jogging skeleton in front of a sign that read "Evergreen Jogging Path" and lobbied city officials to establish a different kind of pathway. With the election approaching, Pacheco persuaded the City's Parks and Recreation Department to provide $800,000 in funding for the lighting and the rubberized path, which would have the added environmental benefit of reducing stormwater runoff. Dedicating the running path with the name RIP (Run in Peace), coalition members were enthusiastic about their results. "It takes a village to improve the community," one of the path advocates, resident Nadine Diaz, said, recalling how, as an infant, she was "pushed in a stroller as her grandmother walked around the cemetery." "It took many, many people to volunteer their time and to advocate," Diaz said of the efforts of the coalition. The process itself served as Diaz's village, and the organizing became the vehicle that created community in an area that was filled with community meanings that needed to be recognized and renewed.[3]

Two years later, in May 2005, Antonio Villaraigosa, fresh off his successful campaign for mayor (after having previously defeated the incumbent councilman, Nick Pacheco, in 2003), assembled his transition team. Villaraigosa had become a magnet for the national media, boldly speaking of a Los Angeles that was an inclusive, multicultural city, where the concept of community extended to all groups. Villaraigosa's own story—he had an abusive father; was raised by a single parent; had been a drug abuser, student rebel, union organizer, and progressive elected official; and now was the mayor-elect of the second-largest city in the country and would be the first Latino mayor in Los Angeles since 1872—was continually evoked by the media as a story of redemption and renewal. His success also seemed to capture the idea that Los Angeles was capable of embarking on a new and exciting experiment in community building. Though he liked to call himself a visionary, Villaraigosa was also a shrewd politician who recognized that forces and conflicts that were capable of tearing apart the city's different groups were close to the surface.[4]

Villaraigosa offered one piece of advice to transition team members. Go see the film *Crash*, he advised, suggesting that the message of this recently released film about a multiethnic Los Angeles filled with anger, conflict, and occasional breakthroughs could serve as a cautionary signal about the dangers confronting his new Los Angeles. Eight months before achieving its high-profile, Oscar-winning status, *Crash* was seen as conveying an updated ethnic and racially tinged version of L.A.'s noir story that could be found in Nathaniel West's *Day of the Locust* or the dark films set in Los Angeles in the late 1940s and early 1950s. But Villaraigosa's fears about a real-world ethnic and racial divide were not imagined. Indeed, literally days after his inaugural, violence broke out between African American and Latino students at an inner-city high school, and Villaraigosa rushed to the school to continue to voice his message about inclusion and the need to recreate a sense of community based on difference as opposed to the traditional message of community based on sameness.[5]

The city that the film *Crash* portrayed, in its often over-the-top story line of individual conflict, was a place where differences predominated and could lead to a breakdown in civil society and the loss of a sense of place. For Villaraigosa, the film was less a realistic portrayal than "a catalyst for important conversations among Angelenos about issues of race and ethnicity." Differences, as the new mayor also argued, further provided an opportunity to demonstrate that community could be available even in the most intense and diverse urban environments, where residents are continually pressed to find a common language—and share common spaces. This was a difficult though compelling message. If community is where community happens, then community in a city like Los Angeles needs to be built on these kinds of shifting foundations and has to be reinvented as it—and many cities like it—continue to experience tension and change.[6]

Locating Community

While *nature* could be considered one of the most complex words in the English language, *community* might be as difficult to define. That difficulty is reflected in the changing and uncertain ways the term is used and acted on as well as how its meanings have changed over time. In *Keywords*, Raymond

Williams argues that *community* has been used to identify "the sense of direct common concern" (for example, a community of values or a community of interests). On the other hand, Williams states, *community* also represents "the materialization of various forms of organization, which may or may not adequately express this [sense of direct common concern]."[7] Such materialization might be in the form of groups (such as community-based organizations or CBOs), places (such as geographic neighborhoods), or activities or institutions (a school or a farmers' market or a community garden). When a community is seen as representing an economic or political system or form of governance, it is usually distinguished by size, referencing a local as opposed to a regional, state, national, or global body or set of relationships. For example, a community food system might have a direct farmer-to-consumer relationship that distinguishes itself from an increasingly global-oriented food system, where food may travel thousands of miles from where it is produced to where it is consumed.

During the past three centuries and particularly during the past sixty years, the idea of community has been challenged and, like the concept of nature, has often come to signify a sense of loss. This loss of community has been especially associated with the idea of modernization, including the rise and enormous growth of the cities, the increased power of the state, the new technologies that have transformed organizations and institutions, the entrenchment of commodity- and market-driven influences over cultural and social activities and relationships, and the impact of globalization on literally every aspect of daily life. The concept of community permeates the literature about society and social organization and is strongly associated with the rise of capitalism and urbanization (and more recently suburbanization). The shifting meanings of community have been reflected in Fernand Tonnies's classic distinction at the turn of the twentieth century between gemeinshaft (communal ties reflecting small, homogeneous communities) and gessellschaft (association through market relationships reflected in mass heterogeneous societies) and in Louis Wirth's 1938 influential essay on "urbanism as a way of life" (contrasting the unity of the rural community with the city's "relatively large, dense and permanent settlement of socially heterogeneous individuals"). Richard Sennett has argued that the gemeinschaft/gesselschaft distinction has been used to reference a "contrast in space between villages and cities and within cities between homes and streets, little cafes and big cafes, a knot of neighbors and a large crowd. The more enclosed and inward in each is supposedly the more sociable." Urbanization, in this context, becomes almost by definition "disruptive of communal patterns of social life." More recently, Robert Putnam's discussion of the loss of community in the United States as the reduction or absence of "social capital," Robert Bellah's arguments about the central importance of community in an atomized and alienating society, and numerous other discussions and arguments suggest the continuing power of the concept and its continuing associations with loss and identity.[8]

Similar to the discussion about urban nature, the question of community is often posed in nostalgic terms to identify that sense of loss. The New England town hall meeting gives way to the multimillion-dollar sound-bite electoral campaign, bowling alone replaces bowling clubs, and the family farm and the home cooked dinner give way to industrialized agriculture and fast food. Much of the loss is seen as inevitable—a condition of the modern age, where the metropolis, the market, and globalization become fixtures that have transformed the past, are embedded in the present, and represent the future. To struggle for community, like the notion of fighting to preserve nature, is seen as defensive and even futile. Even the market and globalization's biggest fans worry about this loss of a local autonomy and sense of community. "The tide of history continues to move toward more market organization," former U.S. treasury department secretary and Harvard university president Lawrence Summers commented to a reporter at the 2006 World Economic Forum at Davos, Switzerland. "[But] for this to work out well," he added, "there must be more attention to local disintegration even as we work toward global integration."[9]

The search for community that is associated with the desire to establish a sense of unity based on a homogeneous group, neighborhood, or set of values can also take on a reactionary and even a racist or chauvinist character. "Racism, ethnic chauvinism, and classic devaluation," Iris Marion Young writes, "grow partly from a desire for community; that is, from the desire to understand others as they understand themselves, and from the desire to be understood as I understand myself. In the dynamics of racism and ethnic chauvinism in the United States today, the positive identification of some groups is often achieved by first defining other groups as the other, the devalued semi-humans."[10]

The search for community—in the face of market forces and the overwhelming nature of the urban and industrial transformation throughout much of the twentieth century—has led to efforts at exclusion and community separation that Iris Young warns about, such as the restrictive covenants that became a prominent part of the early twentieth-century Los Angeles housing market. But the search for community has also been associated with community building and efforts at inclusion rather than separation. The settlement house movement at the turn of the twentieth century up through World War I, for example, established what could be considered a social democratic impulse associated with building new community institutions through neighborhood, workplace, and cultural organizations. In the nineteenth ward of Chicago where Hull House was located, the neighborhood (which also included sweatshops inside overcrowded tenement houses) was a victim of the urban and industrial forces that had degraded the conditions of working-class and immigrant life. The Hull House reformers set out to establish a "higher civic and social life" by helping stimulate civic and class organizations, ranging from cultural clubs to unions for women workers. One of the leaders of this social democratic impulse, Florence Kelley, utilized a strategy of mapping the neighborhood by canvassing the houses, streets, and living conditions of everything within a square mile emanating from Hull House, which was itself seen as the center as well as the stimulus for the community building and community empowerment exercise. The mapping, later published under the title *Hull House Maps and Papers,* was itself a tool for advocacy, documenting the degrading conditions of daily life and establishing a new sense of community.[11]

By the 1960s, the desire to locate community reasserted itself in the form of new social movements, political coalitions, and struggles to create new sources of power and social change mechanisms. During the 1970s and 1980s, new community movements also emerged regarding housing and environmental issues that provided new types of institutions (such as community development corporations) and agendas for change (such as the antitoxics focus of an emerging environmental justice movement).

The rise of the environmental justice movement is significant in relation to this question of what constitutes a community and how the approach to community is reflected in ideas and action. As I discuss in *Forcing the Spring*, during the 1970s and 1980s environmental groups tended to divide in relation to geographic focus (national and global versus local and regional), organizational framework (professional, staff-based, expertise-driven versus ad hoc, single-issue, place-based), and demographics (white middle-class, male-led groups versus lower-income people-of-color groups, often led by women). The local, ad hoc groups—which by the 1990s came to see themselves as part of a better-defined and more comprehensive environmental justice movement—established a strong commitment to the idea of community, often in the form of battling against further hazards being imposed on their local areas, including homes and schools where local residents would be subject to those hazards. The groups defined themselves as oppositional movements (arguing against particular hazardous facilities or negative land uses that were planned for their areas) while simultaneously asserting the primacy of community, place, or home. Similar to the approach pursued by the Progressive Era reformers one hundred years earlier, environmental justice advocates turned to the strategy of mapping hazards and assets in their local areas in relation to a proposed facility to better mobilize as well as to establish a type of community identity.[12]

By the late 1980s and early 1990s, a shift began to occur among the local environmental justice groups, due in part to the preponderance of the dynamics of race and class that were visible in nearly all the battles that had erupted in local areas. Race, in particular, came to be seen by many of the groups as the dominant factor influencing the location of hazardous facilities and related negative land uses, whether waste site, freeway, or polluting industrial plant. By the convening of the First People of Color Environmental Leadership Summit in 1991, the language and arguments of the local groups were less about locating and protecting communities than the civil rights arguments about the need to struggle against the discrimination that communities were subject to.[13]

These civil rights arguments, embedded in the language of environmental racism, were limited to the extent that they were often focused on challenging or removing a particular facility or hazard rather than broader community change. Yet these local areas were themselves subject to multiple environmental hazards as well as a wide range of social, economic, and cultural problems that so heavily impacted the quality of daily life. Seeking to broaden the focus, key figures in the movement began to assert the importance of *justice* as the governing metaphor for their movement, allowing the groups to expand their agendas for change. The nature of the environmental discourse also changed. The importance of community in the context of making places more livable was reasserted, and more equitable access to such key needs as housing, education, jobs, health, parks, and transportation was made a key goal. Place-based identities became less the idea of protecting communities from a particular hazard and more the idea of transforming them to become more livable and just places.[14]

While spatial or place-based identities related to a specific geographic area remained important for environmental and social justice groups that defined themselves in community terms, the location of community was as much a matter of *what* established the basis for that community as to *where* it was located. The concept of community still resonated for those focused on issues of justice, often referencing groups, places, goals, and values that resided at the margins of power. A community politics came to be seen primarily (though not exclusively) as a politics of resistance and social-change advocacy rather than a politics of resistance to change.

Such a politics took on a more reactive nature in the desire to keep others out, as Iris Young argued, through an agenda of exclusion, protection of homogeneous neighborhoods or local areas, or enforcement of immigration regulations. Gated communities represented one type of action aimed at exclusion, as did fences along the U.S.-Mexican border or policies that denied residents a driving license, a residence, or access to higher education because of immigration status. Opposition to affordable-housing projects, homeless shelters, transitional housing, or commercial development that could undermine a purely residential neighborhood also became the focus of divisive community actions. These "not in my backyard" arguments were based on a language of "no growth" and hostility to development. They muddied the notion of a community politics and forced a debate over whether community was best represented by homogeneity or diversity. These distinctions were particularly compelling in urban settings where major demographic shifts, massive migrations, and sharp class, racial, and ethnic divides prevailed. Locating community in the city became a challenging proposition in the context of a global economy and a political, social, and cultural order that threatened the very places, groups, and values where community was being sought.[15]

Urban and Suburban Places

The nineteenth-century industrial city, with its "dark satanic mills" and its housing that "sustained disaster [and] degeneration and disablement of lives," as William Blake and H. G. Wells said of their London, helped frame the consideration that urban life was devoid of community—a perspective that survived through much of the twentieth century and continues into the present. Yet already by the late nineteenth century and into the first two decades of the twentieth century, a wide range of social movements—from the progressive reformers, municipal socialists, and the new City Beautiful, garden city, and regional planning groups—sought to identify changes in urban life that reestablished community identities through a process of social, civic, and economic betterment. The brutal conditions of daily life and absence of any countervailing power in the city "was never the whole story," as Raymond Williams put it. The powerful reform impulses in that period were urban based, including the struggle for the vote, the expansion of educational opportunities, the struggle for cultural amenities such as libraries and parks, and the active growth of a municipal as well as civic culture. Beyond the civic reformers, there was also, Williams argued, "the growing organization of the working class itself; the great civilizing response to industrial tyranny and anarchy; the creation of the unions out of the network of urban friendly and benefit societies, and beyond this expression of a new and active neighbourliness, the vision of mutuality as a new kind of society."[16]

The civic and community impulses of the Progressive Era and the New Deal period of the 1930s provided contending visions about the social organization and political and economic conditions in neighborhoods, civic institutions, workplaces, and cities and regions as a whole. Even in Los Angeles, so often characterized as a model of metropolitan development that assumed the absence of community and civic *urban* identities, an active and vibrant set of social movements established a vision of L.A. as a place that stimulated new democratic ideals and "an alternative vision of social justice and an agenda of workplace reform, civil rights, gender equity, a healthier environment, and more livable communities," as Mark Vallianatos, Regina Freer, and Peter Dreier, and I put it in our book, *The Next Los Angeles: The Struggle for a Livable City.*[17]

The Boyle Heights neighborhood in East Los Angeles, where Evergreen Cemetery is located, provides a key illustration of community identities in Los Angeles that were based on inclusion—hetereogeneity rather than homogeneity—and a vibrant civic culture, even as the metropolitan area took on its contemporary form. As George J. Sanchez has described, throughout the 1950s Boyle Heights was a showcase of new immigrants, active civic and political life, diverse cultures, and the struggle to maintain a community of diversity. In 1954, the magazine *Fortnight* characterized Boyle Heights, which was then engaged in a bitter fight over the location of the Golden State Freeway, as one of the most "ethnically dynamic, religiously and politically tolerant, and community proud" communities in the country with its "open air markets, street sellers, Mexican, Jewish and Chinese restaurants, Spanish and Japanese movies, Russian and Turkish steam baths (one of which is operated by an Irishwoman) and Buddhist temples." The area represented "a United Nations in microcosm," as the L.A.-based publication *Frontier* headlined its portrait of Boyle Heights the following year.[18]

Those articles were published at a time, however, when housing developers were actively breaking up neighborhoods through racial exclusion and public-housing programs were under siege and being reconfigured in ways that undermined rather than strengthened community. The concept of a "blighted" neighborhood—which was the term that was applied to Boyle Heights by developers, appraisers, politicians, and planners—became synonymous with the notion of an inclusive, heterogeneous neighborhood ("honeycombed with diverse and subversive racial elements," as officials with the Home Owners' Loan Corporation put it in justifying their "low red"—that is, redlining-related—rating for Boyle Heights). On the other hand, community values and economically stable neighborhoods were associated with areas where "the population is homogenous," as the HOLC system identified its "A rating" criteria. Nevertheless, even as Boyle Heights changed into a predominantly Latino immigrant neighborhood by the 1980s and 1990s, a powerful sense of place still survived, even as other changes (such as the significant increase in fast-food restaurants, freeways that bordered the neighborhood, and the decline of affordable housing) affected the area as well.[19]

Los Angeles ultimately established multiple identities, each involving a different sense of community. The "gaudy, flamboyant, richly scented, noisy [and] jazzy," place that Carey McWilliams wrote about in the 1920s when he first encountered the city stood in contrast in that same period to the place that elected a former Klu Klux Klan member as mayor and where several neighborhoods panicked over the possibility of racial inclusion and passed some of the strongest restrictive covenants in the nation. By the 1920s, Los Angeles was confronting the growing significance of the automobile in transforming the sense of place and the relationships within and between neighborhoods, places of work, and commercial centers. One 1929 University of Southern California study, for example, argued that as the automobile reduced spatial distances between places, it established a broader Los Angeles rather than community identity.[20]

By the early 1940s, however, planners were already complaining that Los Angeles—thanks in part to the automobile, a new imported water supply, and a pattern of development that was transforming the patterns of urban life—was reinventing the concept of sprawl and its relation to the urban core. "Large sections of the city are now deteriorating as people move out to the suburbs in search of space, order, and quiet for their homes," complained one 1941 statement of planners. L.A.'s business and political elite, many of whom had played a pivotal role in encouraging rapid sprawl, worried about the lack of a center and a strong civic identity. They focused on the need to revitalize a downtown area that had lost its preeminent role as the cultural and economic heart of the city and the region. But by the mid- and late 1940s, the changes began to appear irreversible. The earlier suburban-industrial clusters that had contributed to the changing identities of the region were rapidly giving way to the automobile suburbs farther to the east in the new bedroom cities of the San Gabriel Valley, to the north in the San Fernando Valley, to the southwest in places like Lakewood (which itself became the prototype of a city without control of its own resources and infrastructure), and farther south in Orange and San Diego counties beyond an earlier generation of the "black gold" working-class suburbs that had been spawned by oil development. These new suburbs were similar to many of the other suburban developments that were springing up around the country in the post–World War II period, but the rapidity and extent of the suburban development in Los Angeles cemented its reputation as the capitol of sprawl, with the pervasiveness of automobile use a major factor in the strategies for post–World War II real estate subdivision.[21]

The development of the interstate highway system in the late 1950s and 1960s, with its enormous reach both within urban areas and through its role in extending the boundaries of the metropolitan region, further reinforced these patterns that changed the face of Los Angeles and of urban and suburban America itself. The reach of the highway and the automobile also meant that entire new subdivisions in undeveloped or agricultural lands could be pursued, offering residents, some of whom were escaping the urban core, opportunities to construct new kinds of places in the suburbs. Moreover, these were places, it was hoped, that could also reinvent the sense of community.

Many of the debates that have now raged for fifty years about the rise of the suburbs have focused significantly on the question of community. The defenders of the automobile suburbs, such as Robert Bruegmann, have argued that these new residential communities are far "cleaner, greener, and safer than the [urban industrial city] neighborhoods their great grandparents inhabited" and that they also have "a great deal more affluence, privacy, mobility, and choice." Other defenders, such as Joel Kotkin, have argued that suburbs, owing to their low density and single-family home arrangements, have created strong community bonds. "Although widely vilified, embracing a suburban growth model does not have dystopian consequences," Kotkin argues. "New single-family oriented suburban communities can be developed intelligently, with green space around them and a small, denser core for residents who will want that style of living."[22]

Complementing the idea that the automobile suburbs provide a connection to nature, developers have also emphasized the importance of the small-town feel of community, utilizing such names as "The Village" for their subdivisions. "Remember the street you grew up on? Where neighbors knew neighbors," one Internet ad for an automobile suburb in Illinois declared, asserting that this suburb represented "a traditional hometown for families who aspire to the good life." L.A.'s automobile suburbs have often prided themselves on their "traditional Americanism," as Ronald Reagan put it in speaking of the new suburb of Temecula located in the Inland Empire in the area between Orange, San Diego, and Riverside counties. In his 1983 speech, Reagan highlighted Temecula's "can-do" volunteer spirit as essential to its small-town character, although the small-town community ambience of this suburb, which incorporated in 1989, would soon be transformed. This included its rapid population growth, increasingly congested freeways for commuters traveling in multiple directions, and endless malls, as well as a massive

Indian casino at the suburban city's edge. Between 1990 and 2005, Temecula became in fact one of the fastest-growing areas in the country, expanding from 27,000 to 93,000, while housing prices, ordinarily one of the most important attractions of an automobile suburb, climbed to a median sale price of $465,000.[23]

Critics of the automobile suburbs have argued that instead of creating a renewed sense of community, the new subdivisions have undermined civil society and civic institutions through their reliance on more impersonal institutions such as shopping malls and big-box stores, their lack of public space and pedestrian friendly streets, and their daily life requirements that center around the automobile for shopping and commuting to the job. Long commutes (due in part to what planners have characterized as a jobs and housing mismatch) have meant less time for civic engagement. This in turn has undermined the institutions central to the sense of place and connection to community. The reliance on the automobile and the layout of the automobile suburbs also reduces opportunities for physical activity such as walking and biking, with public-health analysts linking sprawl to weight-related health problems. Suburban establishments like big-box stores such as Wal-Mart, for example, have been organized to discourage anything other than cars from entering their spaces.[24]

Such car-centered places and land-use outcomes thus have both environmental and social consequences. They contribute to major air pollution concerns in what had once been undeveloped or less developed areas or agricultural regions such as the Central Valley in California (now one of the most polluted air basins in the country), and they add to the cultural dislocations and suburban anomie that have been discussed and portrayed in the media and academic literature for more than fifty years. Suburban development, in turn, has benefited significantly from key federal, state, and regional policies in housing, transportation, and other infrastructure issues such as water development. Moreover, continued growth of the suburbs seems likely to continue well into the twenty-first century, due in part to the continuation of many of those same policies and related issues of urban decline based on the lack of affordable housing, the loss of an urban manufacturing base and simultaneous rise of a low-wage service economy, and the continuing stresses of urban life.[25]

New Urbanist Communities

The suburban desire for community has combined with the increased recognition in recent years of certain environmental and social advantages of denser urban places to lead to the rise and celebration of a new urbanism, an approach to planning that planner and journalist Bill Fulton has argued "captured the imagination of the American public like no urban planning movement in decades." Featured on the covers of *Time*, *Newsweek*, the *New York Times*, and *Atlantic Monthly*, among multiple publications, the new urbanists, inheriting in part the mantle of the garden city advocates, have sought to develop a concept of community in the context of denser, more walkable, mixed-use developments. As planners, the new urbanists have been vocal advocates of a new type of residential community. As developers, they have, in several highly visible settings, been able to design a number of planned town sites that have become important landmarks of this new approach to the urban, suburban, community connection.[26]

Evoking the idea of the traditional neighborhood—with its small-town character, mixed uses, and emphasis on the ability of residents to walk to the town center—the new urbanists have sought to blend the traditional and the contemporary features of urban and suburban design, often characterized as "neotraditional." Andrés Duany and Elizabeth Plater-Zybeck, town planners who have been two of the new urbanisms's key players as well as cofounders of the Congress for a New Urbanism, have laid out a series of design concepts and codes that represent a template for the new urbanism approach to development. These include a discernible town center; schools and playgrounds accessible within a short walk from the cluster of residences that radiate from the center; residences that include apartments, row houses, and individual homes; retail and commercial outlets either in the center or at the edge of the town that provide jobs for residents and meet their shopping needs; parking relegated to the rear of homes and buildings; the use of traffic-calming strategies such as narrow streets to deemphasize the role of cars; greening strategies such as tree plantings that also provide a pastoral feel to the development; and a form of self-government to address issues affecting the maintenance and character of the neighborhoods. One prominent new urbanist or neotraditional planned town, Celebration, has, as its town seal, a cameo of a little girl with a ponytail riding her bike past a picket fence, trailed by her dog, and the town seal is "emblazoned on everything from manhole covers and light poles to coffee mugs and golf towels." While seeking to capture this nostalgic desire for a "return of community," as Vincent Scully puts it, the new urbanists have also placed their approach within the context of the need to address urban decline, "placeless" sprawl, environmental deterioration, and loss of agricultural lands and wilderness, all as part of one "inter-related community-building challenge," as described by the mission statement of the Congress for a New Urbanism.[27]

Two of the most recognized of the new urbanist community building ventures have been the Disney Corporation–funded town of Celebration, about twenty miles southwest of the Disney World complex in Orlando, Florida, and Seaside, the first of the new urbanist developments, on eighty acres along the Panhandle coastline in Florida, which was conceived as a new type of "American town," by its designers and most visible promoters, Duany and Plater-Zybeck.[28] Celebration's houses first went on sale in 1996, three decades after Disney purchased 27,000 acres of land in Orange and Osceolo counties and created the Rock Creek Improvement District to establish an ability to plan and develop its properties without public review or intervention. This included 10,000 acres of land that was considered ideal to build a new town and that was located off a new exit on the interstate. It was conceived as a signature development evoking Disneylike associations of the "Main Street" environment of its theme parks. In this Disney Company and new urbanist partnership, Disney provided the funding and then hired a series of high-powered architects and designers to model this neotraditionalist town.

Celebration developed all the accoutrements of the new urbanist philosophy: gridded streets, homes built close to sidewalks, a main street filled with retail stores that are walking distance from homes, and pedestrian and bike paths, golf courses, tennis courts, and a health center. Celebration, its owners proclaim, has "the look and feel of a warm and friendly hometown." But critics of the Celebration town argue that the town has contributed to the sprawl in and around Orlando; that it was not a diverse community (93 percent of the residents are white, according to the 2000 census, and, as its Disney-hired chronicler commented, a "solid 80 percent Republican"); that the developer retains control of unsold plots and further development opportunities rather than the community as a whole; and that Celebration's small-town feel is commercialized and Disney-like ("community as commodity" as Alex Marshall puts it) rather than representative of the type of diverse community promoted in the new urbanist literature.[29]

Similar to Celebration's carving out of a new urbanist town from undeveloped land, the town of Seaside was transformed out of undeveloped land inherited by Robert Davis from his grandfather. Along with Duany and Plater-Zybeck, Davis sought to create a newly constructed town that would convey the message of "civilized livability" with pedestrian-scaled streets, native plants and absence of lawns, accessible beach pavilions, and pastel-colored, Victorian-inspired cottages. Considered by its promoters to be "the most successful example of neo-traditional town planning in existence today," Seaside was designed and built with the premise that "America's eighteenth- and nineteenth-century towns remain great models of urban coherence and felicity," as *Time* magazine's contributing editor for architecture and design argues in one of several volumes dedicated to this new urbanist icon. On its twenty-fifth anniversary, the *New York Times* called Seaside "America's most imitated town."[30]

Located on the Gulf of Mexico, Seaside became not simply a successful elaboration of new urbanist principles but a commercial success and an inspiration for a number of other new urbanist town developments, both in Florida along the Panhandle and in other suburbs, university towns, and a few urban housing clusters around the country. Seaside's success helped reinvent the Florida panhandle, which according to Florida new urbanists had once been "derided as the 'redneck Riviera'" but now "assumed a far more sophisticated identity as other New Urbanist towns grow up on the sugar-sand dunes of the Gulf of Mexico."[31]

The model for the town in the film *The Truman Show* representing a self-contained new urbanist type ideal yet inherently flawed community, Seaside today has become a major tourist destination, a second home for wealthy residents from places like Atlanta, and one of the wealthiest and most rapidly escalating real estate markets in the country. Even more than Celebration's "community as commodity," Seaside could be better described as a real estate agent's Fantasyland. Houses that might have sold for $100,000 in the 1980s became valued in the millions of dollars less than two decades later. At the same time, the heavy tourist appeal of Seaside transformed the town's pedestrian-friendly signature commercial and retail center into a congested thoroughfare, with cars parked throughout the town while "SUVs clog streets designed for bicycles and feet," as a 2005 *New York Times* article pointed out.[32]

Seaside's commercial success has heightened the new urbanist reputation, and the Congress for a New Urbanism's annual meetings have become major gatherings of architects and planners, public officials, real estate developers, and assorted policy makers. The 2005 Congress meeting in Pasadena, California, for example, was keynoted by Los Angeles's new mayor-elect Antonio Villaraigosa, who told the audience that he embraced the new urbanist agenda in his approach to planning issues, a vision he has characterized as "stylish density." After Hurricane Katrina, Mississippi governor Hailey Barbour turned to the new urbanists to help redevelop the coastal towns along the Gulf of Mexico, while the Louisiana Recovery Authority hired several new urbanist firms, including Seaside's builders Duany and Plater-Zyberk, to develop "a comprehensive regional vision" for the areas outside New Orleans.[33]

However, new urbanist critics have pointed out that the combination of enhanced market performance with a nostalgic-type neotraditional emphasis on community shuts out those who cannot afford the new urbanist dream.

Moreover, some of the Gulf Coast rebuilding plans of the new urbanists, not too dissimilar to the ideas underlying the Florida Panhandle ventures, have come to be associated with fantasylike development scenarios, including refurbishing casinos. At the same time, they have the capacity to reinforce rather than challenge the radical post-Katrina demographic shifts that severely disadvantage the most vulnerable, low-income residents, particularly people of color.

The new urbanist approach, though, is not monolithic. Some of its advocates and practitioners have focused on urban core issues as well as equity and diversity needs. For example, the concept of transit-oriented development (identifying commercial and residential developments along transit corridors) assumes new urbanist goals, including reducing reliance on the automobile. Similarly, the location-efficient mortgage concept (also favoring reduced distances among work, home, and transit) has sought to use a new urbanist appeal within a housing-market context. Yet the critics of new urbanism have been most vocal about the failure and, in some cases, the reinforcement of a race and class divide and the market orientation or more exclusive nature of the new urbanist development.[34]

One of the more compelling of the approaches distinct from yet also drawing on new urbanist ideas have been the groups associated with a "Latino new urbanism." Latino new urbanists (planners and social- and environmental-justice advocates alike) link the value of density to more diverse and cross-class settings—"*gentification* rather than gentrification," as one of the Latino new urbanists has characterized it. Instead of seeking to develop an alternative to the dense urban core neighborhoods in places like East Los Angeles, the Latino new urbanists have promoted some of the characteristics of the barrio, particularly where immigrant communities have developed, as the best illustrations of a new urbanist vision of a built environment conducive to community life. Far more than their Seaside-like counterparts, these barrio or immigrant neighborhoods, the Latino new urbanists argue, directly exhibit the kind of pedestrian-oriented, compact living, and active and vibrant street life celebrated in the new urbanist philosophy.[35]

The Latino new urbanist approach complements the arguments of those who have challenged the increasing privatization of urban space. The City Repair organization in Portland, for example, has sought to create a type of design art as part of the streetscape to serve as a gathering place and for traffic-calming purposes. City Repair artists, Latino new urbanists, and other urban street-life advocates have focused on how the role and reach of the automobile in urban and suburban life have a contributed to this loss of public space. The French philosopher Henri Lefebvre characterizes this triumph of the automobile in the modern city as hastening the shift from public space to a market-oriented "commodified space." Instead, Lefebvre asserts, people have a right to the city—to establish public spaces, the engine for a new type of sustainable economy, the fulcrum for cultural diversity, and the place where community becomes a force for social change. "The right to the city is not merely a right of access to what already exists," David Harvey argues, "but a right to change it."[36]

The right to the city, a rich and suggestive concept that Lefebvre introduced in the late 1960s following the student and worker demonstrations of May 1968 in Paris, has stimulated an expansive literature on the cross-currents within and among evolving metropolitan regions. Harvey, for one, further argues that instead of the notion of the city "as a thing," urbanization *as a process* provides a better framework for understanding those cross-currents, including the huge migrations that cross borders into the cities that take on a life of their own. These ideas contrast both with new urbanist villages that exclude rather than include through market forces and with the new suburban advocates like Joel Kotkin and David Brooks, who suggest that their new suburbia-blend of the automobile suburb with a new urbanist village—the "go-go suburbia," as Brooks characterizes it—establishes "market-tested cohesive institutions to counteract the segmenting and niche-ifying forces of the age." This struggle to achieve the right to the city and the possibilities of urban transformation rather than the pull of the market and the nostalgia for homogeneous cities and suburbs effectively becomes a struggle about what constitutes community.[37]

The Power of Place

The concept of place also figures prominently in the new urbanist literature and has (like its counterpart concept of community) multiple reference points and different meanings. The connection to place, for example, has represented efforts to exclude when associated with a particular geographic setting. These could include the gated homes in upper-middle-class urban neighborhoods; the suburban subdivision that is celebrated by the "new suburbanists" as "a new sort of landscape that is neither city nor sprawl" but that lacks the ethnic and income diversity of the urban core; or even, despite their intent, the new urbanist villages like Seaside that consist almost exclusively of expensive homes. Places that take on a homogeneous character often put up walls, figuratively and sometimes literally, to maintain the identity of the community within that geographic place.[38]

But a connection to place can also provide a way to open up those locations, to establish a more direct engagement with "the commonplace and the value of the everyday," and to "become actively engaged in their care," as the English organization Common Ground said of its goal to link a sense of place to an environmental and social-justice awareness. How the connection to place gets situated and defined can also establish a more democratic connection to the past, as Dolores Hayden has argued in her book *The Power of Place* and in the advocacy work she undertook through the organization with that same name. "If Americans were to find their own social history, preserved in the public landscapes of their own neighborhoods and cities, the connection to the past might be very different," Hayden argues. One of the goals of the Power of Place organization in Los Angeles, for example, was to recognize the places associated with Biddy Mason, including her gravesite in Evergreen Cemetery, as a marker in understanding Los Angeles's own diverse and complex history.[39]

This ability to "nurture citizens' public memory" by extending the sense of place to working-class neighborhoods and public spaces, ethnic experiences, or women's history makes a connection to urban spaces and to the city itself and challenges the notion of walled-off places and racial, class, and ethnic divides. These connections to place ideally establish what Iris Young calls "the un-oppressive city" where the varied ways that urban life is experienced come to be accepted and ultimately enhance the notion of the right to the city.[40]

Public spaces such as parks and farmers' markets help illustrate the recognition of such differences within this broader, more democratic connection to place. Urban parks not only provide a diverse and complex way to establish nature in the city but also demonstrate how the use of the park may provide varied connections to place, particularly in cities like Los Angeles that have rapidly shifting neighborhoods and populations. Policies around parks development, however, have not always reflected the "right to the city" notion. The development of parks as recreational places during the Progressive Era was pursued in part to accommodate the needs of urban immigrant and working-class populations for public spaces and in part to assimilate those groups through a more generic notion of "citizenship." Parks programs, especially organized sports and related recreational activities, were established to contain and ultimately eliminate differences among groups.

Nevertheless, the different uses of public spaces persisted, even as the amount of public land dedicated to parks and recreation and the ability to maintain them suffered significantly. Research by UCLA urban planning professor Anastasia Loukaitou-Sideris, for example, identified a range of uses for urban parks in Los Angeles among African Americans (social gathering place), Latinos (active uses, especially recreation for families), Asian Americans (quiet contemplation), and Anglo (picnics, playgrounds, clean areas, and a sense of order and beauty). Some parks could accommodate such differences, although park planning tended to try to advantage one use over the others, often defined as the debate between passive versus active uses of park lands. Loukaitou-Sideris argues that where differences are accommodated, urban parks reestablish a sense of place in urban inner-city neighborhoods. The distinction in part resides in how urban parks—and more broadly, the urban environment in neighborhoods—come to be designed and constructed. Utilizing the language of environmental psychologists, Loukaitou-Sideris further argues that this connection to place can be distinguished between environments that encourage social association and interaction (including urban parks and pedestrian-friendly streets) and environments that prevent such associations (including car-dominated streets and landscapes).[41]

The reemergence of urban farmers' markets in the 1970s provides another striking example of the desire for public and community spaces in the urban environment and the potential (though not always the practice) of accommodating differences in the right to the city. Up through World War II, most farmers' markets were centrally located in or at the edge of the central business districts. They functioned primarily as distribution points for regional farmers to move their goods by trucking their produce for direct sales to consumers as well as to retail and wholesale buyers. Farmers' markets declined significantly after World War II as industrialized agriculture, long-distance food systems, and new policies designed to standardize the sale of produce (particularly size and packaging) rapidly transformed the produce market. These changes coincided with the growth of the supermarket industry in the size of stores, numbers of stores, and increased product availability. While supermarkets were able to increase the availability of fresh produce, they also encouraged the shift to standardization and more durable products with a longer shelf life that in turn reinforced the trend toward an industrial agriculture and longer-distance transport of farm produce. This period from the 1950s through the 1970s became the era of the "hard tomato," the shift towards highly processed foods, and the rapid rise of the fast-food restaurant. It also became the era when supermarkets began to abandon urban core areas to establish new, larger outlets in suburban areas, particularly the automobile suburbs where the markets located off the freeway exit and parking lots took up more land than the store itself.[42]

The increasing abandonment by supermarkets of inner-city neighborhoods was primarily responsible for the revival of urban farmers' markets in the mid- and late 1970s. Though still subject to some of the limitations imposed by the standardization rules, this new generation of farmers' markets was established by antihunger advocates who were able to obtain exemptions from the rules on the basis that they were providing a service for low-income residents who otherwise would not have had access to inexpensive fresh produce. In Los Angeles, the first of the new farmers' markets was organized by the Interfaith Hunger Coalition in the city of Gardena in the southwestern part of the county, whose low- to moderate-income residents included an equal mix of Asian, Latino, African American, and Anglo households. The second market was organized in South Los Angeles in what was then a predominantly low-income African American neighborhood south of USC. These markets provided fresh, moderately priced produce for city residents as well as income for farmers from nearby counties, many of whom were not able to effectively participate in the long-distance marketing that characterized the dominant industrial and increasingly global food-production system.[43]

From this modest renewal as a low-income fresh-food access strategy, farmers' markets began to take off in the 1980s and 1990s. This was due to a number of factors—the increased sales of organic produce, the growing dependence of small local and regional farmers on direct sales as their primary economic strategy, the growing interest in healthy and fresh food, and, perhaps most significantly, the recognition that farmers' markets were *socially beneficial* as one of the few public spaces in the city that provided important opportunities for social interaction. Different cities and neighborhoods welcomed new markets precisely because they were seen as "community building" or at least community enhancing. Farmers' markets became "community events," as one researcher put it. However, farmers' markets were also critical as economic enterprises for the participating farmers, and those that flourished or at least survived did so in part because of the price the farmer could command as well as the volume of sales. As a consequence, by the 1990s, farmers' markets had shifted from their low-income fresh-food access goal to become a thriving middle-income urban amenity that celebrated a connection to place and that supported local and regional farmers who were dependent on direct marketing sales.[44]

Although criticized for their evolution into a type of niche market for middle-income and upper-income residents, farmers' markets still attract and serve multiple publics. They are valued as a source of fresh and diverse (and seasonal) food and as public places that accommodate different kinds of residents and shoppers. While many of the markets attract sizeable numbers of Anglo middle-class customers, several of the low-income markets continued to operate (including in Los Angeles the two original markets in Gardena and South L.A.), while a number of mixed-income markets have flourished. These markets are often located in what can be called "bridge communities"—markets that border different neighborhoods that vary demographically and by income but that attract residents from all of the surrounding neighborhoods. Aside from differences in income and ethnicity, farmers' markets have also become places that attract the young and the old, weekly shopper and browser. Despite only limited support from funding or other public-policy mechanisms, farmers' markets in Los Angeles and other cities have become a symbol of the right to the city. They are democratic, public spaces that recognize and welcome differences as well as common experiences—farmer and shopper, Spanish speaking and English speaking, and consumers of Asian vegetables, habanero peppers, kale, and fingerling potatoes alike.[45]

Community Connections

The Media

When I first arrived in Los Angeles in 1969, I quickly discovered that the *Los Angeles Times* and its owners, the Chandler family, had long assumed the role of economic and political power broker for their city. Beyond that role, the paper and the family had also long sought to shape and thereby represent the community identity of Los Angeles itself through its coverage, its civic roles, and its boosterism. Yet the history of the newspaper and its publishers had yet to be told in a full-length biography. What better way to learn about the region and the ways that it was changing could there be, I convinced myself, than to tell that story?

As my colleague Irene Wolt and I began to research that history, we began to feel, as historians often do, like explorers. Unearthing documents and behind-the-scene activities from interviews helped us understand the paper and the family's assumption about their role in shaping Los Angeles. In one instance, we arranged to speak to Norris Poulson in the automobile suburb of Tustin in Orange County, where he had retired. At one point in the interview, Poulson, the one-time congressman who had successfully run for mayor of Los Angeles in 1953, stopped the conversation to rummage through some files at the back of his house. He came back chuckling, waving a letter and suggesting we read it and then take it back with us to Los Angeles to use for our book. The letter was addressed to Poulson ("Dear Norrie," it began), and it was written by Norman Chandler, the *Times* publisher, on behalf of a self-selected group of key business figures who saw themselves as the city's pre-eminent power brokers. This group wanted Poulson to run for mayor, and the letter described all the perks that would be awaiting Poulson, including "a Cadillac to strut around in." "There really wasn't much choice in the matter," Poulson told us.[46]

This power-broker function, masked as community booster, remained prominent at the *Times* for another decade, and it was a role that was not dissimilar to that played by other newspapers that had also been dominant in their region's market. While newspaper competition in those markets had also provided a modest array of different perspectives about a region's identity, reflecting the different audiences that the papers appealed to, by the 1960s and 1970s such competition had significantly dwindled. Many cities were left with either a single major metropolitan newspaper or a joint publishing arrangement that also tended to diminish the differences in coverage and outlook. Yet concentration of ownership establishing major newspaper chains as well as cross-ownership between different media diluted the role of particular newspapers in providing an identity for any particular urban area. As media conglomerates grew in size and reach, the role of a newspaper in shaping a regional and community identity also began to wane.[47]

This loss of local or regional identity could be seen in Los Angeles, where the *Los Angeles Times* and the Chandler family witnessed a change in how the paper viewed its role as the region's voice. That change had already begun to occur during the 1960s and 1970s as the Times Mirror Company itself became a diverse company (that included other newspapers as well as book publishers and television stations in other cities) and the *Los Angeles Times* sought to establish itself as a prominent national and international oriented paper. While we were researching our book on the *Los Angeles Times* and the Chandlers during the 1970s, a common complaint we heard from some of the staff was the paper's decreasing ability to cover Los Angeles issues, local neighborhoods, a changing regional economy, and the civic and cultural identities in the city that included major demographic changes. During this period and the next three decades, the paper experimented with various methods to establish a parallel, albeit more limited, community identity, including zone editions in different regional areas (such as the San Fernando and San Gabriel Valleys), a separate edition in Orange County, and regional bureaus within the overall metropolitan region, but these methods achieved limited results that eventually led to their abandonment. By the time the company and the paper were sold to the Chicago-based Tribune Company in 2000, the change in ownership came to represent more the culmination than the occasion for this shift in role away from city builder and regional voice.[48]

Even the issues that affected the vast metropolitan area that stretched across city and county lines, watersheds, and the changing suburban and undeveloped land boundaries were not adequately covered. For example, when I became a member in 1980 of the board of directors of the giant water wholesaler, the Metropolitan Water District, an agency whose decisions had powerful land-use and regional-development implications, I discovered almost no *Los Angeles Times* (or other media) focus on the MWD. Moreover, the common assumption within the MWD about the role of its public information office was that it should keep stories out of the media, an approach that was easily accomplished as the regional identity of the paper became blurred and coverage shifted away from this regional role. The change of ownership to the Tribune Company did not alter the nature of this failure of coverage and the absence or decline of community or regional identities. It only reinforced the idea that a *Los Angeles Times* was simply part of a corporate entity that appeared to have no more organic connection to Los Angeles other than through its name and its fading history.[49]

Los Angeles Times coverage, similar to coverage in other metropolitan newspapers and the various chain outlets, also helped delimit the nature of the debate about national and global issues. While the newspapers argued that they provided an objective and unbiased assessment of events as they unfolded, their primary frame of reference for coverage—their use of sources (predominantly people in power) and their acceptance of dominant themes (for example, the triumph of the market and of globalization)—became the basis for identifying the range of perspectives about those events. Broadcast media have tended to be even further limited in how events have been covered and how perspectives and debates are framed, due in part to how the technology has been utilized (it usually is focused on particular settings and sources such as Congress, the White House, or the military) and how audiences have been solicited and defined. Local or regional media outlets (both television and radio) have been even more focused on delivering audiences to their primary constituency (advertisers) that preclude a more open-ended or diverse frame of reference about what constitutes an issue and how a community might be defined and covered. For television and radio, a demographic doesn't reference a particular group with a community or social identity as much as a group of potential consumers evaluated for their willingness to buy any set of particular products. Community in this instance continues to represent a market orientation.[50]

During the 1960s and 1970s, the rise of new social movements associated with such issues as the war in Vietnam, civil rights, the role of women, and environmental issues helped stimulate the search for an alternative media that would challenge the type of coverage (and absence of coverage) and the civic, cultural, and social voices that had been left out of these dominant media frames. This alternative press sought to position itself as a community voice, partly by defining community as representative of various disenfranchised groups and alternative perspectives. Challenging the notion of objective assessment as disguising the support of the existing sources of power, the alternative press sought to establish an advocacy role for the communities that it was hoping to represent and ultimately empower.[51]

By the 1980s and 1990s, however, this type of advocacy voice in the United States either had become muted through its own commercial orientation or eventually disappeared as the alternative press struggled to survive in a changing political environment. However, by the new century, new alternative press strategies began to emerge through the use of the Internet and a growing international network of independent media voices that first emerged with the meetings of the World Social Forum in the 1990s. The concept of advocacy reemerged, although some of these press strategies were also seized by conservative media outlets led by Rupert Murdoch's media empire. While the events of September 11, 2001, established a focus on news as patriotic duty and the Iraq war created what some critics argued were "weapons of mass deception" on the part of the media, the events of Hurricane Katrina demonstrated that even an established medium (such as the *New Orleans Times-Picayune*) had the capacity to redefine its role as a community voice seeking to reestablish community in the context of the need to rebuild the city. "What's wrong with subjectivity in journalism anyway?," one *Times-Picauyune* reporter asked in the wake of Hurrican Katrina. "It works in other Western democracies, where most newspapers openly stake out some wavelength along the political spectrum. It's intellectually honest. And it harkens to the best traditions of advocacy journalism."[52]

The Hurricane Katrina events also coincided with the revival of a new media-reform movement that had been stimulated by a proposed action by the Federal Communications Commission to allow for more cross-city and cross-media ownership, diluting whatever remaining independent local and regional outlets that could still be found among broadcast media. Furthermore, the desire for change could be seen in the proliferation of voices through the Internet, the push for low-power radio, and the increasingly influential role played by new kinds of interactive media, including the blogosphere. While blogs provide a venue for a wide range of participants—including senators, corporate executives, and military supporters but also individuals and groups without power and with alternative perspectives—they are especially able to diversify the nature of information by employing such features as discussion boards, live e-chat, and social networking tools that make them far more interactive and constituency-linked than traditional media. These "netroots," as they call themselves, have become especially noteworthy as an example of a new media that gives access to groups that have no place within and through the established media frames. Despite their limits—in fully reaching across class, racial, and ethnic divides, in providing misinformation, and in their lack of accountability—the netrooters and other media reformers are nevertheless creating the potential for a renewed debate about the role that the media should play in defining community and the community interests that they should serve.[53]

In addition to the netroots, the increasing prominence of ethnic print and broadcast outlets has created linkages for and among constituencies that would not otherwise be available through the established media. For example, the mobilization of immigrant-rights demonstrators in the spring of 2006 highlighted the community-linked role and growing audiences of Spanish-language media, including, for example, Spanish-language station KSCA-FM in Los Angeles, which has the largest audience of any radio station in the region. The immigrant-rights demonstrators have also effectively used the Internet to make information about events available and largely bypass efforts to gain access to the large established media such as the *Los Angeles Times* or the major television broadcast stations. Ultimately, the size, scope, and intensity of the 2006 demonstrations took the established media by surprise and underlined their lack of connection to the extensive community mobilization that had taken place.[54]

Higher Education

Questions concerning public roles and community connections, including how ideas and information are conveyed, have also preoccupied higher-education institutions. The relationship of these institutions to their surrounding communities, often characterized as the town-gown divide, has long been a subject of debate and tension. The debates over the university's public role have extended to questions about what the nature and purpose of academic discourse are and whether what happens inside the academy and how research and the educational processes are conducted have any reference to nonacademic settings or influences. The idea of a separate and discrete ivory tower culture or of an academic freedom that reflects a process of inquiry and activity protected or set apart from the outside world has often failed to acknowledge nonacademic influences regarding how the university functions. This includes funding sources (whether public or private) for research and institutional operations and the role of trustees (who are themselves often representative of business, political, and cultural elites) in shaping decisions. It is also reflected in how academics themselves constitute a source of power and influence *in* a community and the larger society.

Part of the notion of the insular academy can be traced to the development of the research university in the late nineteenth and twentieth centuries and the degree of specialization and organization by discipline that created an academic language that is largely inaccessible to outside publics. Research universities in this period also sought to isolate themselves from the "turmoil of the city," as Johns Hopkins University president Daniel Coit Gilman argued in an 1898 discussion of university issues. This approach contrasted with the Progressive Era orientation of some higher-education institutions and their leaders who defined the city as a laboratory for study and a place to engage in social-reform and social-welfare activities.[55]

However, the direction of the university during the twentieth century emphasized the *expertise* function of the academy, an approach that further reinforced the growing specialization and discipline-based focus that underlined academic research, writing, and public activity. Even many of the Progressive Era academics pursuing a service orientation or public role promoted the notion of discipline-based expert knowledge as the basis of the university's role. As such trends intensified during much of the twentieth century, the notion of a public role for academics diminished. Public intellectuals, as

Russell Jacoby has argued, came to be seen as limited to those outside the university, given the inaccessibility of academic language and ideas.[56]

Yet in the post–World War II period, universities were becoming increasingly engaged, not so much in relation to a public role but in pursuit of the agendas of key government and corporate sponsors. Massive new funding streams for university researchers, particularly in the sciences and engineering fields, provided academics with a direct stake in and influence around such issues as the cold war, the space race, the green revolution and transformation of agriculture, and the introduction of thousands of new chemicals. Particularly during the post–World War II period, when substantial new funding was distributed by the government (especially the military), by corporate sources, and by private foundations, academics became junior partners of the major national and global economic and political players in setting agendas and exercising power.[57]

This form of engagement with power contrasted with the dynamics of the town-gown relationships in urban core settings. By the 1950s and 1960s, higher-education institutions located in these urban areas—particularly those adjacent to depressed or vulnerable neighborhoods, such as the University of Pennsylvania in Philadelphia, Columbia University in New York, the University of Chicago in Chicago, Connecticut College in Hartford, or USC in Los Angeles—were becoming increasingly preoccupied with the higher-education/community relationship. This was underlined by the uneven power balance between a richly endowed institution that was seeking to carve out a safe haven for its students and faculty and neighborhoods that were subject to real estate manipulations, gentrification, and other impacts (such as traffic and air pollution) because of the presence and the actions of the college or university. Universities became landlords, real estate blockbusters, community economic development players, and politically influential entities that sought to guide development scenarios in the areas immediately adjacent to their campuses. In the process, universities established a type of de facto urban agenda—one of protecting their own interests in the face of urban blight while neglecting or in some cases extending the widening divide between neighborhoods (and within neighborhoods) as well as the increasing income, racial, and ethnic divisions that characterized every metropolitan area.[58]

The decline of neighborhoods adjacent to inner-city campuses contributed to the beginnings of a change in perspective among university administrators who were fearful that prospective students might shy away from the urban location of the university (a situation particularly pronounced in Los Angeles after the 1992 civil disorder). A number of academics also began to argue that the university needed to engage with those without power as opposed to its engagement with power, which characterized so many of the academic research agendas. Some of the change came to be reflected in the growing demand for "community-service learning," less a pedagogy than a notion that the university had a responsibility to provide services to groups and communities that were otherwise limited in resources.[59]

Though popular among students who joined service organizations and clubs, community-service learning has been perceived by most key institutional players as a "soft" form of academic activity that contributed little to academic discourse and negatively impacted tenure and advancement among its faculty advocates. Organizations that formed during the 1980s and 1990s, such as Campus Compact, sought to revitalize the service tradition of the university under the new banner of service learning. Such an approach, while considered a marginal academic activity by many of the faculty, became popular among administrators, especially for urban schools, as a way to demonstrate university commitment to improve blighted neighborhoods, including those adjacent to the campus.[60]

As the service-learning movement struggled to gain more acceptance as an integral part of the academic system for teaching, some academics, particularly in the social sciences, health professions, and environmental fields, began to develop a more comprehensive and research-oriented focus that shifted the framework for academic activity from the notion of *service* (which often implied a nonequal relationship between the service provider and the client) to a *partnership* concept (which proposed a more equal, interactive relationship) that blurred the lines between researcher and researched. These partnership approaches were rooted in two important research traditions. The first was the action-research model first developed in the 1940s by Kurt Lewin, which sought to involve people affected by a problem to engage with researchers in practical problem solving. The second model, participatory-action research, had its roots in the 1970s popular education movement associated with Paolo Freiere and other civil-society movement advocates. These critics of the "colonizing" approach associated with research in developing countries also sought to "break the monopoly over knowledge production by universities," as Meredith Minkler and Nina Wallerstein put it. Most of the activist-oriented participatory-research models were established outside the university, particularly in developing countries, though a handful of community-education centers in the United States, including the Highlander Center in Tennessee, were also involved in seeking to empower community members as knowledge bearers and potential researchers. Partly influenced by this action and community-empowerment approach, a new approach towards research within the university also began to develop in the 1990s. New research orientations, such as participatory-action research and community-based participatory research, began to take root and were responsive to the more assertive role of community-based organizations.[61]

This new scholarship of engagement, as the late Ernest Boyer called it, sought to define the academic's role as a social-change agent, "one dedicated not only to the renewal of the academy but, ultimately, to the renewal of society itself." These new research orientations were focused on changing the power dynamics between academics and those who were subjects of the research. They also recognized that the traditional methods of research that were based on unequal power relationships were also contributing to the mistrust of the researcher—and the university—as being less interested if not hostile to a social-change agenda. By reorienting the nature of the research as a shared process (in its design, methods, and implementation), a redistribution of the balance of power "between the observer and the observed" could take place, providing new sources of knowledge and a different orientation toward community. The biggest stumbling block remained how academics, like journalists or other professionals in fields like the environment, came to see their own role—as observant outsiders or as participants in the process of creating change and establishing new ways of thinking about and acting as members of a shared community.[62]

Nature and Community on Santa Catalina Island

Callie White, a student of mine who was born and raised on Santa Catalina Island, twenty miles off the coast near the Los Angeles harbor, had a dilemma. A strong environmentalist, she shared some of the goals of the Santa Catalina Conservancy and had interned with this organization, which controlled nearly 90 percent of the land on the island through the trust arrangement of the Wrigley family, which had bought most of the land on the island in the 1920s. Like other land conservancies, the Santa Catalina Conservancy was most interested in protecting the island's ecosystems. However, its approach more often than not failed to include the issues and needs of the island residents, many of whom were dependent either directly or indirectly on the Conservancy for their own livelihood.

As an environmentalist, Callie was sympathetic to some of the Conservancy agenda. As an island resident, she was concerned about the unequal power relationship and the unwillingness of the Conservancy to listen and learn from Island residents' own understanding of what in fact represented a complex relationship between nature and community on the island. As a student who decided to undertake a major research project on these issues, Callie sought to establish a research partnership with both Conservancy members and island residents, exploring an agenda for change that, she felt, could ultimately benefit both. She characterized her approach as "place-based research" and discovered how valuable resident and Conservancy insights could be in addressing—and potentially overcoming—a nature and human community divide.[63]

As I observed and commented on Callie's project, it struck me that the opportunity to discover the link between nature and community in this isolated island offshore from the city suggested how such links in the far more complex and implacable setting of Los Angeles itself could be pursued as well. While such community-researcher approaches still needed to be more fully explored, the opportunities were available to create agendas to promote the idea of a more livable, sustainable, and just place for nature and community alike in a city that had long been considered hostile to both and that increasingly defined itself as a global city without a connection to that sense of place.

3

Water for the City

> Los Angeles joyously announces to all the world that she has just begun to grow; that she has hitched her wagon to a star, and nothing can stop her progress; that she is making preparation to take care of all the people who are looking longingly toward the shores of the Sunset Sea and are coming to make their homes in this blessed land.
>
> —"*Viva Los Angeles,*" Los Angeles Times, *September 7, 1905*[1]

Water Rebel

When the meeting of the fifty-one-member board of directors of the Metropolitan Water District came to order in the board room at district headquarters on Sunset Boulevard just north of downtown Los Angeles, I was not at all prepared for what was about to take place. It had been about six months since my December 1980 appointment to the MWD board to represent the city of Santa Monica, one of the original members of this regional water organization. Despite my previous research and writing about water issues in the West, I hadn't yet fully appreciated the culture of the organization and the role it played in the region.[2]

The Metropolitan Water District of Southern California is the most important organization dealing with water and land-use issues in Southern California as well as a powerful force influencing water issues in California and the West. MWD board members and staff liked to say that the organization had only twenty-seven customers (now twenty-six), which were its member agencies, referring to the different cities, smaller water districts (themselves mostly composed of cities within a particular subregion or water basin), and one county entity (the San Diego County Water Authority, representing all of San Diego County). MWD is a wholesale rather than a retail organization, managing and allocating two major water supplies—one that imports water from the Colorado River (through the Colorado River Aqueduct) and one that transports imported water from the Sacramento River (through the California Aqueduct). MWD does not deal directly with those who use or consume water. Instead, retailers (like the city of Santa Monica and the City of Los Angeles' Department of Water and Power) provide water to their customers. The number of board members allotted to each agency is based on property tax valuation. When I joined the board, Santa Monica, as a small city, had one board representative, while the City of Los Angeles had eight, and the San Diego County Water Authority had six.[3]

I had already been tagged as controversial at the time of my appointment. In the late 1970s, the city of Santa Monica, a coastal city with a population of about 85,000 at the time, experienced escalating housing costs and a protracted struggle around rent control. The City Council was split among three progressive council members who had been elected on a pro-rent-control slate and three other members who included someone who was concerned about environmental questions and felt that water was an important area that needed attention. City officials and council members felt that Santa Monica's previous representative to the MWD board of directors had not been forthcoming with them about MWD issues and simply went along with the policies that were adopted, more than willing to participate in an organizational culture filled with perks and self-importance, given the enormous budget and reach of this giant water organization. Chris Reed, the conservative environmentalist on the Santa Monica City Council, convinced the council to remove its MWD board representative and to look for someone who would be more communicative and more willing to question the process and substance of MWD decision making. As a Santa Monica resident who had written about water and politics in Los Angeles, I had originally been solicited for the position by Dorothy Green, one of a handful of environmental advocates in Los Angeles who was focused on water issues. The board position seemed like an invaluable and unusual opportunity to learn about and weigh in on questions of water and politics in Southern California from a unique vantage point. After some behind-the-scenes maneuvering, including by MWD officials, a majority of the City Council members agreed to appoint me.[4]

Joining the MWD board of directors felt like a culture shock. I was thirty-six at the time, and by joining the board I lowered the average age of its members to sixty-seven. The District staff had as many as three nurses on call at board meetings in case of a health emergency. The board meeting room consisted of an elevated three-tier horseshoe-shaped seating arrangement for board members who faced each other. They spoke into microphones that they could switch on and off to get permission from the board chair to speak. Seating for the public was arranged in rows extending to the back of the room in a manner that suggested that members of the public were outsiders looking in at a meeting that seemed almost private. Most of those in attendance were staff from the member agencies or consultants and construction and engineering specialists who were there because of a particular issue to be addressed. There were no press representatives, no elected officials, and hardly any community members in attendance (with the exception of one elderly man who saw himself as a watchdog but who was barely tolerated and was often smirked at for his difficulty in using the appropriate technical terms and mastering the obscure language of water policy).[5]

One of my first lessons was to learn the significance of the words *water industry*. This rather obscure term, which was used primarily by those who considered themselves to be a part of it, referenced the various groups, both private and public, that met, deliberated, made water-related decisions, and pursued water projects that, in the context of the western United States, primarily meant the development and transport of water from where it was abundant to where it was scarce. Here were the members of a public agency who did not distinguish their own public role from the goals and approaches of the array of private interests who were also assumed to be card-carrying members of the water industry. These private-interest groups included the real estate developers who needed the water to ensure development; the financial, construction, and engineering interests that built the water projects that delivered the water; the agricultural interests around California and the West that secured the giant share of water supplies flowing through the West; and the water lawyers and lobbyists who drafted the rules and established the language for laws and policies that established the ownership rights to the water and governed how water could be mined, transported, delivered, and used.[6]

The MWD's key role was its ability to secure distant supplies of imported water for a region that had long since associated the need for new water sources as the basis for its continued growth. The issue of the moment was a referendum to be voted on the following June that would prevent implementation of recently enacted legislation calling for a new conveyance facility, the Peripheral Canal, that would bring additional water from the Sacramento Bay Delta, through the State Water Project's California Aqueduct, to the large farms on the west side of the Central Valley, over the Tehachapi Mountains, and into urban Southern California. To the MWD board and staff, a referendum vote against the Peripheral Canal threatened to undermine the painstaking negotiations that had led to a classic water deal, not dissimilar from the original legislation that had established the State Water Project in 1960.[7]

Before joining the board, I had written about the Peripheral Canal issue as illustrative of how imported water sources had made possible the development scenarios that had expanded the boundaries of urban Southern California during the previous four decades. After joining the MWD board, I became its only member to oppose the canal. Even more concerning to the

MWD board and staff, I had publically voiced my opposition, undermining the assumption about a Southern California consensus regarding the *urgent necessity* of making available this source of additional imported water.

While the Peripheral Canal raised a series of complex issues about water-management strategies, environmental protection, and water-quality considerations, the argument about the imported-water link to growth and development was most compelling for its proponents. Without additional imported water, board members argued, Southern California was in danger of facing serious decline. "We'll see people needing to drink water out of their toilet bowl," one board member exclaimed during a debate I had with him at a community forum. On the other hand, the successful completion of the Peripheral Canal constituted what MWD board chair Blais called "the final solution."[8]

As the board members took their seats for the meeting, I happily waved to my in-laws who were seated in the back of the board room. They were visiting from their small town, Dundee, Illinois, and did not expect what was about to occur. Then the chair, a big, gruff in-your-face type of man named Earle Blais, called on one of the senior members of the board, a major real estate developer named Preston Hotchkiss. I realized afterward that this was a prearranged maneuver. Both Blais and Hotchkiss were among the board members who were most furious about the board's loss of consensus on the canal and the challenge to the imported-water link to growth and development. "We have a board member who has a conflict of interest," Hotchkiss began, arguing that since I was writing about water issues, including the Peripheral Canal and its urban expansion implications, I was deriving income and therefore acting in an illegal manner. Blais then opened the floor to a series of attacks by board members who argued that I was a "water rebel," someone willing to undermine the very health and vitality of the region through its water lifeline. Even one of the L.A. board members who considered himself a long-standing progressive berated me with a reference to Woody Guthrie. "He loved water projects," the board member said of Guthrie's series of songs about Bonneville Dam during the New Deal.[9]

Although the board members threatened to take immediate action on my presumed conflict of interest, this never came to pass. Although I was just one of fifty-one members on the board, there was no longer a consensus in the region. The argument about the continuing need for imported water was not quite as fear-inspiring and motivating as it had been on countless other occasions in Southern California's history of support for new water development. It also became clear to me that the board's sense of violation was due to the fear that the consensus on water development was coming to an end. The water industry itself was no longer a monolith, whether in relation to those who developed and delivered water to the cities and the farms (like the MWD) or to those who wanted to keep the water off the land (like the Army Corps of Engineers and the flood-control bureaucracies that had long controlled the flow of the L.A. River). To change the terms of debate was perhaps the greatest threat for a water industry that was beginning to experience duress despite its long-standing dominance of all matters pertaining to water. The very assumptions about how to manage and control its flow—whether for a city like Los Angeles, a region like Southern California, the lands and places west of the 100th meridian, or the regions from the Mississippi River to the Gulf of Mexico—were now being challenged. Water had become contested, and a new type of language about water and what it represented had also begun to emerge.

The Culture of Water

If *nature* and *community* are complex terms whose meanings reflect the ways in which they are talked about, designed, managed, and socially constructed, then *water* also has its own wealth of complex meanings and social contexts. Water can be seen as healing. It can be portrayed as powerful and destructive. It is known to flow towards the sea but also uphill towards money. It has a life of its own but has also been transformed into a market commodity. It is considered both public good and private resource. It ultimately has strong cultural associations and rich historical allusions, but it is also deeply embedded in political and economic agendas. By recognizing that a culture of nature can influence how issues about nature in the city are defined and adopted and that the language about community can influence how community-based issues are addressed, then, similarly, it is important to explore the ways in which language and perceptions about water are framed and how such a culture of water can influence how water issues are considered and pursued.

When I joined the MWD board, for example, I was continually struck by how often the term *wasted* was applied when discussing a water source that was not captured for its social and economic uses. For the California water-industry participants, the water that flowed out to San Francisco Bay through the Sacramento Bay Delta was wasted insofar as it did not become a water source available for uses elsewhere through the State Water Project or the federal Central Valley Project. For water users from Nebraska to the Texas Panhandle who depended on the Ogallala Aquifer as their primary source of supply, rainfall that percolated into the ground but was not then pumped up to be used as a source for agriculture, mining, or municipal and industrial uses was wasted rainfall. In contrast, environmentalists argued that water that did not replenish aquifers or restore ecosystems in wetlands, marshes, or the Sacramento Bay Delta was water subject to a system of control and management that harmed nature. And for the flood-control engineers, rainfall in areas where the river flow would spill over its banks and threaten the development alongside the flood plain was dangerous and threatening if the river was not controlled by such appropriate measures as channelization, dams that served as flood basins, or other engineering strategies.

The focus on the quality of the water—whether rivers or streams, groundwater or surface water, drinking water or water used for irrigation—has also influenced the way we think about, describe, and act in relation to water issues. The industrial city of the nineteenth and early twentieth centuries, for example, transformed urban rivers and streams into highly polluted discharge outlets for the untreated wastes flowing from industrial and urban activities. This in turn created a language of water as contaminated and suspect. While twentieth-century water policies began to focus on ways to reduce or mitigate those practices, the language about urban water, particularly the above-ground sources of water that flowed through the city and remained subject to various discharges, still reflected the historical legacy of pollution. A distinction came to be made between drinking water (assumed to be safe after the introduction of chlorinated treatment strategies early in the twentieth century) and the water subject to discharges that flowed in streams and rivers and into the sea. Rivers and streams were nature's sewers, the chemical industry would still argue as late as the 1950s, asserting that the flow of the water would dilute and essentially eliminate any pollutants in the process. Those arguments largely prevailed, and discharge policies did not become significantly altered for another several decades. But they also reinforced the popular notion that untreated water, especially urban water, was not fit for drinking, swimming, or even fishing. Dramatic events during the 1960s, such as the increasing eutrophication of Lake Erie and the burning of the Cuyahoga River in Cleveland, sparked by discharges from the chemical and steel plants that bordered the river, highlighted the problem. But these events, by their magnitude and visibility, also helped change the nature of the discourse and contributed to an emerging era of environmental action. "Clean water" became a goal and a framework for policy as well as an important shift in the culture of water itself. Water needed to be protected, not simply used or made available for dumping.[10]

This shift in the discourse about clean water was further influenced by two significant discoveries during the 1970s—the health impacts related to disinfection by-products in drinking water and the growing recognition that numerous groundwater wells around the country had become contaminated by different chemicals and pollutants from various sources, such as industrial discharges, sewer system leaks, and landfills. Chlorine, itself a chemical that had altered the public perceptions about water as fit to drink, began to be seen as a part of the problem rather than just the solution. New research indicated that, when combined with organic matter, the chlorination process produced by-products that, at high enough levels, could contribute to various kinds of cancers, thus muddying the understanding of what was safe about the water coming out of the tap. The groundwater contamination issue also undermined a core meaning associated with water—namely, that once in the ground, the earth itself would act as a filter against unwanted contaminants from urban, agricultural, mining, and industrial activities. The discovery of contaminated wells in the late 1970s and 1980s heightened the fears that these same agricultural, industrial, mining, and urban-related practices were transforming a pure substance into something harmful. While a complex system of risk assessment and regulatory action came into play to address these discoveries, market forces also went to work. A dramatic upsurge in private bottled water sales was stimulated, feeding on the fears about water quality and the taste and odor problems that had become prevalent in a number of treated-water systems, despite bottled water's prohibitive cost in comparison to tap water and the lack of any regulatory review that could answer to the quality of the bottled water as well.[11]

The shifting language of water in this period also extended to the debates about the policies and economic designs on water as a source of supply, a form of transport, or a resource that needed to be managed and controlled. When Jimmy Carter did battle with some key water-industry interest groups early in his presidency, he argued that much of existing water policy was dictated by the interest groups, their congressional allies, and the federal and local agencies—the classic iron triangle—that set the terms of water policy. But to Carter, who likened himself to an engineer in the form of a reformer, the iron triangle represented disjointed, poorly conceived, and inefficient projects and use of public funds. The dictates of keeping the water off the land in the East and bringing it to the land in the West had formed the basis for how the big bureaucracies like the Army Corps of Engineers and the Bureau of Reclamation functioned, and some of the environmental and regional consequences of those policies resulted less from a condition of nature (flooding, drought, amount of rainfall) than the desire of the water agencies to meet the needs of their core constituencies.[12]

But despite Carter's failure to fully implement reform, the iron triangle involving various water-industry interests was no longer in full control. When

Ronald Reagan replaced Jimmy Carter, and James Watt, the scourge of environmentalists, replaced the friendlier Cecil Andrus as secretary of interior, the water industry breathed a sigh of relief. But rhetoric notwithstanding, funding and political support for many of the water projects by the 1980s was no longer available, and a stalemate ensued. The stalemate, moreover, reflected both policy differences and differences in language and the culture of water. Water might be wasted if it didn't become available as a supply source, the water-industry interests argued, but the counterarguments were increasingly compelling about the need for water as a source for environmental renewal or that water needed to be clean and safe. This difference in language was true as well when it came to the concerns about flood hazards. While the engineers from the water agencies spoke of the need to tame and straighten unpredictable and dangerous rivers through their flood-control approaches, the environmentalist counterarguments spoke of respecting the rivers by working with nature through wetland protection, land-use approaches, and other strategies that maintained the ability of the land to absorb the flood waters. Similarly, the debates about the nature and extent of regulations related to drinking-water supplies were often framed by questions of perception and language. For example, how safe was safe when it came to pollutant levels, and did risks that were involuntary require a more stringent approach? By the twenty-first century, these differing approaches came to represent both contested language and stalemated policy, a culture of water that was itself a source of division.[13]

Like nature, water could then be considered a type of *hybrid*, "neither purely natural, nor purely a human product," as Maria Kaika put it. With the rise of urban and industrial society, water increasingly came to be defined as a "materially produced commodity," a resource to be manipulated and controlled for transport, irrigation, mining, real estate development, or drinking water. Much of this history of water development in the United States occurred through the actions of public entities working with and often on behalf of private interests—the water-industry model. The more recent shift toward a model of privatization, both in the United States and abroad, has further reinforced what could be characterized as a commodity-based culture of water.

But water is also seen as a part of nature, albeit a form of nature whose meanings become socially constructed. Rainfall, rivers and streams, lakes, oceans, and water in the ground and above the ground are, like nature, talked about, perceived nostalgically, made a part of amusement park rides, seen as life giving, or feared as a potential hazard in the form of floods, tsunamis, or drought. In this way, water can be seen as helping establish and constitute a community-based culture of water. A river or a lake can help define a region, while the concept of *watershed,* an increasingly popular concept of water management, assumes community functions by linking a river system to a specific area of land and its various ecosystems and human and nonhuman populations. The public-trust doctrine, an approach to water law that gained favor in the 1980s and 1990s, sought to apply the community value of water to key environmental battlegrounds, such as the conflicts over the city of Los Angeles's claim on water from the Mono Lake Basin. The community-value-of-water concept, however, has also been used to justify the transfer of water from its area of origin to its ultimate destination, in some cases thousands of miles distant. Thus, various societies and communities (particularly in the western United States, where water availability has been limited and contested) have been organized according to differing interpretations of how water should be managed and controlled. The communitarian Mormon villages in the Great Basin Kingdom of the Mountain West in the nineteenth century, for example, were organized around water as a shared resource, but Los Angeles was a city and a region whose scale and capacity to grow became dependent on locating ever more distant sources of water.[14]

Whether water policies are established based on water's commodity associations, on its community association, or on some combination of the two can tell us about the societies that are influenced and shaped by those policies. The culture of water has, in turn, influenced the nature of the debates that have emerged about the kinds of policies to be adopted, including in those urban settings where water has been and continues to be a critical factor in how the cities have grown and changed.

Bringing Water to Los Angeles

The story of water for Los Angeles has been told often and in a variety of ways—in the story line for films, novels, and short stories; in newspapers; and in the histories, institutional analyses, and popular literature about the region. It includes the story about the Los Angeles River, efforts to control the river's turbulent ways, and the search for and the capture of distant sources of water for development, beginning with the Owens River watershed in the Sierras and extending to the Colorado River Basin and Northern California's complex river systems, which drain out to the Pacific Ocean through San Francisco Bay. Los Angeles's story is frequently described as illustrative of the urban assault on nature as well as the city and region's imperial ambitions. In the telling of this story, a kind of Los Angeles exceptionalism is also assumed.

Yet L.A.'s water story is far from unique. The story of urban development in the early nineteenth century in places like New York and Philadelphia references the key role of capturing and importing water from distant places as essential to the growth of those cities. Such efforts not only laid the groundwork for the expansion of the cities but served as a kind of water-development template for other cities in the decades to come. As the cities grew, two key factors stimulated the search for new water—the limits of existing sources to accommodate expansion and the increasing contamination of existing sources that plagued nineteenth- and twentieth-century cities and suburbs alike.[15]

The transformation of urban rivers in places like Chicago also predated the massive changes that began to be instituted during the late 1930s for the L.A. River. In the 1890s, Chicago officials successfully altered the flow of the Chicago River in response to contamination from urban and industry sources along the river's edge. In the process, Chicago exported its pollution problems to downstream communities while remaking the river itself. Like the Chicago River, urban rivers throughout the country underwent a policy of containment and reconfiguration to address problems of flooding, inappropriate land uses, and problems of pollution that undermined recreational and drinking-water uses. Similar to the L.A. River, the urban rivers, streams, and creeks in cities like Salt Lake City, Providence, and San Jose were turned into unattractive flood channels. And similar to the events that unfolded in Los Angeles, an urban river-restoration movement emerged during the 1980s and 1990s to contest the strategies and seek to revive, albeit in a new form, an attractive and "living" stream, creek, or river.[16]

These urban water strategies and conflicts are national in scope, and their details depend on each region's geography, history, politics, and land-use patterns. Nevertheless, Los Angeles does present important stories that are both distinctive and representative about the desire to bring water into the city as well as efforts to keep water off the land in the city. This is due partly to L.A.'s reputation as a water imperialist and as an urban environmental dystopia. But it is also due to the stories of renewal and reinvention that have also characterized this region.

The story of water *for* Los Angeles and *in* Los Angeles is complex and sometimes elusive. For example, the assertion that L.A. is desert area, totally dependent on imported water, is misleading and masks the importance of the different players and decisions that have influenced water and land-use development in the region. Average rainfall in the Los Angeles basin is now calculated at about thirteen inches a year, which in turn qualifies the area as semiarid at its edges but less so along the coastal plain, where a Mediterranean climate prevails. Moreover, *average* rainfall is itself not reflective of the mean: the amount often varies widely year to year, decade to decade. Long dry periods might be immediately followed by heavy rainfall and rapid surges of the local rivers and streams. The Los Angeles River gained its early reputation as placid but also mysterious owing to these abrupt shifts in rainfall—small trickles, varied vegetation, and then sudden tempestuous flows. The course of the river was erratic, directions changed, and some of L.A.'s earliest

Anglo residents worried that without historical memories of the river poor decisions about what and where to settle might be made.[17]

The combination of uncertainty about local sources, pressures to expand commercial activity, population surges, and the unpredictability of the weather led to the quest, as early as the 1880s and 1890s in Anglo Los Angeles's infancy, to find new water sources. This also meant relying less on the L.A. River, the San Gabriel River to its east, the Santa Ana River to the south, the water that percolated into the ground, and the various aquifers that were located throughout the region. The citrus groves that were planted in the late nineteenth century, particularly after the introduction of the navel orange from Brazil in 1873 and the subsequent planting of lemon and grapefruit trees, created a citrus culture and industry to the east of the city that eagerly sought a readily available water supply. This came initially from the surface waters of the San Gabriel River and was followed by wells that were drilled to tap the local groundwater. When electric motors and gas engines became available towards the end of the nineteenth century, groundwater production increased significantly, particularly for irrigation purposes, not only in Southern California but throughout the country.[18]

The railroads and various commercial and retail interests led by the *Los Angeles Times* went on a promotional spree during the 1880s to attract new residents to the city, using the weather and the attractions of the landscape of the San Gabriel Mountains, the Arroyo, the Los Angeles River, and the Pacific Ocean as their calling card. The region did experience an extraordinary boom in population, less so in terms of commercial and industrial activity, although agriculture continued to flourish. But bust followed boom, also creating a cyclical development that further skewered planning beyond the assumption that a cycle—primarily associated with rapid increases in population—would continue.

For the *Times* and its various allies, including the powerful railroad interests, this growth cycle represented two important opportunities—(1) new real estate development possibilities, particularly in less developed areas as long as core infrastructure needs (such as water availability) were met and (2) population growth based on attracting the right kinds of newcomers, which could also be used for political advantage, particularly by undercutting a resurgent labor movement and its Socialist Party allies. In both instances, securing a new water supply seemed imperative, since the type of land development and population growth that L.A.'s leading boosters were promoting could have been limited or redirected if local water sources were exclusively utilized. "If L.A. didn't get the water, it would never need it," William Mulholland, the leading promoter of the city's bid for a new water supply and the new Water Department's chief engineer, would often comment.[19]

The efforts to secure that supply by obtaining the rights to Owens River water and the subsequent authorization and construction of the L.A. Aqueduct, which was completed in 1913, have been the subject of multiple histories of water and the city, as well as novels, dramatizations, documentaries, and films. The story is filled with intrigue, shady maneuvers, and dynamite. The project's critics have focused on the plight of the Owens Valley residents who lost access to their own local water source and on the cabal associated with the L.A. real estate syndicate that secured the water for its development to the northeast of the city and thereby made enormous profits. The project's defenders have argued that the L.A. Aqueduct was the region's first great infrastructure project that made it possible for the city to grow and prosper. Defenders of the Owens River project also pointed to the huge vote in 1905 in favor of authorizing the funds to build the L.A. Aqueduct as the first indication of popular support for what would become, in the years that followed, the continuing need to obtain water to expand Los Angeles. "Owens River is ours, and our business is to hustle and bring it here, and make Los Angeles the garden spot of the earth and the home of a million contented people," the *Los Angeles Times*, the aqueduct's leading booster, crowed about the 1905 vote.[20]

Water for Annexation

The story of the L.A. Aqueduct is the story of how an imported water supply defined and shaped the city. Between 1905, when funds to construct the aqueduct were authorized, and 1913, when construction was completed, a series of debates unfolded about the purpose of the new water supply. Was it to meet existing needs that would allow for modest and staged growth, to meet future needs that would accommodate more rapid economic growth and population increases but within existing boundaries, or to utilize the water as a basis for population growth, economic development, *and* geographic expansion? Related to those debates were arguments about labor issues (whether L.A. would be a closed-shop versus open-shop city, including whether the workforce in the new industries would be unionized) and about population issues (whether Los Angeles would be a "city of scale" with just modest increases in population contained within existing city limits, as the Socialists and their labor-movement allies argued). This contrasted with the idea of an expanding metropolis whose slogan "Watch us Grow" emerged by 1907 as "the legend emblazoned . . . over store and trolley lines, which extend far out into the country," as City Beautiful chronicler Dana W. Bartlett put it. The disposition of Owens Valley water became the basis for resolving the outcome of those debates. "In a land where water means so much, where it has the power to transform the desert into a garden," Bartlett said of that moment, "who can picture the beauty and greatness of the future City of the Angels when this gigantic scheme becomes an accomplished fact?" "Los Angeles is the Mecca of the whole countryside from the mountains to the sea," the *Los Angeles Times* declared in a similar vein six months after the bond election. "Every crossroads village within twenty miles of our outposts knows of the river of delight that is leaping from the snow-capped Sierras, bound oceanward! And each seeks to share the golden glory of the flower and the fruit and the vine that will strew its course."[21]

The sharp political divide in Los Angeles, including the debates about the aqueduct water, continued for a short period after the aqueduct's completion, but a shift began to take place with the advent of World War I as the patriotic fervor associated with the war undermined the position of the antiwar Socialists. By 1915, the issue of a water-stimulated expansion of the city had been resolved in favor of extending the boundary lines of

Los Angeles. The city aggressively marketed its new water supply to both unincorporated areas and existing cities, arguing that growth in those areas would be limited without the imported water from Los Angeles but that growth could occur by making those areas part of the city of Los Angeles. A new set of campaigns about whether to annex to L.A. ensued in these areas. These campaigns mimicked in part the earlier debate in the late nineteen century to get towns and unincorporated areas such as Highland Park to annex to Los Angeles to obtain the city's surplus water supply from the Los Angeles River, which the city controlled but had no right to sell outside its boundaries. These new campaigns around the L.A. Aqueduct water were also hotly contested with overblown rhetoric and dire warnings, and most areas decided to annex, leading Los Angeles to grow to four times its size in a little more than a decade. However, some cities like Santa Monica, Pasadena, and Beverly Hills resisted the push to annex, arguing that while the availability of imported water was attractive, they wished to preserve their autonomy as separate municipal entities.[22]

During the 1920s, to use Tom Sitton and William Deverell's suggestive title to their 2001 book on L.A., Los Angeles was a "metropolis in the making," a city that was "population mad, annexation mad, and speculation mad," as a newly arrived Oliver Carlson called Los Angeles in a famous comment about the period. This new and more expansive boom was stimulated by the emerging oil and entertainment industries and supported by available water and cheap power, with the continuing supply of additional imported water an essential ingredient of the boom philosophy. But with the annexation strategy running its course, a coalition of nine cities (eventually thirteen), including Los Angeles in conjunction with the other cities that had resisted annexation, weighed in on securing the next major imported water source from the Colorado River, anticipating another cycle of imported water linked to urban expansion. Achieving a regional consensus, in this instance, was compounded by divisions among some key players as well as competition for supplies among the seven states constituting the overall Colorado River Basin area. The *Los Angeles Times* publisher, Harry Chandler, wanted to protect his large landholdings on the Mexican side of the border, fearful that an "All-American Canal" project associated with the plans for Colorado River development would reduce water availability for his 400,000-acre ranch. Arizona also feared that California's desire for Colorado River water (which included the agricultural areas to the east of Los Angeles in the Imperial and Coachella Valleys as well as the urban areas in the Southern California coastal plain) would undercut Arizona's own claims on the water, creating a bitter dispute over water allocations that would not be resolved until nearly four decades later in a critical 1963 court ruling.[23]

The desire of most of the local elites in those cities and in the region to obtain another round of imported water remained constant, however. Even sworn political enemies worked together to establish a Southern California claim by getting federal approval to construct a huge dam site in Black Canyon within Nevada near the California and Arizona borders (which became the Hoover Dam) and, subsequently, a Colorado River Aqueduct to carry water into California for its agricultural and urban users. At the same time, the cities coalesced into a new entity, established in 1927 as a special district by the state and organized the following year as the Metropolitan Water District of Southern California. The MWD, rather than the city of Los Angeles, would now be the instrument for expansion, with the Colorado River Aqueduct "planned not as a Los Angeles product, but as a Southern California enterprise, not on the basis of meeting immediate needs alone, but on the far broader basis of insuring for generations to come an adequate water supply for the region as a whole," as the MWD's first annual report put it.[24]

The decision to secure a new imported water supply from the Colorado River on a regional basis coincided with the push to reconfigure the Los Angeles River as a flood-control channel. Already with the development of the L.A. Aqueduct and the changing patterns of land development in the city and the region, the L.A. River had lost some of its significance as the key supply source for Los Angeles. Pumping water from the groundwater basin replenished by the L.A. River (as well as the aquifer fed by the San Gabriel River) continued to represent the major source of water in the city and in much of the region, but the focus on securing imported water was seen as essential to the future growth of the city and particularly the region, since the city of Los Angeles's geographic boundaries were now set with the end of the water-for-annexation policy. Annexation to the MWD became critical for this new water entity to survive and eventually establish its claim as the driver of growth and expansion for the region.[25]

When the Colorado River Aqueduct came on line in 1941, economic and population growth in Los Angeles and the metropolitan region had significantly declined during the 1930s, and the MWD found itself with a source of imported water but not enough buyers, similar to the situation that the city of Los Angeles had found itself in 1913 when the L.A. Aqueduct had been completed. The situation was further compounded by the city's decision to also expand its own imported water supply by constructing a second aqueduct link to the Mono Lake Basin adjacent to its Owen River supply and by further increasing the pumping of groundwater from the Owens Valley, decisions that further secured the city's reputation as a water imperialist. Therefore, the MWD faced a double bind—reduced growth, which dampened demand, and a major customer that was increasing its own source of imported water, coveted in part because of the gravitational flow downhill from the Owens Valley that enabled the city to obtain a parallel source of cheap hydroelectric power. As a result, the MWD established its own annexation policy, less a policy of water blackmail as the city had pursued between 1913 and 1928 than the carrot of subsidized annexation fees (we've built it, and now we want you to come). The MWD then became more willing to annex *non-urban* areas to the east and south of the MWD service area (and eventually to its west), despite the water district's own founding documents that defined it as providing water for municipal and industrial uses in the urban areas within the Coastal Plain. These annexation policies included the absorption of the San Diego County Water Authority in 1946 and several of the agricultural areas—the original citrus belt—in the San Gabriel Valley, which had begun to experience the post–World War II transformation of agricultural and undeveloped land into Southern California's automobile suburbs.[26]

By the early 1950s, the MWD itself had become less a supplemental imported water supplier for the developed urban areas than a water broker playing a powerful role in changing the land patterns in the region. This revised role was codified in a document known as the Laguna Declaration. Adopted at a December 1952 meeting of the MWD board of directors, it had been stimulated in part from a debate about some additional annexations involving the cities of Pomona and Ontario and the question of securing a third major imported water supply (the second for the MWD) for the region. The declaration, a short mission statement, argued that making the MWD's imported water available "to meet expanding and increasing needs in the years ahead" would be accomplished by identifying the MWD as the logical and exclusive entity to secure and manage water "from other sources as required in the years ahead." Underlying the declaration were the discussions about constructing a California Water Project to secure water in northern California from the Feather River (a tributary of the Sacramento River system), through the Sacramento Bay Delta, to the west side of the Central Valley, and into the Southern California region. There was also talk about the need to secure other, even more distant and massive imported water sources, such as the Columbia River in the Pacific Northwest and even the Yukon River in Alaska. By announcing a desire to pursue, control, and manage these new sources of water, the Laguna policy statement assumed that new areas outside its presumed service area could now annex to the district since the MWD was asserting it could provide water no matter where the location or the amount of the request. A continuous cycle had been established of securing new imported water to expand the urban and suburban land base, which in turn stimulated the need to secure another imported water supply for another round of expansion.[27]

These discussions were taking place in the midst of the massive expansion of the region owing to population growth, economic changes, and, in relation to the MWD and the question of water, the patterns of land development and suburbanization that would make Los Angeles and the Southern California region the poster child of urban to suburban to even further suburban transformation. Southern California was becoming not only a growing region but an "exploding metropolis" or as the *Los Angeles Times* put it "a metropolis of the Pacific Littoral," a region (like the newspaper) that could "think big" by expanding in every direction and "which gives promise of being the center of tomorrow's universe." New cities—based on subdivisions that were made possible when areas annexed to the MWD or its member agencies and thus could obtain imported water—sprung up overnight. By the early 1960s, more than eighty cities had been organized in Los Angeles County alone, and dozens more were forming in adjacent counties that now constituted the six-county MWD service area. Although the MWD had not been entirely successful in obtaining its role as the exclusive Southern California contractor and supplier of the State Water Project (San Bernardino and some of the communities in the San Gabriel Valley opted for their own contractual relationships with the State Project), it had become the dominant force in obtaining new water and linking it to the patterns of development.[28]

This defining role was reflected in two types of policy approaches that emerged during the 1950s and 1960s. The first involved the preeminence of imported water as providing the margin for new growth, as opposed to the long-standing role of local water as the primary supply source, particularly groundwater. That policy shift was also made possible since groundwater basins were beginning to be adjudicated between its different users and thus managed in a manner that limited the amount of water to be used to prevent too much pumping or an overdraft of the basin. While these legal arrangements were establishing a more efficient and rational policy to prevent any overdraft, the adjudications also made possible the push from the MWD to pledge new sources of imported water to meet any and all anticipated growth, thus linking any new growth to the availability of imported water rather than relying on local water. Second, by establishing explicitly that link, annexation to the MWD helped inspire or reinforce the kind of crazy-quilt pattern of development in the region where new growth often leapfrogged over existing boundaries of development. This pattern created what the MWD called "windows." These were places not contiguous to existing annexed areas to the water district, which also skewered the boundaries of the wildland/urban interface.[29]

During my tenure on the MWD board of directors during the 1980s, window annexations were frequent but not discussed. They often were simply placed on the agenda as part of the consent calendar. MWD board and staff assumed that continuous growth in the region would eventually close any specific window that they created, even as new ones were opened beyond the existing boundaries of the district. "We are not involved in land-use decisions," the MWD board and staff would assert, while its decisions weighed heavily on the land-use patterns in the region. Thus, while the new freeways built in the 1950s and 1960s helped stimulate the development of the automobile suburbs, water availability made it possible to build in areas with no groundwater or with limited groundwater and no existing water infrastructure. The new developments needed to be able to assure their future residents that water was plentiful and secure, that lakes and fountains could be constructed, and that limits on use would not become a part of the culture of water for those newly organized areas established beyond the urban edge.[30]

During the 1960s, as these imported-water policies linked to a mission of expansion seemed to be most firmly in place and the region continued to grow, an air of unreality about what was possible and what the costs might be also seemed to prevail among Southern California's water-industry leaders and elected officials. Even though the State Water Project had yet to be completed (the link to the MWD service area would become available only in 1972), a series of proposals about building new imported-water projects continued to be floated, many of them at enormous scale, with an assumption that demand based on growth would continue to drive water policy. One of the areas and water sources under consideration, the Pacific Northwest and the Columbia River, were taken off the water industry's agenda when key figures such as Senator Henry Jackson (D-Wash) inserted language in 1968 water-related legislation that forbade the Bureau of Reclamation to undertake studies, over a ten-year period, of the feasibility of a Columbia River–related imported-water project, a provision that continued to be extended until the proposal eventually faded. That failed to stop the champions of big water projects, however. Another proposal, put forth by the Ralph M. Parsons Company, a Pasadena-based engineering company that sought to demonstrate its capacity to handle massive projects, suggested that the area of origin's water source could extend beyond the Pacific Northwest to British Columbia to serve the water needs of the Great Lakes region, the rapidly growing southwestern cities of Phoenix and Tucson, and the leading customer for imported water, the continually expanding Southern California region. Even the fanciful concept of towing icebergs from Alaska to Southern California appeared from time to time and, like the Pacific Northwest and British Columbia proposals, continued to have advocates into the twenty-first century. When I first joined the MWD board, I received in the mail a large poster of the Alaska iceberg that called for a "2,000 plus undersea aquaduct [sic] transporting pure Alaskan water from the abundant rivers of Southeastern Alaska to the arid Southwestern United States" at a terminus at the port of Los Angeles. A statement highlighted on the poster from then Alaska governor (and former secretary of the interior) Walter Hickel proclaimed, "Big projects define a civilization. So why war—why not big projects?"[31]

But by the late 1960s, the opportunities to develop these types of massive imported or transbasin water projects were no longer readily available due to high costs, environmental factors that were becoming more prominent, and regional conflicts, particularly growing opposition from the area-of-origin regions such as the Pacific Northwest and, as emerged during the Peripheral Canal fight, Northern California. The Ralph M. Parsons Company's British Columbia project, for example, never received serious attention owing to costs (the Parsons Company's original price tag of $200 billion seemed to increase by another $200 billion every ten years) and environmental constraints (one water-industry figure in Colorado mockingly suggested that the project's environmental-impact report might take forty years to prepare). The growth and expansion scenarios in Southern California also began to change. While the rapidly suburbanizing, urban edge areas of San Diego, San Bernardino, and Riverside counties (designated the "Inland Empire" by its boosters) continued to grow and expand with new subdivisions and water hookups, some of the automobile suburbs in Los Angeles and adjacent counties grew in population but also began to become denser. The massive demographic changes in the region during the 1970s, 1980s, and 1990s, especially in Los Angeles and parts of Orange County, which witnessed a huge new influx of population from Asian and Latin American countries, further reinforced and extended the trend toward denser urban and suburban areas, even as the creation and expansion of yet another generation of new automobile suburbs persisted. Water-use patterns also began to change, with increases most pronounced in the high-growth areas, while the original member agencies of the MWD, including the City of Los Angeles, began to experience a leveling off of water use. Inter-regional conflicts, including tensions between Los Angeles and San Diego, over such issues as funding mechanisms, annexation fees, and other user subsidies, while always present, had tended to be minimized during the quest for new imported water sources.[32]

Even more than the imported water issues, water-quality problems that emerged during the 1970s and early 1980s revealed a water industry, including in Southern California, at its most defensive—insecure about the future of its water supply and unable to establish an effective language to dismiss the public's growing water-quality concerns. Dismissing the fears about the toxic substances found in its water sources as well as its treatment approaches, water-industry figures argued that the trace levels of the substances identified in the water were insignificant and that the problem was due as much to the introduction of the more sophisticated technology that became available during the 1970s since it could detect even more minute trace levels of substances in the water. To the water agencies, public fears about drinking-water quality appeared at first to be more of a public relations problem (but one that would continue to grow) since potable or drinking water constituted a small percentage of overall use. But the magnitude of the problem soon became apparent with the loss of groundwater supplies. Wells were shut down, and regulations regarding risk levels pointed to the need to change to a more expensive treatment approach, such as a granular-activated carbon- or ozone-based system, given their major water-supply and -cost impacts. But the inability to demonstrate an active approach that indicated that the quality of the water constituted a core part of the water agency mission undermined some of the credibility of the water industry and fueled the rapid growth in sales of a privatized drinking-water supply in the form of bottled water and home filter systems.[33]

The Peripheral Canal debate further compounded the Southern California water industry's dilemma. Some of the advocates for the canal had included a handful of environmental-oriented officials in the State Department of Water Resources within Governor Jerry Brown's administration. These officials began to raise the idea during the referendum campaign that the package had important water-quality benefits for Southern California. This was due to the reduction of organic matter if the imported Northern California water were to flow around rather than through the agricultural lands in the Sacramento Bay Delta. That, in turn, could potentially reduce the amount of water-quality problems related to the disinfection by-products created by the chlorination process in treating Southern Californian's drinking water. Yet to raise such water-quality considerations also undercut the argument of the water industry that water quality was not a problem. Since environmental opponents to the Peripheral Canal were also focused on the water-supply issues and their environmental impacts in the delta and in relation to Southern California's cycle of expansion, the water-quality issue never gained traction during that election.[34]

The Peripheral Canal debates also were revealing by indicating the decline of the historic water-industry position about imported water, the complexity and uncertainties about the future of water policy, and the limits of the environmentalist positions about water and growth. During the 1980s, a couple of new initiatives were proposed to address the shortcomings of the State Water Project but without success, eventually leading to a new type of bargaining process that brought the Bay delta–focused environmentalists to the table and culminated in the establishment of a joint federal-state entity called CalFed. An uneasy and often tentative series of water-*management* rather than water-supply discussions and proposals were now on the agenda, particularly forcing the Southern California water interests to revisit their dependence on imported water from big-water projects as a way to manage its supplies and best accommodate new growth. The focus of environmentalists on limiting new imported-water supplies as a way to manage growth—if they don't have the water, then maybe they won't come—failed to address how growth should be managed and the urban landscape redefined. Nor did it address the foundation of water policy and the mission of the water agencies that helped frame those policies. These questions had yet to be fully explored during the period when I served on the MWD board during the early and mid-1980s. By the time I left the board in December 1987, they could no longer be ignored. The water agenda could no longer simply be defined as a matter of ensuring, to use Dana Bartlett's City Beautiful evocative phrase, the continuous transformation of desert into garden. Questions regarding water origins, water quality, best uses of water, and best water-management policies—as well as the purpose and goals of the water agencies providing those supplies—had become part of the water for the city debate.

Privatizing Water

While private interests have played a major role within the water industry, the water-delivery and -management systems that were developed in Los Angeles and most other urban and regional areas have been publicly owned and operated. This occurred despite periodic efforts to privatize urban- or regional-based water utilities that date back to the efforts of a Wall Street banking combine (that had included Alexander Hamilton, Aaron Burr, and DeWitt Clinton) to privatize New York City's water operations in the late 1790s. But the early private water operations were often inefficient, were not often upgraded, did not provide adequate service, and functioned more as capital pools than water utilities. The shift to public ownership of the water utilities during the nineteenth century reflected the interest of many urban areas to identify and tap new large-volume water supplies, much of which involved the purchase and delivery of distant sources. In contrast, private owners had been more reluctant to make the investments and operate the kind of large-scale system that would have been required. As a result, from 1800 to the end of the nineteenth century, public ownership of water systems jumped from 6 percent to 53 percent, while as many as forty-one of the fifty largest cities became publicly owned. In the case of Los Angeles, the development of the Owens Valley water supply was similarly made possible in part by the city's purchase of the local private water company. This contrasted with the Progressive Era arrangements of the late nineteenth and early twentieth centuries, which had led to the development of state-regulated private or investor-owned electric utilities whose investments and rate of return were determined by state utility commissions in exchange for the utilities' investor-owned ownership structure. Los Angeles, with its publicly owned electric utility (made possible by the hydroelectric generation from the Owens Valley project), was the exception.[35]

The city of Los Angeles's Water Department and, in particular, some of its key water commissioners operated in a manner similar to the water-industry model where public goals and private development interests tended to converge. L.A.'s electric utility, established in 1911 through a city charter amendment, became, on the other hand, far more aggressive in defining its mission as a public enterprise and service provider. Political coalitions consisting of Socialist Party public ownership advocates such as City Council member Fred Wheeler and Progressive Era reformers such as John Randolph Haynes fought hard to keep the electric utility publicly owned and publicly accountable, a battle that also took place in a number of other Southern California municipalities such as Pasadena, Glendale, and Anaheim in Orange County. The successful battle in Los Angeles to establish public ownership of electric power (in contrast to the failure to municipalize San Francisco's electric operation related to its Hetch Hetchy Dam project) was due in part to the contentious debates about how water and power affected the development of the city itself. Although combined in a single entity called the Los Angeles Department of Public Service (later the Department of Water and Power), its two entities—the Water Bureau and the Power and Light Bureau—were for a time distinctive in mission and operational priorities. "Los Angeles business interests did not look as kindly on public power as they did public water," a DWP history coyly put it, and even after Southern California Edison's distribution system in the city limits was purchased in 1922, which allowed the public utility to significantly expand its reach, those debates had not been fully resolved. The power side of DWP, in particular, continued to function as a semi-autonomous political machine, jealously guarding its service-provider role (which included low rates, efficient delivery, and a desire to expand its capacity to accommodate growth). It played a key role in the development of Hoover Dam, for example, by guaranteeing power purchases generated by the Dam to repay the federal government's costs for the project.

At the same time, on the water side, particularly with the development of the Metropolitan Water District as the main instrument for regional water development, the Department of Water and Power tended to play less of a leadership role around regional water issues while still pursuing its primary goal of obtaining water for the city, as opposed to the region. While the DWP was seen as the prototype of the water imperialist (a legacy of the Owens Valley project and its subsequent expansion in the 1920s and again in the 1940s with the Mono Lake project), it was the MWD—less visible, not as directly accountable, and more in tune with the patterns of regional development—that fit more directly with the water-industry model that often blurred the lines between private and public.[36]

As the water industry began to lose some of its ability to influence water policy (particularly with the decline of the big water projects in the 1980s) and water utilities at the municipal level and regional water agencies like the MWD began to receive more visibility and public attention (due in part to the growing concerns about water quality), the debates about water began to reference the contrasting choices between its commodity and community values. The push for a more private or marketable water supply and delivery system as well as changes in the ownership structure provided one possible direction—the goal of privatizing water—in places like California but also in cities and regions around the world. In the wake of the Peripheral Canal referendum, the concept of privatization and water markets began to have some appeal to Southern California's water interests. Several of the MWD directors, for example, began in the 1980s to promote the idea of outsourcing some of the MWD operations to private companies. But it was the issue of water markets—transferring water from one entity to another on the basis of a market transaction—that emerged as the most significant example of this new interest.

For MWD, the water-markets issue had been particularly contentious owing to its prominence during the Peripheral Canal debate. In the course of that election, one of the key environmental groups opposing the canal, the Environmental Defense Fund (now Environmental Defense), argued that a transfer of water through a market arrangement between the Imperial Irrigation District (IID) and MWD could potentially represent the amount of water that would otherwise be obtained by the additional water from the State Water Project made available through the Peripheral Canal project. The EDF focus was on leakage from the pipelines and poor conservation practices among Imperial Valley agricultural interests and the Valley's water agency, the Imperial Irrigation District. If MWD would pay for conservation improvements, the Environmental Defense Fund argued, then the water that was saved could in turn be reallocated to Southern California. But both MWD and IID officials strongly objected to the idea, arguing that the complex agreements established to divide California's share of the Colorado River water precluded any such arrangement. The political concern that such a proposal could undermine Southern California's argument about its desperate need for additional water from Northern California remained a critical subtext to the debate over whether and how such a market transaction could occur.[37]

After the 1982 referendum election, the Peripheral Canal's resounding defeat, and other subsequent unsuccessful legislative initiatives to secure additional Northern California water, the debate about water markets became far more open-ended and uncertain. Positions now varied between and among water-agency officials, agricultural interests, irrigation districts, urban development interests, water entrepreneurs, and other key water-industry figures and water officials. Soon after the 1982 election, the San Diego County Water Authority, the largest user among MWD member agencies, began to argue that future water supplies to meet its anticipated growth were now in jeopardy and that the Laguna Declaration promise of meeting any future anticipated demand would be subsumed by the District's policy on how to allocate *shortages*, a policy that potentially favored the city of Los Angeles and, the San Diego board members argued, harmed San Diego's interests. San Diego now expressed an interest in a water-market deal with the IID and its farmers and even prepared to build a separate pipeline to make it happen if MWD would not cooperate by making available its own delivery system. MWD continued to argue forcefully against such a deal, concerned that this separate water-market arrangement not only would undercut its own position but would ultimately undermine the complex agreements that had led to the Laguna Declaration in the first place—centralized control in exchange for always meeting future demand.[38]

Though MWD's opposition to a separate San Diego/IID arrangement continued through the 1980s and 1990s, its own position on water-market arrangements (or water transfers between districts, as MWD officials defined the concept) began to evolve as well. The most vocal proponent of water markets within the MWD, a former RAND Corporation economist who became the District's assistant general manager, helped steer MWD toward a more nuanced position that largely coincided with the Environmental Defense Fund's initial focus on conservation measures or water banking-related mechanisms as part of an interdistrict water transfer. Beginning in the late 1980s, MWD began to explore a number of interdistrict transfer agreements, including one with IID that essentially reversed its earlier position, as well as arrangements that included keeping agricultural lands fallow during dry years or using wet-year, dry-year cycles to store and transfer Northern California water to urban Southern California.[39]

While a few of these arrangements were successfully concluded, the MWD and other water interests became more open to the idea of water transfer, and water-market arrangements significantly increased the interest in and potential shift toward the privatization of water and related land- and water-rights speculation. A new set of players, including several European companies that were at the forefront of the global push for water privatization, began to carve out a role in the fast-evolving water-markets situation. In the Imperial Valley situation, the Bass brothers, Texas investors and speculators, purchased water rights through sizeable agricultural land holdings in the Imperial Valley in anticipation of a possible San Diego/IID arrangement. With the deal held up owing to intraagency as well as legal obstacles, the Bass Brothers decided to sell the water rights for stock ownership in the U.S. Filter Corporation, a major water-privatization conglomerate that was the largest seller of water-treatment equipment. Other water and land entrepreneurs, such as the politically connected and big campaign donor Keith Brackpool, eagerly sought to position themselves through land purchases and related control of and access to water rights and water-storage capacity that could be used as bargaining chips in future water-market deals. The concepts underlying how water should be viewed and the speculation associated with water markets—a more explicit version highlighting the commodity value of water—reinforced a long-standing southwestern adage that land was worth only the rights to the water that attached to the land.[40]

In the last few decades, private water companies like U.S. Filter (first acquired by Vivendi and later sold to the electronics conglomerate Siemens), with its holdings in wastewater treatment, chemical manufacture and treatment, and bottled-water production, among dozens of other products and services; large French conglomerates like Suez and Vivendi (now Veolia Environnement SA); and the British company Thames Water (itself owned by a German conglomerate RWE) aggressively sought to take over public water operations in hundreds of countries, including the United States Large conglomerates like General Electric and Siemens and chemical industry giants like Dow also aggressively entered this expanding market. This in turn inspired a cottage industry of consultants and academics who promoted private ownership of this long-standing public resource. Globalization-promoting institutions like the World Bank also linked their own economic policies, including loans to developing countries, to the potential transfer of water operations to private entities. Some of these privatization schemes subsequently became national scandals, such as the water-privatization programs in South Africa, shepherded in part by the World Bank, that were aimed at making South Africans pay for the full cost of running water. However, this privatization scheme instead caused literally millions of poor people to have their water supply cut off because of inability to pay, forcing them in the process to get their water from polluted rivers and lakes and ultimately leading to South Africa's worst cholera outbreak in its history.[41]

Unlike the water-industry concept that assumed a public role in the delivery and operational side of water systems and that relied on public financing of the large imported water projects that had made possible urban expansion and changes in agriculture, the privatization advocates sought to minimize if not entirely eliminate the public role. This potentially exacerbated issues of access, equity, and the ideas and values associated with the quality and availability of water for urban residents and farmers alike. Nowhere were these issues more apparent than with drinking water, as the increased presence of private companies in this area threatened to continue undermining the public's confidence in the municipal and regional water agencies. Sales of bottled water and water-filter systems rose at a phenomenal rate in the 1980s and 1990s and into the twenty-first century. This increase in demand could be attributed in part to two core concerns—(1) taste and odor issues that plagued the water agencies due in part to their reliance on older chlorine-based treatment processes and (2) fears of the public about the quality of the drinking water coming out of the tap. The defensive posture assumed by the water agencies about these water-quality issues were effectively exploited by the private water firms, which at times engaged in misleading marketing practices or failed to inform the public about the purity of their product. For one, bacterial contamination at room temperature was greater in open bottled-water containers than in tap water, an issue that was not addressed by the bottled-water companies. Second, a significant percentage of the bottled-water systems used the same water sources as the tap water provided by the agencies. In some cases, the taste of the water was simply improved through a more advanced treatment of those same water sources, even as marketing claims were made about a more pristine water source. In one particularly egregious example in which a company claimed that its source was "spring water" (with a label on the bottle picturing a lake and mountains), the water source was a well located in an industrial facility's parking lot near a hazardous waste dump. Moreover, the well periodically recorded pollution levels above U.S. Food and Drug Administration standards due to the industrial chemicals contaminating the groundwater, as a study by the Natural Resources Defense Council on the bottled-water industry noted.[42]

The increased sales of bottled water also created a significant equity issue associated with the price differential between bottled water and tap water. The water delivered by the water utilities was in effect a public good that represented a tiny portion of a water bill. The amount of water used as drinking water by a household, for example, was minimal compared to other residential uses, such as for lawns or plumbing. Thus, the unit cost of the equivalent of a half gallon of drinking water from the tap cost a resident less than one twentieth of a cent, even among those Southern California agencies that were dependent on long-distance imported water sources. This compared to the cost of a half gallon of bottled water at a supermarket, which might be priced at $2.19—or more than a thousand times the cost of public water. Some estimates of the price differential indicated that the cost of bottled water could rise to as much as ten thousand times the cost of tap water and would even be more expensive than milk per fluid ounce. The equity problem was further exacerbated by the aggressive marketing in low-income communities about the safety of the tap water, an issue that also resonated for immigrant communities that had experienced significant water-quality problems in their home countries. Although bottled-water users crossed class, racial, and ethnic lines, studies indicated that taste tended to be the significant factor among high-income consumers, while quality remained the dominant concern among low-income consumers.[43]

As the drinking-water-quality issue became more pronounced during the 1980s, I was struck by how the Metropolitan Water District board and staff addressed the issue. The jump in sales of bottled water and water filter systems was assumed to be limited to a marketing problem requiring better explanations about the safety of the tap water. There was still enormous resistance to making any change with regard to treatment options to address quality concerns (a position that eventually changed for the MWD in response to federal regulations regarding disinfection by-product standards). Nor was there much focus on improving the quality of the water at its source by adopting a prevention-oriented watershed management approach—by reducing, for example, pesticide runoff or other pollutant discharges into source waters. The bottled-water emphasis on taste also received little attention other than the use of chemicals to mitigate the most obvious of the taste, odor, and appearance problems of the water.[44]

During my MWD board tenure, the bottled-water issue, even more than the debates about imported water, revealed an uncertain and defensive water industry, including but not limited to MWD and other water agency participants. At a 1985 retreat to discuss water issues in Park City, Utah, sponsored by the Sundance Institute, the head of the water side of DWP and the general manager of MWD both acknowledged that they had been unprepared to address the growing public concerns about water quality and the new kinds of regulatory and staffing requirements those concerns had generated. The DWP had been further embarrassed when newspaper articles pointed out that a large majority of DWP employees drank bottled water, an issue that continued to plague DWP as late as 2005, when it was revealed that DWP was serving bottled water at several of its events and locations.[45]

The water-quality and bottled-water controversies also touched directly on the issue of privatization and public roles. The bottom line for bottled-water companies, as with other privatization enterprises, was economic—rate of return, market share, profit margins. Even the arguments about greater efficiencies or quality products by privatization advocates were based on market-related arguments that highlighted the commodity value of the water. The mission of the public agencies in the water-quality area was more obscure as long as the primary focus remained the water supply and growth and development goals of the agencies. The participation of the public agencies in the water-industry network of players had long been tied to this overarching goal of identifying and making available water for development. With the rise of the concern about water quality, the water agencies found themselves with potentially competing purposes—and claims on the resources available to accomplish those purposes. At the same time, it increased the public visibility of the agencies and their decisions as well as how they operated. As that happened, it brought the question of water's community role to the forefront as well.

Alternative Scenarios

Soon after I joined the Metropolitan Water District board in December 1980, I began to meet with a network of water activists who had come together to oppose the Peripheral Canal. The water-activist group was as concerned with the elusive and unaccountable nature of water decision making and its land-use implications as the lack of water planning for watershed approaches and reductions in demand. However, with a powerful set of Southern California interests in support of the Canal (ranging from the Chamber of Commerce, nearly all elected officials, the *Los Angeles Times* and other Southern California media, and all other fifty members of the MWD board), the water-activist network was generally perceived as marginal and not likely to influence the outcome of the election significantly. That perception turned out to be inaccurate. Operating like a kind of "behind-the-lines" guerilla force, the Southern California water activists were more influential than expected. Change was in the air, and the ability of that cadre of water activists to influence the election became an illustration of that change.

Since that early 1980s shift in water politics, the MWD, the DWP, and other water agencies have witnessed the development of a number of programmatic initiatives and alternative scenarios about their role, mission, and organizational focus. At the same time, some of those same water agencies have resisted the pressures for change, and the water-industry culture in many ways has persisted. While some of the old-guard leaders of the agencies are now called "water buffalos," the changes remain contested, and the old mission and focus survive in part.

Nevertheless, despite the resistance of the water buffalos, new players and a changing environment for water issues have appeared in Southern California and other parts of the country. For example, a couple of years after I left the MWD board, I helped establish in 1991 a new network of environmentally oriented water-agency board members and staff in California. Instead of a handful of activists at the margins meeting in the living room of one of its participants, those who became involved in this new network ten years later had emerged as "water leaders." The new network called itself POWER, or Public Officials for Water and Environmental Reform, and included board members from the Los Angeles Department of Water and Power and MWD, the East Bay Municipal Utility District (EBMUD) in Northern California, and even some board members in the fast-growing suburban water districts who recognized the need to reexamine the expanding supply and development-oriented approach of the agencies. The focus of POWER was to identify those alternative scenarios, open up and make more transparent the public water agencies, and explore issues like watershed management, integrated resource planning approaches, normal and dry-year conservation strategies, and other ways to manage demand. Some of POWER's participants also advocated a new "landscape ethic" to complement the idea of a new "water ethic" that could govern the strategies and programs of water agency and water user alike.[46]

Two of POWER's key players who had been among the earlier group of water activists seeking to develop alternative scenarios were Dorothy Green and Tim Brick. Dorothy Green was a self-taught water and environmental advocate. A housewife living in Westwood, she first became involved in water issues during the late 1970s during the L.A. DWP's continuing battles with environmentalists and Inyo County residents regarding Owens Valley and Mono Lake issues. She subsequently became engaged in the Peripheral Canal battles, where she helped convene the group of water activists engaged in that campaign. Both passionate in her advocacy and hungry to expand her knowledge about issues and strategies for change, Green became a kind of doyenne of the water activists. She gathered in her living room the different participants involved in the planning sessions, campaigns, or actions designed to challenge the prevailing wisdom around water policy. In the process, she helped initiate a number of key water-related groups and networks. This included the Heal the Bay organization that addressed ocean pollution and runoff from urban and industrial sources; Unpave L.A., a collection of advocates who focused on landscape and hardscape issues and who popularized the term "landscape ethic"; and the San Gabriel and Los Angeles River Watershed Council organization, a gathering of environmental advocates, water agency officials, engineering types, and even staff from the Army Corps of Engineers and L.A. County Public Works Department who had long been regarded as hostile to any approach other than the laying of concrete. By the late 1980s, Green's philosophy had evolved, and she was arguing that key environmental approaches, such as a focus on watersheds, were ultimately "win-win" strategies since they were efficient, had a sound technical basis, and could meet multiple goals, whether flood management, recreational opportunities, or environmental benefits in the case of the L.A. River and San Gabriel River watersheds. For Green, the development of the POWER organization provided a potential venue for exploring such approaches while including former antagonists within the water industry as part of a new dialogue. Along those lines, Green, who had been appointed by then Mayor Tom Bradley to the L.A. DWP Board in the early 1990s, helped organize an annual conference under the auspices of POWER. The conference brought together the different players both inside and outside the water industry in a context of dialogue rather than action, a process that had also begun to take place with the establishment, in the post–Peripheral Canal election era, of the California Bay–Delta Authority or Cal Fed, a collaboration of twenty-five state and federal agencies to address Sacramento Bay Delta issues.[47]

Similar to Dorothy Green, Tim Brick has gone from water rebel to water leader, suggestive of the changes in the water industry. An old friend and long-time activist, Tim participated with me in a number of anti–Vietnam War and community-based organizing initiatives in the early 1970s. He subsequently became actively involved in energy and water issues related to the Los Angeles Department of Water and Power and then joined me and others in the campaign against the Peripheral Canal, where he became the coordinator for the opposition campaign in Southern California. Like myself, Tim joined the inner sanctum of the water industry when in 1985 the city of Pasadena appointed him as its representative to the Metropolitan Water District board of directors. The two of us overlapped for a couple of years until I left the board in December 1987. Continuing to raise issues about the need for new approaches, Tim emerged on the board as a key advocate for more integrated water-resource planning and the importance of watersheds in identifying how to approach water-policy development. Tim also continued to consider himself an activist and subsequently became a key player in his own community regarding the Arroyo Seco subwatershed, which fed into the Los Angeles River. In that capacity, he became the project director for Pasadena's Earth Day Festival and as the founder and director of the Arroyo Seco Foundation. Then in October 2006, Tim accomplished what would have seemed inconceivable to those of us who had met as a group of water activists twenty-five years earlier. With the longest tenure on the MWD board and while continuing to preach the language of watersheds and resource management, Tim Brick pulled off an upset when he got a majority of the board—including the Los Angeles and San Diego delegations—to support him as chair of the board of MWD. The election clearly took a number of water insiders by surprise. But it also opened up the possibility that water policy at the MWD could be potentially aligned with new agendas for greening the region and making water politics itself a more visible and accountable arena for action.

Tim Brick's election was not entirely a surprise to those who had been part of the process of trying to bring about those changes. The introduction of alternative scenarios, the changing faces among water-agency participants, and the development of an alternative language of water had impacted a number of the key water players, including MWD. For the first time since the

1950s, the District had decided, even prior to Tim Brick's election, to change its mission statement to include an environmental component. It had established an integrated resources planning initiative that actively pursued ways to stretch existing supplies and focus on the demand side of water use. It had explored the possibility of linking its imported water supply in conjunction with the groundwater basins in the MWD service area. It had developed a series of landscape and native-plant programs, including a "California Friendly Landscapes" funding initiative to award cities in the region that introduced low water use and native-plant landscaping. And it had also established a Community Partnership Program with small grants for community-based organizations partnering with local agencies that had been the brainchild of Adan Ortega, an MWD assistant general manager who had previously served on the Heal the Bay board of directors.[48]

By 2007, MWD had gone through changes that had significantly altered some of the ways it pursued its goals, although it had yet to fully alter several of the goals themselves, a challenge not only for Tim Brick and Dorothy Green but for a water-activist community that had helped change the nature of the water debate. That challenge included the shifting argument that obtaining an ample supply of water for the city remained the essence of water policy and that the commodity value of water still continued to trump its community value. The alternative scenarios that had become part of the dialogue around water had forced the agencies to respond to new initiatives regarding protection of water sources, managing the water supply differently, assuming that conservation could play a significant role in stretching supplies, and focusing on watersheds—the key to the community-value-of-water approach—as a basis for the development of new kinds of policies. And while the scenarios affecting water for the city had become a place for conflict as well as new visions and dialogue, the question of water in the city emerged as the fiercest of the battlegrounds around water. It was here that the opportunities to construct a new language and new politics of water that had once seemed least likely were now becoming most available.

4

Re-Envisioning the Los Angeles River

The Explorer and the Poet

How does one discover the Los Angeles River? In 1985, two people—a poet and a journalist who called himself "the Explorer"—set out to identify not only where the L.A. River could be found but what it represented.

The journalist, *Los Angeles Times* writer Dick Roraback, was given the task to write about a river that few in Los Angeles knew even existed. The structure of his twenty-part series, published over several months, was based on Roraback's journey from the river's mouth to its source. This reverse process of discovery was designed in part to answer such questions as "What is the L.A. River?" and "Where does it begin?" The articles were part comic relief, part ironic juxtaposition. They were written in the third person, through the voice of the Explorer, a narrator whose mocking tone reflected the intent of the series.[1]

Roraback began his series at the mouth of the L.A. River near the ocean in Long Beach. Because it has a soft bottom area due to the high level of the groundwater table, this short stretch of the river was not fully channelized, and the Explorer happily reported that he had succeeded in finding an actual river. His happiness, however, was short lived. As he heads upstream, the narrator comments ruefully: "For a last time, the Explorer looks south, at the real Los Angeles River. A heron-like bird, easily four feet tall, stands motionless in the stream, graceful and haughty." This glimpse of nature in the city quickly fades as the bed of the river changes to flat concrete and passes through the old industrial section southeast of downtown Los Angeles, consisting of some of the densest communities in the region. Here the Explorer sees the river as a "desolate vista, a wasteland . . . just a threadbare coat of unspeakable slime." Drawing on river analogies and poking fun at every opportunity about the degraded river and its nearly nonexistent flow, the Explorer wonders whether he has discovered " 'Old Man River' in drag' " or the " 'Beautiful Blue Danube' in a mudpack." When he finally reaches what he assumes is the river's source, the Explorer complains that he never really did find what could be considered a river. "It hasn't any whitecaps. It hasn't any fish," the Explorer laments. "Just to see one ripple would be my fondest wish." Instead, he moans, the L.A. River "just hauls its load of sad debris from the sewage pipes to the mighty sea." "Ooze on, L.A. River, oooooze on," Roraback urged as he concluded his ironic homage to this bleak part of the Los Angeles landscape.[2]

While Dick Roraback was pursuing his Explorer role, Lewis MacAdams, a poet, activist, and one-time board member of a Northern California public utility district, had decided to make his own statement about the L.A. River through a type of performance art. With two other friends, MacAdams, who settled in Los Angeles a few years earlier, borrowed a pair of wire cutters, went to one of the fences that had a "No Trespassing: $500 fine or 6 months in jail" sign, and defied the warning by cutting a hole in the fence to enter the river. In an act that was part theater and part action designed to spur organizing, MacAdams descended to the river bottom at a location just north of downtown Los Angeles close to where one of the river's other soft-bottom areas gave way to a concrete channel. He then proclaimed that the river still lived below the concrete. "We asked the river if we could speak for it in the human realm. We didn't hear it say no," MacAdams would later comment on his act on a number of occasions. By cutting through the fence—one of many fences "that had kept people away from the river for decades," as he put it—

MacAdams and his two friends declared that their intent was to open the river. Later that night, he did a show at the Wallenboyd Theater that he called "Friends of the Los Angeles River." He painted himself green, wore a white suit to represent William Mulholland, and then turned himself into various river animals like a hawk, an owl, a rattlesnake, and a frog. "I was really out there," MacAdams later told *The New Yorker.* A *Los Angeles Times* reviewer agreed, wondering whether his performance was really supposed to be a put-on, like an Andy Kaufman event. "With friends like MacAdams," the *Times* critic declared, "the river needs no enemies."[3]

Following his panned performance art action, MacAdams and his co-performer activists Pat Patterson and Roger Wong established over the next several months a new group they called "Friends of the Los Angeles River" (FoLAR). For this tiny band of activists, the initial goal of the organization was to focus on language and symbols by insisting that the L.A. River was indeed a river. MacAdams, whose activist roots were more bound up with his identity as poet and affinity for imaginative 1960s-style protest than any specific environmental or river-advocacy lineage, tended to attract like-minded artists, planners, architects, designers, and neighborhood activists in this quest to "bring the River back to life," as he wrote in a letter to the editor of the *Los Angeles Times* in response to Dick Roraback's series. "[The River] doesn't have to be treated like a sewer," MacAdams complained in the letter. In response, Roraback happily placed a FoLAR bumper sticker on his car, getting in turn skeptical stares and one suggestion that the bumper sticker ought to have a subtitle: "Sons of the Ditch."[4]

For Roraback and his readers, the L.A. River, at best a memory of nature in the city, was symbolic of its absence in L.A. This "alleged L.A. River is finally a joke," Roraback said of his journey, passing through what he characterized as the river's "comical crawl." For Lewis MacAdams, the discovery of the river was just about to begin. It would transform him in the process into the organizer and the poet laureate of the river, prepared to undertake what he characterized as his "forty-year art project" that would hopefully lead to the river's own transformation.[5]

REMAKING THE RIVER

When Dick Roraback undertook his journey and Lewis MacAdams his performance art, a free-standing L.A. River had become more memory than reality. The L.A. River has, in fact, experienced multiple lives, from the pueblo era to the boom town and eventually to a continually expanding urban land mass. "Making the river a critical part of the landscape made sense in the early days of the little village's history," historian William Deverell wrote of the pueblo period during the eighteenth and nineteenth centuries. "Local knowledge, based on lived experience in the Los Angeles basin, incorporated the river into the rhythms of everyday life. Los Angeles needed no more water than the river could provide, and it was an especially prominent landscape feature along with other local markers such as the Pacific Ocean or the San Gabriel Mountains."[6]

Today, L.A. River restoration advocates seek to invoke the historical image of a free-flowing river filled with those willows, cottonwoods, watercress, and duckweed. The river in turn was able to fertilize a rich soil that was "black and loamy and [capable] of producing every kind of grain and fruit which may be planted," while its "banks were grassy and covered with fragrant herbs and watercress," as its first Spanish chronicler Father Juan Crespi wrote about the river he called Porciuncula in his diaries back in 1769, a sentiment about a lost river often evoked by river advocates. But the river also served the communities that grew up around it. They used it as a source for drinking water and for irrigating agricultural land (including during flood episodes) throughout much of the nineteenth century. In the mid- to late nineteenth century and the early part of the twentieth century, areas adjacent to the river became available as cheap land for railroad yards or for the industrial plants that settled along its edge as part of the East Side industrial corridor. And in 1930, it became the centerpiece of a Chamber of Commerce–commissioned study by Harlan Bartholomew and Frederick Law Olmsted. This study and a subsequent series of articles in 1931 by real estate attorney and city planning advocate W. L. Pollard laid out a vision of greenbelts, parkways, and new park lands that could also serve as flood-control strategies. These concepts were occasionally referenced in later planning and park-development proposals, but they achieved more widespread recognition when the Olmsted-Bartholomew report was discussed by Mike Davis in several of his writings on L.A. during the 1990s and when the report was reprinted in 2000, seventy years after its initial publication, with a new introduction by L.A. historians Greg Hise and William Deverell.[7]

Most significantly, through much of the early twentieth century, as the city began to go through its periodic boom and bust cycles, the L.A. River began to lose some of its visual appeal as anchoring the region's attractive landscape. Instead, it came to be seen as a barrier for existing and future residential and industrial development along its path, owing to its propensity to carry rapidly flowing flood waters during the occasional but fierce storms that periodically occurred. The L.A. River, in fact, was both symbol and substance of Anglo Los Angeles's complex view of its surrounding environment. On the one hand, early promoters touted the "grand views" available when surveying this wondrous region. "A month may be spent in explorations and still fresh beauties found," one 1891 commentator noted, as he viewed the river from his vantage point at Boyle Heights. "East Los Angeles looks like a vast forest or park, so thickly is it embowered in shade trees," he said of this area that would in two decades be reconfigured by its industrial and residential settlements. But already by this period, warnings began to note the river's capacity to descend quickly and explosively and have a significant impact on development. Changes in the landscape, including deforestation of the mountain areas where the river's flow increased in intensity generated warnings that the city was in danger from "a sudden flood delivery of the rainfall from the burned and bare mountains" that would create "a torrent bed running right through the city," as Abbot Kinney, the one-time chair of the California Board of Forestry and the developer of the beach community of Venice, put it in a 1899 lecture at USC.[8]

The focus on the L.A. River also began to change with the construction of the L.A. Aqueduct and the availability of Los Angeles' first important imported water supply. With the L.A. River no longer central to the planning regarding L.A.'s growth—as water *for* the city and its soon to be annexed territory—the River increasingly came to be seen as a hazard rather than a supply source when water flowed *into* the city during major storms. Just one year after the Owens River supply reached its destination point in the L.A. area and while the annexation debates were beginning to heat up, the L.A. River unleashed its own fury in several areas of the city through a series of storms that affected a wide range of areas throughout the county, including the Arroyo Seco in the northeast and in areas to the south of downtown. Although knowledge about the capacity of the L.A. River to flood had been raised throughout the history of the pueblo and even during the early years of the Anglo development, the 1914 floods took the city's political and business elites by surprise. One *Los Angeles Times* editorial argued that the city "never had reason to think that so destructive a torrent could really come rolling down from the hills and through the gully that had been lightly termed the 'Los Angeles River,'" suggesting that the city would in the future be committed to an engineering solution that would prevent future floods. The 1914 floods also led to a wide range of initiatives and a continuing series of debates about how best to proceed, with the newly formed Los Angeles County Flood Control District charged with establishing the organizational infrastructure and technical capacity to try to contain the river's flow in any future storm episode.[9]

But local initiative and expertise were not sufficient to handle future flood episodes, as new development outpaced the city's capacity to establish sufficient measures to limit damage from the river. How much to spend and where to spend it also became concerns as the Depression-era squeeze on local funding created barriers for flood-control measures and also for land-use strategies that could have served as a form of flood management. But cost wasn't the only issue. The patterns of residential and industrial development along the river's path, as well as the loss of farm land and the placement of railroad and utility lines by the river's edge, had severely limited the ability to establish nonengineering land-use approaches such as the Olmsted-

Bartholomew plan that linked parkway development with opportunities to create open spaces and parklands to serve as buffers for flooding, including along the Arroyo Seco and Los Angeles River. But these nonengineering approaches were not considered feasible and also went against the dominant land-use patterns that essentially undercut any alternative strategy for flood management. As a result, in 1934 and again in 1938, two major storms created intense pressures for a far more massive engineering solution and thereby helped facilitate the entry of federal dollars (part of a broader New Deal job-creation strategy associated with public-works projects) to initiate a wide range of construction projects to effectively (and, it was assumed, finally) manage the Los Angeles River to prevent future flooding.[10]

The Army Corps of Engineers, armed with new federal flood-control legislation, defined this new mission as a declaration of war on the river. "Rebuilding this fractious stream presents engineering novelty," one commentator exclaimed, and the Army Corps embraced its role of redesigning a river's course and its very function. From 1938 (when the Army Corps began to straighten the river by widening it, pouring reinforced concrete along much of the river's flow, building and raising several flood-control basins, and taking a series of flood-control measures along the river's tributaries and washes that emptied into the river) through the 1980s (when Dick Roraback set out to find his lost river), the river became transformed into a fifty-one-mile flood-control throughway. This "water freeway" now had a singular purpose: when it rained fast and hard, it sent the flow of water surging out of the San Gabriel Mountains quickly to the sea. Similar to the flood-control projects that also sought to reconfigure urban streams and rivers around the country during this same period (though not of the same magnitude and extent of the intervention required in Los Angeles, given the region's particular geographic, hydrologic, and land-use features), the now channelized L.A. River essentially redefined the urban landscape along a north-south axis. Areas surrounding the river became fenced off, a forbidden territory that effectively belonged to the engineering agencies. For these flood-control managers, this was now "the river we built," as one Army Corps engineer described it, encased as it was in concrete in all but three of the soft-bottom sections of its fifty-one-mile path.[11]

The emergence in 1985 of FoLAR and its advocacy around the need to bring back a "living river" was, at first, barely noticed by the river's primary gatekeepers, the engineer/managers in the L.A. County Department of Public Works and the Army Corps of Engineers who had undertaken the construction of the channel and other flood-control measures. More of a nuisance than a significant challenge to the roles they had assumed as flood-control managers, the river advocates were ignored rather than contested. Three sets of events during the 1980s and 1990s, however, forced the flood-control managers to respond to FoLAR's insistence about a living river, creating what turned into a war of words and competing symbols whose outcome would influence the policy and institutional framework around river management and related land-use issues.[12]

The first event involved the flow of the river itself. In 1984, the city of Los Angeles's Department of Water and Power began operating its Donald C. Tillman Water Reclamation Plant north of the Sepulveda Dam near the source of the river and one of the three soft-bottom areas where the groundwater table was too high to allow for channelization along the river bottom. This tertiary sewage-treatment plant discharged directly into the river. Along with the releases of two other smaller treatment plants south of Tillman, these discharges provided a year-round flow of water for the river. As a consequence, it increased the vegetation growth and habitat along the soft-bottom areas and reinforced the FoLAR argument that, at least in these stretches of the river, visually and functionally (in relation to a renewed ecosystem), the L.A. River had become once again a free-flowing river. Even with the negative symbolism of treated sewage as its water source (a source nevertheless cleaner than the runoff flow, given the tertiary treatment process involved and the large pollutant loads in the runoff), this new river flow reinforced the appeal about a living river. "Come down to the river," became a constant refrain in talks by MacAdams and his FoLAR allies, an action they considered essential to legitimating their argument about the river.[13]

The second event involved a concept put forth by then State Assembly member Richard Katz. In 1989, Katz proposed that the river, much of it channelized and lacking any human contact, could serve as a "bargain freeway" for trucks and automobile traffic. The River Freeway concept was further explored through a $100,000 L.A. County Transportation study that concluded that such a river freeway could result in a 20 percent reduction in congestion for two nearby freeways. Katz, who was a major advocate of water transfers and sought to appeal to environmental groups in his subsequent run for mayor and the State Senate, also spoke of greenbelts, bikeways, and adjacent parks, which were ideas that had been promoted by several of the river advocates. Katz's argument about a river freeway was never seriously pursued and was eventually ridiculed by the media, but it nevertheless created a new kind of focus on the river that FoLAR and its allies were able to exploit. "Why not reenvision the L.A. River as an actual river?" the activists argued in documents and materials they generated, in events they hosted (such as river cleanup days and kayak rides along the soft-bottom areas of the river), and in the increasing number of press interviews and articles that identified their vision of a living river.[14]

The third event involved the protracted battle over the Army Corps' proposal to raise the channel walls in the downstream segment of the river prior to its entering the Long Beach Harbor area. This was the same segment of the river that Dick Roraback had characterized as a wasteland. In 1987, the Army Corps produced an update for its L.A. River master plan for the Los Angeles County Drainage Area (LACDA) that included the warning that disastrous flooding could return to Los Angeles County and that residents that lived along the river, including in low-income communities like Lynwood, Compton, and Paramount, were advised to "take action"—that is, to purchase federal flood insurance. Over the next several years, the Corps proposed a series of measures to address a number of problems that had emerged subsequent to the river's channelization. These included increased residential development along the river's edge, debris flow concerns, and emerging fears about flood-damage insurance for homeowners owing to FEMA's redrawing of its maps that now indicated that certain areas bordering the river, including several of the working-class neighborhoods that had been some of L.A.'s earliest suburban-industrial clusters, would now be considered "flood-hazard zones." Residents of those areas called the FEMA action a "flood tax" and strongly opposed any effort to prevent the Army Corps from completing LACDA. Having asserted that neighborhoods bordering the L.A. River, particularly those downstream in Dick Roraback's wasteland areas, required protection from a hundred-year flood (an assertion that had triggered FEMA's redrawing of the maps), the Corps plan included widening the channel, modifying bridges, and, most controversially, constructing new parapet walls from two feet to as much as eight feet higher than their existing height.[15]

Almost immediately, the Army Corps' LACDA proposal became a flash point for river advocates who sought to challenge the Army Corps' approach while identifying their own alternative plans to contain the L.A. River as a counterpoint to the raising of the walls. The credibility of the river advocates was also subsequently enhanced by their participation in a Los Angeles River Task Force, established in 1990 by L.A. Mayor Tom Bradley to "articulate a vision for the river." Through the early 1990s, the LACDA fight became the centerpiece of the debates over the future of the river. On the one hand, the Army Corps and the L.A. County Department of Public Works argued that the LACDA proposal was simply an extension of their mission as flood-control managers ("to keep water and people apart," as one Army Corps official argued) and that to call the L.A. River a "river" was a misnomer. The L.A. River, they declared, served just two purposes—to keep flood waters from destroying property and lives ("a killer [that was now] encased in a concrete straight jacket," as one water-agency publication put it) and to serve as an outlet for discharges from the sewage-treatment plants. FoLAR countered with its plan that included the twin concepts of "restoration" (tearing up the concrete where feasible) and "flood management" (including creating some new parkland and implementing some land-use approaches).[16]

In one memorable encounter, described by Blake Gumprecht in his history of the L.A. River, Jim Noyes, the chief deputy director of the L.A.

County Public Works Department, got into a sharp exchange with FoLAR's MacAdams over what term to use when describing the L.A. River. Each time Noyes used the term *flood-control channel* as part of a presentation he was making, MacAdams would interrupt to declare "you mean *river*". This happened again and again, with Noyes insisting on using the term *flood-control channel* and MacAdams interrupting each time to assert "*River*!" MacAdams later recalled the incident as turning "really ugly," with Noyes becoming more and more furious. "I saw him a couple of days later," MacAdams told Gumprecht, "and he wouldn't even speak to me." Despite FoLAR's newly established visibility and the increased interest by policymakers and the media in a "renewed" L.A. River, the County Board of Supervisors in 1995 voted four to one to allow the LACDA plan to proceed. Given the changing discourse around the river, the focus on the limits of traditional flood-control strategies, and concerns about community blight, the LACDA plan was modified to the extent that in some areas levees were raised and walls were not built. In addition, adjacent bike paths were maintained, and more plantings and vegetation were added to counter the "urban blight" and "wasteland" characterization of the engineering approach. And although the river advocates had lost a battle, they continued to make inroads in how the L.A. River was to be defined and how it might ultimately be managed and renewed. "I always saw [the LACDA fight] as a symbolic issue," MacAdams recalled, arguing that the war of words was in fact "a battle over the definition of the river, and what the river is going to be."[17]

CREATING A COLLABORATIVE

In the fall of 1998, Lewis MacAdams came to Occidental College to speak to my Environment and Society class about L.A. River issues. In the course of the discussion with students, MacAdams commented that a key dimension of FoLAR's approach was changing the image of the river, an image, MacAdams speculated, that had been shaped in part by its portrayal in Hollywood films. MacAdams suggested that research was needed regarding how such images and the language about the river were framed, how the management of the river had evolved historically, how it could be reengineered differently, and how planning could help reconfigure the river and the lands adjacent to it. "Today, the research regarding the River serves as barrier rather than opportunity for renewal. Can you help make that renewal possible?," MacAdams challenged his audience.[18]

MacAdams was also interested in a possible partnership with our Urban and Environmental Policy Institute (UEPI). It seemed to me to be a good fit. I had recently come to Occidental after more than fifteen years at UCLA in the urban planning department. Having parachuted into UCLA (my first relationship was as a client for a graduate student group project analyzing the institutional issues related to the MWD), I had sought to carve out a role as researcher, educator, and activist and to continue writing about the environment, history, and development of Los Angeles and the resource issues (like water and energy) that had shaped the major urban centers in the West. In 1991, I had cofounded, with John Froines and David Allen, an interdisciplinary environmental center called the Pollution Prevention Education and Research Center (PPERC) that linked the departments of chemical engineering, public health, and urban planning. PPERC had two distinctive components—(1) a research and teaching aspect that drew on the technical and research capacity of the different disciplines to address urban and industrial environmental issues and (2) an "action research" program that was designed to help develop new public policies and establish linkages with key stakeholders, including community-based organizations. With a focus on the work environment, industry issues, and urban daily-life concerns, PPERC, owing to its interdisciplinary and action-oriented nature, operated as a "small c" interdisciplinary center at the edges of the University of California's complex system of research centers and departmental silos. Located on the west side of Los Angeles in a wealthy neighborhood, most UCLA researchers tended to focus less on neighborhood and regional issues and more on national and global questions, a research orientation that began to change when the 1992 civil disorders in Los Angeles sent shock waves through the region.[19]

When the opportunity arose to move to Occidental in 1997, I was happy to do so, given the nature of the appointment and the college's mission, location, and focus on Los Angeles. Occidental is a small, diverse liberal arts college located in the Eagle Rock neighborhood of northeast Los Angeles, not far from the area where the L.A. River begins to enter downtown L.A. and where Lewis MacAdams had in 1985 cut his hole in the fence to go down to the river. With my new position defined as part teaching, part research, and part community engagement regarding environmental issues in Los Angeles, I was able to reestablish several of the projects of the Pollution Prevention Center at Occidental. With colleagues from Occidental and managers from several of our UCLA projects who had also joined me at Occidental, we decided to expand the center's focus and rename it the Urban and Environmental Policy Institute (UEPI). UEPI became a visible example of Occidental's own commitment to community engagement and "learning by doing." Thus, while UEPI was located within an academic program (urban and environmental policy), it was able to strengthen and significantly expand its community emphasis. It also sought to define itself as a multifaceted, social-change-oriented institute that provided a place where faculty, students, organizers, community partners, researchers, and policy analysts could collaborate. Its mission—"to help create a more just, livable, and democratic region"—became the backdrop for bringing together community groups and researchers. In this way, UEPI came to be defined as part academic center with strong community ties and part community-based organization with a strong research and policy-development capacity.[20]

A major gap in UEPI's approach, particularly given its new location, was the absence of any research, educational work, or policy and program development regarding the Los Angeles River. MacAdams's challenge was timely: L.A. River renewal advocacy had reached a critical point. FoLAR and its allies had been able to challenge the prevailing engineering language regarding the river to the point that the idea of renewal was no longer marginalized, even if initiatives to change the approaches to river management still appeared out of reach. A series of gatherings and charrettes at four different sites including Chinatown and Boyle Heights called The River through Downtown had produced an exciting mix of proposals and early outreach activities concerning land adjacent to the river, particularly an undeveloped parcel just north of downtown across from Chinatown known as the Cornfield and a second site to its north called Taylor Yard that the railroads had used as rail yards. The river advocates had also initiated their Down by the River programs, such as their annual river clean-up program (La Gran Limpieza) and their monthly walks and carefully staged kayak rides and had received increasing attention in the press and vocal support from various officials and policy makers. Important allies, such as the Trust for Public Land and the two major land conservancies in the region, the Santa Monica Mountains Conservancy and the Coastal Conservancy, sought to extend their agenda and their operating boundaries and focus on opportunities to purchase land adjacent to the river for park development and to increase awareness about the river's history and possible rebirth.[21]

A number of important policy initiatives were also coming to the fore. This included then State Senator (and long-time activist) Tom Hayden's legislation calling on the MWD to address watershed issues in its service area, including the L.A. River watershed. Another piece of legislation, then under consideration, would establish a new San Gabriel River and Lower Los Angeles Rivers and Mountains Conservancy that could address issues north and east in the San Gabriel Valley watershed as well as the area south of downtown Los Angeles where the LACDA fight had been most pronounced. While impressive in their potential reach and the fact that they were under consideration at all, given the apparent quixotic nature of the quest for river renewal, these initiatives nevertheless constituted a kind of scattershot approach to river renewal. Underlying these efforts, moreover, was the continuing "discourse battle," still far from resolved, pitting the language of river renewal against the sixty-year history of flood control and its own language of danger and hazard. Despite the successes of the river advocates in shifting some of the terms of the debate about river management, for most Angelenos the L.A. River, while perhaps a bit more visible than when Dick Roraback wrote his series, was still off limits, a contained hazard that was filled with debris, guarded by barbed wire, marked by graffiti, and inhabited at its banks by the homeless.[22]

The need that MacAdams articulated when he spoke to the students was for not just one more new research or educational program but a way to situate those various policy and discourse battles in their broadest context—as a set of historical, cultural, political, environmental, engineering, planning, land-use, recreational, and physical activity questions. It was important to see L.A. River issues as environmental *and* community-based and to find ways to explore river issues as an integrated set of issues as well as through a more innovative approach from each of those different entry points. To do so effectively (we hoped when we first sat down to talk about a collaboration) would provide a way to reenvision not just the river but the entire Los Angeles region.

By the summer of 1999, we had agreed to pursue an intense and ambitious type of program that would advance ways to reenvision the river and could identify a "right to the river," an idea I associated with the suggestive concept of the right to the city. It was agreed that UEPI and FoLAR would cohost, along with dozens of other community-based organizations, environmental groups, policy makers, and academic participants and cosponsors, a one-year series of events, activities, forums, and research under the heading Re-Envisioning the L.A. River: A Program of Ecological and Community Revitalization. More than forty programs were to be scheduled to capture the multiple ways to explore and analyze and act on L.A. River issues. Programs for the series would include historical, engineering, environmental, and watershed discussions; poetry readings and art installations; and bike rides and walks along the river's edge. There would be a Hollywood Looks at the River forum where a film montage entitled *River Madness*, commissioned for the program series, would be screened. And additional sessions would focus on the the Arroyo Seco, the San Gabriel Valley, and the Southeast L.A. areas of the broader, linked watershed.[23]

Research initiatives included a client-based research project undertaken by a UCLA Urban Planning team working in conjunction with UEPI (as the client) to provide a community and planning profile of the Cornfield area and an assessment of alternative plans to develop it. The battle over the fate of the Cornfield, which emerged as a key subtext of the Re-Envisioning the L.A. River program, would ultimately emerge during the course of the Re-Envisioning program as the first major debate over the fate of the river and its surrounding areas since the protracted battle over LACDA that had occurred during the early and mid-1990s. With the timing right and with a large number of environmental and community-based organizations as well as cultural groups, bike riders, and K–12 principals, teachers, and students signing on to the programs while the two host organizations worked out the details, the Re-Envisioning the Los Angeles River series was launched.[24]

"A Very, Very Pretty Duck"

The opening session of the Re-Envisioning the L.A. River series on October 1, 1999, included talks by environmental officials Mary Nichols and Felicia Marcus. Nichols spoke of the critical importance of forums like the Re-Envisioning series in focusing the attention of policy makers regarding open space and river revitalization as community issues as well as environmental concerns—issues that needed to "cut across geographic, ethnic, racial and class boundaries." Marcus talked of the importance of combining a vision of the river's transformation with the ability to act in a practical and incremental manner. She spoke of the need to establish new management paradigms while recognizing the barriers and gently pursuing a shift in the traditional agendas of the engineers and water-industry and flood-control actors who had managed the river for more than six decades. "We have to learn to speak 'DWP,'" Marcus argued, referring to the obscure and technically oriented language of the Los Angeles Department of Water and Power. This learning process needed to extend to other water and engineering agencies as well. While the river might not be "the [trumpeter] swan in L.A.'s future," Marcus concluded, "it could be a very, very pretty duck," citing writer Jennifer Price's compelling metaphor of a reenvisioned river in an article about the series that had appeared shortly before the Nichols and Marcus talk.[25]

The opening session helped set the backdrop for two of the key programs in the series—the history and engineering-related forums. The History of the River forum included Blake Gumprecht, who challenged some of the assumptions of the river-renewal advocates, suggesting that the river's past had been "romanticized" and that because of Los Angeles's unique environment of mountains, climate, extremely variable rainfall, and the river's propensity to flood, "there were very good reasons that the river was entirely cased in concrete." The second panelist, Jared Orsi, then at work on his own book on the L.A. River, also sought to place in historical context some of the first debates about river management associated with the 1914 floods, arguing that the Progressive Era faith in experts, which included the technical expertise of the civil engineers, was as much a political and cultural choice of what to do with the river as a purely technical one. Orsi suggested that the recent history of river advocacy, however, had brought new people to the table "with different values and backgrounds and talents and skill sets" to be part of the dialogue about the river. The third panelist, William Deverell, who had written extensively about the pre-Anglo and early Anglo development of Los Angeles during the nineteenth and twentieth centuries, spoke of the L.A. River's central place in the history of the Los Angeles Basin and as a critical feature of the regional landscape that was transformed as a flood-control channel into an invisible river that failed to satisfy the dominant aesthetic understanding of what a river should or should not be. The river, Deverell warned, contributed to the kinds of jurisdictional, neighborhood, racial, cultural, and class divides that have characterized Los Angeles. "To overcome those divides and for Los Angeles to build community," Deverell asserted, "the river's a really good place to start."[26]

The history panel, despite the different interpretations of how historical insights informed the current debates, was nevertheless important to the river-renewal advocates. While Gumprecht's position could be interpreted as a defense of the ultimate fate of the river as a flood-control channel, he also argued that other choices could have been made. The city, for example, at the turn of the twentieth century had failed to address its high per capita water consumption (among the highest in the nation). This was caused in part by its lack of water metering, its desire to promote its image as a "garden" based on ample water, and a rate structure that favored wealthier residents who were the largest users of water. In contrast, the prevailing choices included the land-use decisions that had increased the potential for flooding impacts and the push for an imported water source as the basis for the city's expansion. As those choices were implemented and the river evolved into a flood-control channel, its engineer-managers had sought to control how the river was to be managed and had closed off discussion about the history of the river itself, situating it exclusively as a story of periodic floods that required their intervention to finally contain those dangers. By arguing that the river's fate was neither inevitable nor isolated from the choices about the development of the city and the region, it opened up the discussion about the past in the context of the current debates about the future of the river and the region.[27]

The opportunities for new insights and interpretations about the past and for what choices might be available for the future of the river also framed the discussions about the technical and engineering issues in the session entitled Managing the River. Six months prior to the panel, at the dedication of the Heron Gates, a striking sculpture that opened to one of the only foot paths along the side of the L.A. River north of downtown Los Angeles, Harry Stone, then head of the L.A. County Department of Public Works, proposed that the county, for the first time ever, would explore the possibility of removing concrete from the L.A. River. The timing of the Managing the River session, coming shortly after Stone's comments and the earlier bruising LACDA fight, provided the first opportunity to gauge whether the engineering-driven agencies like the Army Corps of Engineers and L.A. County Department of Public Works were in fact ready to undertake a major paradigm shift and, if so, how a project designed to take concrete out of the river could actually be accomplished.[28]

The session indicated not only opportunities for change but the barriers that still remained deeply embedded in the culture and practices of the engineering agencies. On the one hand, the speakers who emphasized the need for a paradigm shift focused on the strategies already available for river and stream restoration, including those successfully implemented at other urban river sites that could inform any L.A. River restoration effort. Emphasizing the need for new management tools, the restorationist-oriented panelists spoke of the need to go "beyond the concrete era," through such varied techniques as revegetation and tree plantings, creating land barriers between homes or businesses and the river that could also serve as open space, and reducing the stormwater flow from the urban hardscape *into* the river through the emerging strategy of stormwater management that addressed a wide range of urban land-use, transportation, and environmental practices. By doing so, perspectives could be broadened beyond the river to the flood plains and the watershed, creating more of a system understanding, a crucial step for enabling engineers and other gatekeepers of the river to think ecologically.

The panelists from the engineering agencies commented that the approaches from the concrete era were still crucial for purposes of flood control, but they recognized that they could be complemented by other kinds of initiatives. Agencies like the Army Corps of Engineers, it was suggested, were slowly transitioning away from their long-term, single-purpose flood-control mission. However, the economic and social costs in any major change of direction, they argued, could be enormous. "Our bias at the field level is trying to give clients what they want, but those would have to be clients able to bring money to the table," commented panelist Eldon Kraft, the chief of the Army Corps of Engineers Los Angeles District's planning section. That bias had been reinforced over the years by the way costs—and benefits—had themselves been calculated by the Army Corps. As panelist Ann Riley, the director of the Waterways Restoration Institute, pointed out, the benefits from restoration projects that opened up new kinds of land uses such as river walks, new park lands, or new kinds of river-connected community identities were excluded from the benefit side of the ledger. Moreover, the Army Corps failed to identify in their calculations some of the long-term flood-control costs such as repair and maintenance of the channels and levees as well as requiring additional structural improvements such as those the LACDA plan had called for. Most important, Riley argued, the Army Corps approach had established a framework where benefits were calculated first for situations that advantaged structural interventions—creating the flood-control channel or raising the walls of an existing channel—rather than helping "a community avoid or prevent a hazardous situation," such as a land-use-based prevention approach through creating land buffers. That, in turn, magnified the economic problems associated with an ecological approach, placing the burden of change on the restorationists rather than on the engineers who had little experience and, for many of the agency personnel, little inclination to assume that a paradigm shift was necessary.[29]

Nevertheless, the Managing the River session did indicate that the engineering agencies were adopting a more defensive posture and that the discourse battles, apparent in the LACDA fights, might be ready to take a new turn, suggested by the forum session itself. While disagreements between the agency representatives and the restorationists were apparent about the scope and scale of change, there was also agreement that change was increasingly likely. Such changes were not only an issue of particular projects, such as an initiative to take out some of the concrete at Tujunga Wash in the northern stretch of the river in the San Fernando Valley. They also included changes in the culture of the agencies themselves, which would determine whether a new generation of environmental and community-oriented civil engineers would come to the fore, would be skilled in both ecological and watershed approaches, and would have the ability to work with and respond to community insights and concerns.[30]

The notion that changes were possible in managing rivers and streams, including opportunities for restoration, was elaborated in another session on the San Gabriel River watershed area. Located to the east of downtown Los Angeles, flowing through some of the original citrus belt lands and several of the bedroom suburbs of the San Gabriel Valley, and connecting to the Los Angeles River in the Whittier Narrows area before it descends toward the ocean at Long Beach, the San Gabriel River watershed has been a less visible battleground than its Los Angeles River counterpart. While nearly three-fourths of the watershed area represents undeveloped open space, including a wildlife corridor in the Puente-Chino Hills that connects the Cleveland National Forest to the San Gabriel River, it is also an area under increasing pressure from new development. The developed, urbanized sections of the watershed area, particularly the older suburbs that had been developed prior to World War II, have also been experiencing rapidly changing demographic shifts that include increasingly large Latino and Asian populations. The session, an organizing conference with the title Re-Envisioning the San Gabriel River, took place at a moment when new constituencies and coalitions had begun to form to protect open space and to highlight and enhance parts of the river and connecting streams as community and environmental assets. Interest in the issues of the session had further been fueled by the passage of legislation that had created the San Gabriel and Lower Los Angeles Rivers and Mountains Conservancy (RMC), the sixth conservancy in California and the first fully urban conservancy.[31]

While sharing some of the same language about ecological restoration and protecting open space, the political dynamics around San Gabriel River watershed issues have also been suggestive of the potential for broadening the appeal for river renewal and a place-based politics. The RMC legislation, for example, had been introduced by Hilda Solis, who was then state senator (and subsequently elected to Congress) from the San Gabriel Valley city of El Monte and who had also served as the cohost of the Re-Envisioning the San Gabriel River conference. Solis was then emerging as a major political figure and coalition builder who was linking community, environmental, and social and economic-justice issues and identifying how such issues crossed ethnic, racial, and class lines. Solis symbolized an emerging Latino voice in the community, environmental, and social-justice link, including its L.A. River and San Gabriel River dimensions. A new type of open-space and public-space advocacy was beginning to be associated with this new Latino environmental voice that connected access to open space and recreational areas with urban quality-of-life concerns. The San Gabriel River conference and subsequent organizing events, which brought together a wide range of traditional environmental groups as well as some of these new players, helped jump start organizing and create visibility around San Gabriel River issues, including Solis's proposal to study the feasibility of turning the San Gabriel River watershed into a national park. It also brought visibility to the idea of a new urban environmental politics, a possibility reinforced several months later with the large majorities provided by Latino and African American voters in passing two park and water bonds. Voter interest was in turn helped by earmarking urban park acquisitions and urban watershed initiatives as part of the bond funds.[32]

While the session on the San Gabriel River suggested new types of political coalitions, the barriers for restoration and change seemed far more formidable once the river made its way south to the areas where the walls had been raised. To reenvision the channelized river in this edgier landscape seemed a near impossible task, although the importance of broadening the community base for river renewal was compelling. In recognition of that need, FoLAR collaborated with a team from Harvard University's Department of Landscape Architecture to extend the vision of river renewal to the channelized segments through downtown Los Angeles toward the L.A. city limits at Vernon. The charge of the Harvard team was to consider how river managers might be able to "bring back the habitat, clean the water, and make it a natural amenity, while maintaining flood protection," with the goal of transforming the river "from today's poor joke into the centerpiece of a great city."[33]

The limits of the Harvard study were also apparent, as indicated by a Re-Envisioning community forum cohosted by the City of South Gate that sought to focus on the south of downtown segment of the San Gabriel River. In advance of this program, an Occidental College student, who had grown up and gone to school in South Gate, undertook outreach to high school students in the area who were largely unaware of the existence of the river at the edge of their city but who expressed interest in such ideas as creating a new park along the river or painting murals on the channel walls. The high school students, however, had no interest in the community forum since they perceived it to be "a city thing." Poorly attended, the forum suffered from a disconnect between the reality of a concrete river, a visioning process that had yet to establish connection to the needs of the communities south of downtown, and a political culture that reinforced the disconnect between the city and its residents. South Gate was then in the midst of a protracted political conflict involving corrupt City Council members and a city official who sought to run the city as a kind of fiefdom. The Council members were eventually thrown out of office in a recall election, and the political *cacique* was convicted of looting the city's meager resources, suggesting that change in the southeastern segments of the L.A. River, among the densest areas in the region and the site of the original suburban-industrial clusters, also required a change in the political culture as well as a connection to the right to the river.[34]

The Re-Envisioning series further sought to increase the focus on the L.A. River through a series of walks and bike rides along the river's path, as these activities were emerging as an important dimension of river advocacy. Prior to the Re-Envisioning series, which sponsored the first bike ride along the L.A. River, most bike advocates had largely ignored issues that had to do with the river, while river greening advocates had just begun to consider the value of a bikeway along the river's edge. The first bike event helped catalyze the link between river and biking advocates. Many of those who participated had never been to the river previously, and the experience identified how a bikeway along the river could have enormous appeal and value. As a result, quarterly bike rides along different stretches of the river were subsequently organized. Similarly, the monthly Walk along the River events—part exploration, part organizing—were inspired by Lewis MacAdams's mantra to "come down to the river." "We're so used to going so fast through L.A. that we don't notice places," FoLAR organizer and bike advocate Joe Linton said of these exploratory activities." "You see different ways that nature in the city can be found, but that's not all. You discover pathways and neighborhoods and streetscapes that you never knew were there before. You just have to get out of the car, get up on a bike, or learn to walk and slow down."[35]

To see the L.A. River in new ways required a cultural process of reinvention—whether through poetry, photography, art installations, or film. The Re-Envisioning events included an evening of poetry reading and song that suggested why the river was a place requiring discovery. For several of the poets, the river remained obscure and difficult to know, a place without an easily defined beauty or grace but a "*recovering* body of water," as poet Amy Gerstler wrote, that was part of a process that needed to be taken "one day at a time." The photography of the river also provided different kinds of landscapes. The "natural" images along the soft-bottom areas suggested the recovery of a free-flowing river, and the images of the predominantly industrial and intensely urban environments were associated with the river's channelized segments.

But the river, as the Re-Envisioning series' event Hollywood Looks at the River underlined, remained fixed less as a landscape of recovery than one of violence and danger. This could be seen in the *River Madness* video film montage of images of the L.A. River in Hollywood films over seventy years, which was produced for the Re-Envisioning series by filmmaker Dana Plays. The montage demonstrated how extensively the L.A. River has been used in Hollywood for thematic and background purposes, including the car race in the film *Grease* and the pursuit of Arnold Schwarznegger's motorcycle along the channel bed in *Terminator 2*. The river has also provided thematic ironies, such as the 1950s film *Them*, where giant irradiated ants come out of the storm drains that feed into the river, and the film *Volcano*, where the concrete channel becomes the instrument to carry the lava flow to the ocean. "Landscapes tell stories, and the Los Angeles River tells a story of violence and danger," filmmaker Wim Wenders declared at a Re-Envisioning series panel of Hollywood filmmakers and film-industry players, commenting on the montage of Hollywood films. The panelists were struck by the number of films in which the river had served as a location and as part of the story telling. Moreover, the panelists were not able to easily identify how the river could be presented in their films other than as the hostile and forebidding place that had been seen in the earlier films. Hollywood's version of the river indicated the continuing challenge of creating a sense of place based on community or environmental renewal, whether for the river or for Los Angeles itself, when the representation of place as isolating, dangerous, or hostile still permeated the culture of the river.[36]

The Battle over the Cornfield

Right after the Re-Envisioning the L.A. River series' opening session with Mary Nichols and Felicia Marcus, the two environmental officials joined with me and a group of FoLAR activists and UEPI staff at an Eagle Rock restaurant to talk about the future of the river. In the course of the discussion, the FoLAR and UEPI participants informed Nichols and Marcus about the emerging conflict regarding the Cornfield site. A year earlier, FoLAR had hosted its The River through Downtown gathering, including a session in Chinatown that had included urban architects and designers, participants from the Chinatown community, and various river advocates. FoLAR had been excited about the community involvement that had emerged from the process, facilitated in part by a Chinatown activist named Chi Mui, who then was State Senator Richard Polanco's Chinatown field deputy. Mui, who had his own activist roots but was not at the time a river advocate or focused on environmental issues, nevertheless came to see the development of the Cornfield—a forty acre site then owned by the Union Pacific railroad near the Chinatown and predominantly Latino Lincoln Heights neighborhoods north of downtown—as a major opportunity to address a wide range of community needs. Through discussions with community members and through the design and envisioning process from The River through Downtown gathering and its aftermath, FoLAR, Chi Mui, and a number of community participants had come up with a plan for schools, housing, bike paths, recreational facilities, and a park, along with a concept of a more extensive riverfront development that could be ultimately tied to the broader vision of river renewal.[37]

But both FoLAR and the Chinatown activists soon discovered that the process they had launched was about to be undercut owing to a very different kind of development proposed by a large developer, Majestic Realty, for thirty-two of the acres of the Cornfield site. Majestic was considered among the most connected and politically powerful developers in the region. It had worked out a deal with Union Pacific (whose owner also had an interest in the development) to turn the site into a new warehouse and light industrial district for manufacturing, food processing, distribution, and international trade that it planned to call "River Station." Majestic also had ties and the strong support of then Los Angeles mayor Richard Riordan, who touted the project as part of his Genesis L.A. economic redevelopment plan for the city that included remaking brownfield sites into economic development opportunities. As a consequence, Majestic's development plan was on a fast-track to get the project through the city's and the federal government's review process.[38]

Neither Nichols nor Marcus had heard about Majestic's warehouse development, which had been pursued largely under the public radar. Despite the growing interest in river renewal, the FoLAR/community plan seemed dead in its tracks, given the forces pushing the Majestic Realty plan. When Nichols and Marcus heard about this possible new battleground, they warned the river advocates that in taking on such powerful interests, they needed, as Nichols put it, to "slay the King, if you're going to win the battle."[39]

The fight over the Cornfield came to represent a new stage in the advocacy for river renewal. Enlisting the support of a wide range of community and environmental organizations, evoking historical and cultural arguments about the significance of the site, and employing a range of legal and lobbying strategies to block Majestic's fast-track to development, the river advocates displayed a new level of sophistication and capacity to act. The Re-Envisioning program also played a role in the unfolding Cornfield dispute. In September 2000, the final event in the Re-Envisioning series, a mayoral candidates' debate about the L.A. River and the urban environment, witnessed animated discussion about the Majestic Realty project and alternative scenarios about the site—as well as river-renewal issues more broadly. During the debate, the candidates either declared opposition to the project or proposed slowing down the fast-track approach for approval of the warehouses. Their positions in turn suggested that the political climate around the project had significantly changed. "It is hard to adjust to the fact that the L.A. River has become a kind of mom and apple pie issue," MacAdams commented to me right after the debate.[40]

The Majestic Realty developers had also recognized the new political climate around the river and started to reposition themselves to gain whatever advantages that could still be exploited if forced to abandon their development plans. In the months following the mayoral candidates' debate, negotiations took place between the developer and state of California officials over the price and conditions of a sale of the Cornfield property to the state. In 2001, a deal was reached, significantly benefiting the developer in relation to the final price but also making available undeveloped property along the edge of the river to be transformed into a state park. A second deal was consummated the following year for purchase of the Taylor Yard, an adjacent tract by the river that had also been targeted for commercial and industrial development by other developers. Organizing around the Taylor Yard issue included a successful lawsuit regarding the city's approval process, as well as mobilization of Latino soccer clubs and community residents, who emphasized the lack of public space and parks available for recreation in the surrounding neighborhoods. With the successful acquisition of the two sites by the state, the river activists began to raise the possibility of eventually establishing a connected park that could stretch from the confluence of the Arroyo Seco and the L.A. River through Taylor Yard and the Cornfield toward downtown Los Angeles. This parkland in turn could enhance the visibility of the river and the potential to develop the kinds of alternative strategies spelled out in the various reenvisioning exercises such as The River through Downtown charrettes and the Harvard design study.[41]

The two acquisitions, along with the Re-Envisioning the L.A. River programs and multiple other events, policy initiatives, media coverage, conferences, and workshops, significantly shifted the nature of the debates and the discourse about the river. One example of this shift in discourse was reflected in D. J. Waldie's changing commentaries about the L.A. River. Waldie, a novelist and city official for one of the cities south of downtown most affected by floodwaters, frequently wrote about the river in the opinion pages of the *Los Angeles Times*. Waldie's *Holy Land*, an evocative memoir of growing up in one of Los Angeles' post–World War II suburbs, had come to be recognized as a classic description of daily life in the new suburbs. Though sympathetic to the search for a regional identity, Waldie, who lived and worked in an area that had been directly impacted by the LACDA fight, had emerged as a kind of spokesperson for the claims of the south of downtown constituencies, including residents fearful of the Army Corps' warnings about hundred-year flood episodes and skeptical of claims about alternative flood-management approaches. In an *L.A. Times* commentary written shortly before the opening event of the Re-Envisioning the L.A. River series, Waldie described the San Gabriel and Los Angeles Rivers as "problematic." "The gated and trespass-forbidden river channels seem superfluous, the ultimate 'no place' in notoriously placeless L.A," Waldie wrote. Reflecting that shift in discourse around the river, Waldie's position also evolved, reflected in another *Los Angeles Times* commentary soon after the Re-Envisioning the L.A. River series had concluded. "As we begin to encounter the river as a place, not an abstraction, we encounter each other," Waldie wrote. "The riverbank is not the perfect place for this meeting, but it's the only place we have that extends the length of metropolitan Los Angeles and along nearly all the borders of our social divides. Think of the river we're making as the anti-freeway—not dispersing L.A. but pulling it together."[42]

A few years later, Waldie, now a champion of river renewal, noted in a 2002 *New York Times* commentary that "recovering parks from industrial brownfields" wouldn't "restore a lost Eden," given that the greening of the Los Angeles River was "a sobering demonstration of the limits of environmental restoration in an urban landscape." But the major accomplishment of the river-renewal advocates had been to help start the process of enabling the engineers, policymakers, and community residents to change agendas and establish those new management paradigms while seeking to reverse what had also seemed to be an inexorable outcome of closeting the river and dividing a city and a region. "It has been the nature of Angelenos to be heedless about their landscape," Waldie concluded in his *New York Times* commentary. "That's changing, because it must, as we finally gather at the river."[43]

Trashed River Movements

Similar to what the Los Angeles River experienced, numerous other urban rivers and streams across the nation during the nineteenth and twentieth centuries experienced land-use shifts and industrial activities that increased problems of water-quality degradation and flooding while isolating the rivers from the communities that bordered them. This, in turn, stimulated the intervention by the engineering agencies like the Army Corps, which, in its most expansive period from the 1940s through the early 1970s, buried many of the streams and remade the rivers as flood-control conduits. Additional land-use changes during the 1950s and 1960s, such as the development of urban highways built along or over urban riverfronts, further intensified the trends towards the abandonment of urban rivers as community places and environmental assets. As these changes occurred and the flood-control engineers took charge, urban rivers became "cut off from the cities they had once spawned," as a joint American Rivers and American Planning Association report put it.[44]

Interest in urban river restoration first emerged during the 1930s. New Deal initiatives, including the Civilian Conservation Corps' stream- and river-restoration projects, pursued a wide range of labor-intensive conservation activities, including the planting of trees, the building of stone retaining walls and masonry seats, and the creation of new public garden and woodland settings in urban areas. These activities were valued primarily for their job-creation purposes although a strong conservation perspective was also housed within the Interior Department, with backing from hunting, fishing, and wildlife groups. Some of the 1930s urban stream- and river-restoration projects, with their emphasis on the use of natural materials, watershed gully repair, and cuttings from plants came to be seen as "the current day's stream restoration technology," noteworthy as well for their "unusual durability," according to Ann Riley.[45]

However, the urban stream and river initiatives spawned by the Interior Department, the CCC, and the Works Progress Administration (WPA) where the CCC was housed were quickly eclipsed by the rapidly expanding role in urban flood-control activities of the Army Corps of Engineers. Armed with new legislation, the Corps was able to extend its activities as a single-purpose agency defined largely by its use of structural engineering techniques for containing flood waters. With the advent of World War II and the fading of the unemployment crises of the Depression years, the Army Corps approach quickly supplanted the conservation-oriented and nonstructural strategies of the CCC. In less than three decades, the Army Corps cut a wide swath through urban areas, including a number of downtown districts, undertaking more than sixteen hundred projects that ended up pouring concrete in nearly 35,000 miles of waterways. "An engineer's project that was not conceived in concrete could not be perceived as a serious, professionally designed project," argued Riley. At the same time, the flood-control strategies did not prevent the problem of flood hazards and in some instances, as a study of the St. Louis region indicated, might have made communities even more prone to flood damage.[46]

Similar to the enormous impact of urban highway development that divided neighborhoods and degraded downtown areas, the urban river and stream flood-control projects contributed to the reconfiguration of whole neighborhoods and sections of cities into bleak new urban landscapes. As a result, new types of coalitions began to form, motivated in part by concerns for economic renewal, particularly of downtown areas, as well as for scenic improvements in and around urban waterways. In places as different as San Antonio, Texas, Chattanooga, Tennessee, and Cleveland, Ohio, this emerging network of downtown boosters and environmentalists identified new river-restoration projects and adjacent developments as a means of attracting new tourist dollars and reviving their downtown economies.[47]

San Antonio River Walk, with its hotels, boutiques, shops, and restaurants along the San Antonio River, became the poster child for the new river-restoration movement. Subject to Army Corps flood-control projects up through the 1950s, the area adjacent to the river was considered dangerous and was even placed off limits to military personnel. But plans in the early

1960s called for a riverfront commercial revival, including an initial design prepared by Marco Engineering Company, a company that had helped design Disneyland. By the 1970s, a number of hotels and other commercial establishments had opened, as bypass channels and diversion projects also came to be completed, turning the River Walk into the symbol of riverfront development as a tourism-driven economic-development strategy. However, a number of other cities and communities, such as Los Angeles, had far more significant barriers for linking river and downtown redevelopment. In some cases, the notion of river restoration as an aesthetic or visual enterprise led to poor choices for restoration strategies, reinforcing the need for a more comprehensive approach to river restoration beyond its aesthetic or tourist appeal.[48]

By the early 1990s, advocates like FoLAR and its other urban counterparts, such as the Friends of the Chicago River and the Friends of the White River, came together under the banner of the Friends of Trashed Rivers. The modest expectations of reform of agencies like the Army Corps during the Carter presidency, which had modified the operating *Principles and Standards* for the Corps to include nonstructural alternatives in the planning process, were basically eliminated during the Reagan and Bush years. The LACDA battles, which began to unfold during this period, paralleled other fierce debates about Army Corps interventions in other cities, even as community initiatives launched around the country began to seek out possible restoration approaches. By the time the first Trashed Rivers Conference was held in 1993 at Fort Mason in San Francisco, more than three hundred people from twenty states, many of them in the process of forming groups like FoLAR, were ready to establish a more formal network of river restorationists under the banner of the Coalition to Restore Urban Waters. Annual Trashed Rivers Conferences during the 1990s helped build and extend this network, and a number of the coalitions moved beyond mobilization and the debates about flood control versus flood management to begin projects designed for "daylighting" some of the buried streams and creating pockets of new life in and adjacent to the rivers flowing through the cities. Some of the more ambitious projects, like Chattanooga's twenty-two-mile RiverPark along the Tennessee River (which also included mixed-use developments, including affordable housing) were seen as key to the reenvisioning of the cities themselves. "From California to New England, cities are re-engaging with their rivers," commented the *Christian Science Monitor* in 2001 about these new initiatives.[49]

These efforts were aimed at stimulating interest in urban waterways as a part of the city rather than cut off from neighborhoods and community life. Part of the motivation continued to be driven by the recognition that river redevelopment enhanced the value of riverfront properties, both residential and commercial, with the potential for a rapid jump in property values and a transformation through gentrification of the adjacent neighborhoods. Even some of the most devastated and polluted areas adjacent to urban rivers, such as the famed Bubbly Creek neighborhood in Chicago, were potential targets for gentrification and skyrocketing property values stimulated by riverfront redevelopment. Bubbly Creek had been the epitome of a polluted area during the late nineteenth and early twentieth centuries. It was bounded by meat-packing plants, garbage dumps, and a vacant land known as the "hair fields," named for the animal hair and other types of wastes from the slaughter houses that were discarded onto the land and laid out to dry. Described by Upton Sinclair in his 1906 novel *The Jungle* and a major targeted area for the social reform movements of the Progressive Era, Bubbly Creek (which had gotten its name due to the bubbling of the waters from the discarded animal carcasses) remained a depressed area through much of the twentieth century, representing a kind of stagnant backwater of the South Fork of the Chicago River. But with the various initiatives for riverfront revival that emerged in the 1980s and 1990s, riverfront properties along Bubbly Creek, as well as along several other riverfront areas, experienced rapidly escalating price inflation, with some Bubbly Creek residential properties selling for more than a million dollars.[50]

The tension for the river-restoration movements was bound up in the ways that it had become connected to the power of the real estate market to transform areas, not just in being part of a downtown revival but also in remaking, through gentrification, low-income and working-class neighborhoods that had been as neglected as the urban rivers themselves. These pressures for gentrification extended to a number of neighborhoods impacted by river-restoration plans, including several of the working-class neighborhoods along the Chicago River, Washington, D.C.'s low-income and largely African American eastside neighborhoods that bordered the Anacostia River, and even Cleveland's Flats neighborhood near where the Cuyahoga River burst into flames in 1969. For the river-restoration movement to achieve a level of environmental and aesthetic change and thereby establish a right to the river, it needed to establish a democratic right that strengthened rather than displaced the residents of river communities and created access for all neighborhoods in the city.

Let a Thousand Wildflowers Bloom

In March 2006, several months after the corn had been harvested and its seeds packaged for distribution to various antihunger, food-security, and community-development groups, my colleague Andrea Azuma and I met with the *Not a Cornfield* artist, Lauren Bon, at her bungalow at the Cornfield site. It was a stunning sight to stand at the north end of the park and look out past the thirty-two acres towards the southern end of the downtown skyline. It had rained earlier that day, and strong winds had cleared the air. You could see a train passing through on the overhead tracks that emerged from downtown's Union Station on its way to the Chinatown stop of the new Metro Gold Line. A truck was moving around the Cornfield site, spraying a green hydroseed mix for wildflowers, creating a green field that looked almost painted and that reminded me of the time I had seen the Chicago River dyed green on St. Patrick's Day.

Five months earlier, when the corn had reached its full height, the view from that same northern end of the Cornfield had created an impression of an improvised landscape, an idea of possible transformation. The view now also suggested change, but it was also dominated by powerful images associated with Los Angeles—the clear air after it had rained, the silhouette from

City Hall and the high rises behind it, the tracks suspended in air passing above the warehouses and through the backyard of Chinatown like a view from the movie *Blade Runner*, although less threatening. All suggested a visual connection to Los Angeles' future as well as its past.

Andrea and I wanted to explore with Bon how her *Not a Cornfield project*—created as a one-time growing-cycle event in a period when the future design of the park was still being explored—had been linked to the surrounding neighborhoods and also fit with the possible future scenarios for the park and the Los Angeles River. Our Urban and Environmental Policy Institute had been commissioned by the California State Parks Department to undertake an environmental impact and "economic benefits analysis" of the Cornfield (whose official title was the Los Angeles State Historical Park, since the park was now formally part of the state parks system) as well as the adjacent Taylor Yard site to its north, which directly bordered the river and was also now part of the state parks system. The Cornfield and Taylor Yard parks were very different from other state parks. Located in the heart of the city, these parks included contaminated land defined as brownfields. They had previously been used as rail yards and had been purchased after protracted community and environmental struggles that were connected to various urban river, open-space, and public-space movements. The idea of an urban state park, let alone a state park on what had been degraded, vacant inner-city land, was difficult to justify for a state parks bureaucracy that was oriented towards rural parks and for the state legislators who funded the acquisition and maintenance of those parks. State parks, officials had long assumed, were supposed to be in pristine areas outside the cities where urban residents could escape to view nature set apart from urban places that had uprooted natural environments. But the passage of bond measures in 2000 and again in 2002 created an important new funding stream for urban parks acquisition, a change made possible by the huge majorities provided in Los Angeles and other urban areas, particularly among Latino and African American voters. In the wake of this new funding stream, the push to purchase particular sites like the Cornfield (as opposed to the traditional listing and ranking of the more pristine sites available for parks acquisition) had been influenced by the new urban environmental and community movements.

The concept of "economic benefits" implicitly assumed increasing real estate values, a first step towards gentrification of the Lincoln Heights and Chinatown areas that bordered the Cornfield. State Parks officials could then use that economic argument to try to access more funding from legislators who otherwise might not assume that park funds should be spent for such inner-city properties. In developing our study, we proposed that "economic benefits" be replaced by the term "community benefits" to help identify the ways that a park provided a type of community identity for the immediate areas adjacent to it as well as for the city itself. This included the health, educational, environmental, and landscape values associated with public space and open space in the city. A community benefit, moreover, could be place-based at the same time that it was associated with a "community of struggle" associated with the movements and the events that had led to the creation of the park in the first place. Rather than a gentrification model that linked river renewal and park development as an economic value for those who could now afford their connection to the park, a community-benefit model established a right to the city and to the river through a connection to the park for the existing Lincoln Heights and Chinatown neighborhoods, for downtown residents, and for city residents as a whole.[51]

But could a community-benefits model prevail? An artist's *Not a Cornfield* project suggested some of the opportunities for—and the barriers to—pursuing that path. While the nine-foot-high fields of corn had provided a type of transformative moment, the wildflowers that bloomed a few months later provided a sea of color and growth throughout the thirty-two acres that became the concluding signage for this temporary, one-year project. The links to the William Mead Homes, the public-housing project in Lincoln Heights, included hiring twenty-five teenagers from the project to act as community docents and holding weekly Sunday storytelling and Friday evening gatherings, also indicating how the park could serve as community space. But the project was unusual in its conception and its funding. It had been made possible by an artist who was herself a member of the family that ran the Annenberg Foundation and that had provided the funding for this one-year initiative that was part public art, part demonstration, and part visualization of the different land uses that could define the park or other lands in and around the River. In doing so, Lauren Bon had also become transformed, redefining her art as type of community engagement, a process she hoped would establish a more permanent connection among those that had participated. Bon had also begun to root herself in the Cornfield area by occupying space across from the park in an underutilized and rather desolate building and by strengthening her ties with the Lincoln Heights public-housing resident groups. The new building was reconfigured into a new community space, designed to host community meetings, presentations, community libraries, and artworks that could further enhance the new types of Cornfield-related identities. "Part of belonging to a community includes the ways one participates in helping build community and creating a sense of place," Bon said of her own journey.[52]

Through our own research for the state parks study, we had also begun to identify other innovative connections and changes associated with the struggle around the park and the efforts to bring about a renewal of the Los Angeles River. In 2005, as the corn reached its height, about a quarter mile away the California Endowment, one of the largest private foundations in California, relocated its headquarters from the suburban Woodland Hills area in the San Fernando Valley to a downtown site near Union Station in an area that bridged downtown with the Cornfield and the Lincoln Heights and Chinatown neighborhoods. In the process, the foundation created a new Center for Healthy Communities and developed a landscaping design for its massive new building and community space that was "inspired by the natural environment of the surrounding neighborhood, which includes the Los Angeles River and the Cornfield State Park," as the new center's newsletter put it. "The L.A. River is a unique natural resource in the heart of the city, and we want to highlight local initiatives that recognize that a healthy river is essential to healthy communities," Jean Miao, a program officer with the

Center for Healthy Communities said of their new center and its location and approach.[53]

To the north of the new California Endowment building, along the route that led to Lincoln Heights and the Cornfield, Homeboy Industries was also building a new home for its multiple operations designed to provide jobs and services for former gang members. Further north, as one entered the warehouse district at the edge of Lincoln Heights and a few blocks from the Cornfield, a large Thai food and crafts wholesale complex called LAX-C had created a bright exterior mural, hoping to attract residents to its food market, restaurant, and other shops, while park advocates explored the idea of a Thai herb garden at the edge of the Cornfield site. Plans for a farmers' market (to be located at the edge of the Cornfield) were also under discussion, and a demonstration urban garden and a series of Friday evening programs would be part of the effort to transform the drug-infested alley that had been converted by the *Not a Cornfield* group into a community space. The Castelar elementary school in Chinatown and a new school to be opened near the Taylor Yard also awaited opportunities for their students to utilize the park site, given the lack of open space in the Chinatown and Lincoln Heights neighborhoods and the interest in the river among teachers, students, and parents that had been stimulated by the long, protracted fight that had led to the creation of the park. And directly across from the park, an artist-engineer had purchased the oldest building in Lincoln Heights and had begun to develop plans to establish yet another public space where his high-tech visualization of the history of the area, which he called *L.A. Remap*, could be screened. In the winter of 2006, a design team was selected by the State Parks Department after a long process of community input and engagement, with the design team's plans promising an array of park designs and configurations of space that sought to establish the Cornfield as the symbol of L.A.'s pursuit of becoming a green—or at least a greener—city.[54]

But the barriers for a community-benefits framework for the park and its connection to the river-renewal agendas remained formidable. The Gold Line tracks had created a divide between the Cornfield and the bluffs that led up toward the historic Chavez Ravine neighborhood where the Los Angeles Dodgers had uprooted homes to build Dodger Stadium. The Gold Line stop at Chinatown provided an important opportunity to develop transit access to the park, but there were still street crossings that had to be established while the City's Transportation Department remained more focused on widening streets for the automobile traffic passing through than on providing for pedestrian needs. Access to the river and to the majestic Broadway Bridge that looked over the river as it headed toward downtown was not readily available since there was no direct route to either the river or the bridge. And the potential for gentrification remained strong, as the largest owner of property in the area explored the development of expensive high-rise condos that could look onto the park and the river.[55]

As we stood at the northern end of the Cornfield, our backs to the river and looking toward downtown Los Angeles where the river would make its way on towards the eastern edge of downtown near the dividing line between downtown and the Boyle Heights neighborhood, the idea of transforming L.A. didn't seem quite so remote even for a city long identified with its freeways and its endless sprawl. Here we were close to the original center of the city, the site of the *zanja madre*, the historic water ditch from the pre-Anglo period that captured the flow of the L.A. River for the lands to be fertilized and the crops to be grown. Give power to the imagination, we thought, even as the cars sped by on North Spring Street on the way into and out of downtown towards the massive complex of freeways that also defined this city and this region. But could the powers of imagination, powers that Lewis MacAdams had evoked when he had made his way down to the river twenty years earlier, also create the conditions for the type of changes on the ground necessary to reenvision not just the Cornfield or the river but Los Angeles itself? And who could help make that happen?

5

Cars and Freeways in the City

Making Sure No One in L.A. Has to Walk.

—Jiffy Lube billboard sign across from several Los Angeles freeways

Sig Alerts and the Rapture of the Freeway

I'm not sure when I first heard the words "Sig Alert," but I know that I first began to focus on what the words meant when I started commuting to Occidental College in 1997 from my home in Santa Monica about twenty miles to the west of the college. Growing up in Brooklyn, New York, I had not experienced automobile culture when I was a teenager. I learned to drive when I was twenty-six, about a year after I arrived in Los Angeles. When I worked as a writer during the 1970s, I did not use my car frequently and never used it during peak hours. Later, when I commuted to the Metropolitan Water District at its downtown headquarters, I occasionally took the freeway flyer bus and also staggered my hours when I did take the car. And when I began to teach at UCLA during the 1980s and 1990s, I rode my bike or took a local bus that stopped about a block from where I lived, which took about the same time as driving my car that four-mile stretch, when I factored in parking at UCLA. When traffic built up on the surface streets near UCLA, I'd be unhappy sitting in the bus, but I was less stressed than I would have been in my car experiencing one of L.A.'s infamous Sig Alerts.

The Sig Alert has occupied a central place in the culture of the freeway in Los Angeles since the mid-1950s. The term was coined by Los Angeles police chief William Parker based on an electronic device invented by Loyd Sigmon, a radio reporter and vice president of Golden West Broadcasters, a local radio and television chain owned by the singing cowboy, Gene Autry. Sigmon developed a specialized radio receiver and tape recorder that could record police bulletins of traffic accidents and emergencies and pass them on to the local broadcast news stations. First used on Labor Day in 1955, the device became an immediate hit with the local media because of the increasing interest of car commuters in traffic conditions. Even its early glitches (one Sig Alert about an accident that called for help from doctors or nurses caused a traffic jam near Union Station while another used by civil defense administrators failed to work because the right button wasn't pushed) didn't prevent a rapid increase in its use. By the 1960s and 1970s, as freeway commutes continued to increase in time and distance and the California Highway Patrol took over the practice of issuing Sig Alerts, the warning had become a part of daily life in the region. Sig Alerts now refer to the worst of L.A.'s notorious traffic jams—those involving "any traffic incident that will tie up two or more lanes of a freeway for two or more hours," as a California Department of Transportation Web site defines it. The term is used so often that it entered the *Oxford English Dictionary* in 1993. Angelenos live in dread of Sig Alerts, and like most people who commute I find myself gravitating toward the radio stations that issue periodic freeway traffic updates every ten minutes, ready to abandon the freeway for surface streets once I hear the words Sig Alert or its equivalent, "The freeway is now a parking lot."[1]

The culture of the freeway in Los Angeles is today dominated by the ever-lengthening car commute. As in nearly every other major-American city or region, commuting distances have increased, and congestion has become an everyday occurrence during peak hours. But for the first three-quarters of the twentieth century, car culture and freeway culture represented the notion of liberation in space and time and also provided a source of power for the user. These ideas were reflected in the rise of the automobile in the 1920s, through the development of the freeway systems in the 1940s, 1950s, and 1960s, and up through the heyday of car and freeway culture in the late twentieth century. The connection of Los Angeles to the car and the freeway is perhaps most notable for a vast infrastructure that became a permanent feature of the landscape and culture of L.A. The freeway experience, Joan Didion wrote in the 1970s, eventually became "the only secular communion Los Angeles has," characterizing the sense of liberation that a driver felt as the "rapture-of-the-freeway."[2]

By the late twentieth century, the "rapture of the freeway" experience had become more problematic, especially in Los Angeles. Long touted for its "comfort, convenience, power, and style," as General Motor's head Alfred Sloan characterized his automobiles in the 1920s, the freeway became the next logical step in linking speed with power. The freeway revolution of the 1940s and 1950s, including its urban dimension, not only liberated the car from its confines on the street (which it had already helped reconfigure) but also extended the transformation of the region that railroad lines and electric interurban streetcars had previously stimulated. The freeway became the new center of the city or rather inexorably linked city and suburb. Moving from one location to the next—from home to work, to stores, to schools, to appointments, or just to friends—became increasingly not just a car ride but a freeway experience as distances between places increased. The freeway, as much if not more than the imported water supply, helped extend the boundaries of Otis Chandler's "supercity." By the end of the twentieth century, Los Angeles had indeed become the proverbial string of freeways in search of a city, "the first major city that was not quite a city," as David Brodsly put it in his classic study of L.A. freeways, reflecting the common perception of Los Angeles.[3]

But as the culture of the Sig Alert and freeway as parking lot lurked ominously as the backdrop to the freeway experience, rapture turned to frustration. Even Los Angeles, freeway builders discovered by the 1970s and 1980s, was no longer in love with the freeway system. The car and the freeway were no longer liberators. They had become daily life necessities with new kinds of stresses. Trucks leaving the Port of Los Angeles carrying goods from the globalization trade, for example, transformed certain freeways into a nightmare experience for communities and car commuters alike. Other freeways, such as the notorious "Orange Crush" interchange where the Route 405 and Route 5 freeways converge as they feed into the automobile suburbs of south Orange County, became known for their constant congestion and not only in peak hours.[4]

This is not just a Los Angeles phenomenon. According to the Texas Transportation Institute's assessment of the hours of delay on highways from 1982 to 2003, dozens of cities like Houston and transit-oriented cities like San Francisco and New York are finding that the time spent on highways has been increasing, creating a "creeping crisis," as the Brookings Institution's Anthony Downs characterized this increasing congestion. The Los Angeles experience has become the national experience and increasingly the experience of other cities around the world. Yet with each new freeway and freeway extension built, cars fill the lanes, and the automobile's power, speed, and convenience seem more diminished with each passing year and with the increasing number of Sig Alerts. An impasse has developed: Los Angeles residents need their cars (if they can afford one), the huge new globalization-inspired movement of goods needs its trucks, and cars and trucks need their freeways. But the cars, trucks, and freeways are no longer fully meeting their intended goals as convenient, rapid, and liberating systems of transport. And given the vast reach of the freeways and the ubiquitous presence of cars in the city, particularly in relation to their voracious appetite for land and the myriad of environmental problems they generate, the idea of reinventing Los Angeles has also meant reenvisioning the connection of the city to the car and its freeways.[5]

The Interurban Interlude

The history of the car and the freeway in Los Angeles and in most other urban areas in the United States is linked to the rise and fall of the streetcar—the electric interurban system that became the first major urban transportation system to emerge in the late nineteenth century. The interurbans were a hybrid that was both a transportation system (that became the primary means of city residents getting from one place to the next) and also a force for development (that influenced the patterns of residential and commercial settlement as the boundaries of the cities and the new suburbs began to take shape). They were initially linked to the railroad companies, which were, particularly in the West, a dominant economic power. The railroads owned huge amounts of land both inside and beyond the city's borders and used their dominant role as landowners to shape the cities as economic way stations and entry points for the movement of goods and people. The interurban streetcars, as electric systems, were also linked to the private electric utilities that provided the power source and the infrastructure that enabled the system to expand.

In Los Angeles, as elsewhere, both the interurban streetcar interests (such as Henry Huntington's Big Red Car company, the Pacific Electric Railway, which was subsequently taken over in 1911 by the Southern Pacific) and the utilities (led by Southern California Edison, which gained the monopoly for service to many of the new suburbs, including the suburban industrial clusters of the 1920s) were influential in creating new subdivisions and helping plat new townships. Similar to the new imported water supplies that traced the patterns of settlement and subdivision, the streetcars, as the source for transportation, and the utility services, as the source for power, were also essential parts of the system of subdivision and urban expansion. Huntington, especially, served as L.A.'s primary de facto land-use and transportation planner up through World War I, extending the interurban service north and east to the foothill communities of Pasadena and Glendale, southeast to Santa Ana and Orange, southwest to Redondo Beach, and north to the San Fernando Valley, among other areas in the region. Huntington then linked his interurban systems with some of the real estate subdivisions he had also helped launch. "It would never do for an electric line to wait until the demand for it came," Huntington told the *Los Angeles Examiner* in 1904. "It must anticipate the growth of communities and be there when the builders arrive." In the process, the Big Red Car system became massive, the "largest intercity electric railway system in the U.S.," according to interurban streetcar historians George Hilton and John Due. During the peak years, 6,000 streetcars each day served over 115 routes, covering 1,000 miles of track and between 520 and 700 miles of service. Along with Pacific Electric's rival, the Los Angeles Railway with its Yellow Car system, interurban service had become fully entrenched in the Southern California landscape. The president of the Yellow Car system called Los Angeles an "electric railway paradise."[6]

While the interurban streetcars laid the groundwork for urban expansion in Los Angeles and the Southern California region, as early as 1920 the automobile began to emerge as their transportation and land-use rival. The early 1920s were a pivotal moment that influenced which transportation system would become dominant and how the car and the land required to support a car-dominated transportation system would play a role in determining future urban expansion. In 1920, production of Henry Ford's new, simple, and relatively inexpensive Model T car significantly expanded the potential for car ownership. At the same time, Ford's rival, the General Motors Company, armed with design and development innovations including styling changes and the introduction of tetraethyl lead in gasoline, was rapidly moving from an "easy sell to a hard sell" marketing approach to make the car become "more and more a part of daily life," as an official with the L.A. Motor Car Dealer's Association put it. By 1920, car-ownership figures were increasing rapidly, and although California still ranked behind a few other states, such as New York and Ohio, in automobile ownership, the *Los Angeles Times* asserted that the automobile was ushering in an era of "almost limitless expansion."[7]

As their numbers increased, cars vied with streetcars for use of the streets, creating dangerous passage ways for pedestrians and creating new bottlenecks, particularly in the downtown area. As early as 1919 in Los Angeles, downtown congestion had begun to cause major delays in interurban service, and the streetcar companies began to contest the car's role in capturing the streets as well as the parking spots that had otherwise been available for streetcar use. In early 1920, the streetcar companies succeeded in getting the Los Angeles City Council to enact an automobile parking ban in the downtown area between the hours of 11 a.m. and 6:15 p.m., a concept that the newly organized automobile support groups saw as "cutting the throat of business." The downtown retailers, the auto interests declared, had become increasingly dependent on their auto-related customer base. To a certain extent, the divide between the streetcar riders and car drivers broke along class lines, but the auto interests had already begun to effectively portray automobile use as a social necessity. Moreover, although the trolleys were able to run on time and downtown traffic congestion decreased markedly in the short period that the ban took effect, many of Los Angeles' key business and economic leaders (like those in other cities such as Detroit and Cleveland, where the automobile was also reshaping the downtown districts) had begun to ally themselves with the emerging role of the automobile. "Southern California throb[s] in unison with the purring motors of its automobiles," the *Los Angeles Times* said favorably when the Council lifted the parking ban.[8]

The problems of traffic congestion—and the contending roles of the interurban streetcars and the automobile—intensified over the next two decades. Transportation analyst David Jones argues that Los Angeles in the 1920s began to resemble other urban areas around the country, with its growth of downtown retailing, central-city manufacturing, and greater central-city density. But Los Angeles' transportation problems increased in this period, even as proposals for the expansion of new transit operations were never pursued, such as the 1925 "comprehensive rapid transit plan" put forth by transportation consultants Kelker, De Leuw & Co. for a subway system, an elevated rail system to relieve the downtown congestion, and longer-term enlargement of the street railway and bus systems. The Southern Pacific's own proposal for a downtown elevated railway, mired in part by competing downtown interests, was defeated the next year in a memorable election that featured the evangelist Aimee Semple McPherson arguing in a radio broadcast a week before the election that L.A. voters needed to "rise up" and "not let this horrible menace be slipped over," while the *Los Angeles Times* urged voters to defeat this "device of the devil" that plagued cities like Chicago and New York.[9]

Despite the failure to expand and integrate rail service as proposed by Kelker, De Leuw & Co, rail and interurban streetcar use during the 1920s and 1930s increased in numbers of passengers, length of travel time, and extensiveness of transit lines and destinations. Despite that growth, however, the railroad companies continued to neglect the passenger side of their business in favor of the more lucrative freight operations, which, in the case of the Pacific Electric system, meant that 75 percent of their cars in service by the mid-1920s were now earmarked for freight rather than transit. Privately owned but publicly regulated, rail service was still the mainstay of the urban commuter but had also become the target of public criticism for its delays and increased fares. When the extensive scandals around the utility monopolies were revealed during the early 1930s and the railroads continued to neglect their passenger service in favor of freight during this same period, public support for the interurban streetcars further declined. At the same time, the promotion of the automobile as central to the Southern California lifestyle was becoming more deeply embedded as part of the region's culture, while also reinforcing and extending the low density and dispersed nature of its population centers. Still, the debate over which transportation system would prevail was far from resolved, and even as the first of the Southern California freeways began to be constructed during the early 1940s, issues such as the financing, design, and purpose of the new roadways and the compatibility of the contending rail, bus, and car systems remained uncertain.[10]

The period from the late 1930s to the end of the 1940s—when the car-dominated, freeway-oriented transportation system ultimately prevailed and major changes in transit service and ownership patterns as well as freeway financing and construction took place—directly influenced that outcome. Up through World War II, the interurban streetcars had managed to keep much of their status as a major transportation service in the region. Even as late as 1945, when the Pacific Electric Railway sold its Yellow Car system to a national consortium known as National City Lines, the company had managed to stay in the black during the war years and was still operating more than 1,000 interurban cars, 373 miles of track, and 421 buses. By the time of the purchase of the Yellow Car system and several other acquisitions by the NCL consortium in a number of cities around the country, however, the interurbans had become vulnerable, a situation reinforced by National City Lines' strategy to eliminate passenger rail service in favor of its new bus systems. This major new transportation system player was backed by the largest automobile, trucking, oil, and rubber interests, such as Standard Oil and General Motors, which became both investors and suppliers. It succeeded in facilitating the shift away from the interurbans. But that trend had already begun to occur in Los Angeles and elsewhere prior to the NCL acquisitions. Even storied lines, such as the interurban service to Alhambra in the San Gabriel Valley, which was first established in 1902, had already been eliminated in favor of bus service prior to the NCL acquisitions.[11]

With the shift from rail to bus and truck during the 1940s and, more broadly, with the increasing prominence of the car and the freeway, the NCL acquisitions eventually came to represent for some not just the occasion but the reason for the demise of the interurbans. When the NCL consortium was eventually charged and successfully prosecuted during the late 1940s for violation of federal antitrust laws (though fines were minimal and no one served time in prison), this shift in ownership came to be seen as a full-blown conspiracy. Whether such a conspiracy was fully or even partly responsible for the dismantling of the interurban systems has since become a major source of debate among transportation researchers and analysts. For some, the court case, which received little attention at the time, was clear evidence of a conspiratorial intent on the part of the investor-supplier partners to shift away from a rail-oriented transportation system. But other transportation historians have argued that the interurban streetcars had become too inefficient and costly and that the automobile and the freeway provided a more convenient, efficient, and pleasurable transportation alternative. Since the shift from rail to bus had already begun to occur prior to these acquisitions, these analysts have argued, such a conspiracy argument detracts from an understanding of the limits of a rail system in an urban area like Los Angeles and the need for the transit companies to transition to buses as a way to survive in a period when the automobile was becoming increasingly dominant.[12]

While it remains debatable whether the intent of the National City Lines consortia had been to destroy the interurban system in favor of motorized transport, it's clear that a confluence of interests had come together that sought to transform the transportation system in favor of the automobile and related oil, trucking, and freeway interests and in opposition to the "mongrel vehicle," as one General Motors official characterized streetcar service in 1953. These linked interests also included the pivotal role of the state highway officials who had little interest in any kind of transit alternative to cars and freeways, whether rail or bus. While supplier contracts, among other aspects of the National City Lines arrangement, clearly warranted the finding of conspiracy, less prominent in the debate about conspiracies has been the importance of the private nature of the ownership of the rail and the bus systems, which effectively identified in the public mind transit as a private business and the automobiles as a public good. Unlike most European cities—where the rail lines were publicly owned and where transit was seen as a type of social service provided by the government, which was also more broadly engaged in transportation planning for a region—private ownership of the rail lines underlined the uncoordinated and patchwork patterns of rail service broadly tied to real estate speculation and development. It also made it easier for the NCL consortia and other private transit operators to preside over the final dismantling of the urban rail systems in Los Angeles and elsewhere around the country.[13]

Attempts to municipalize interurban streetcar service in Southern California and other urban areas was effectively resisted by the railroad companies and their electric utility allies throughout the first half of the twentieth century. Proponents of municipal ownership of both rail and bus lines in

Los Angeles, Pasadena, Long Beach, and other Southern California communities focused on the problems of service and inadequate planning. Municipal ownership proposals began as early as 1899 in Los Angeles, continued through the Progressive Era into the 1920s, and again briefly made their appearance in the 1940s. Attacked by the railroad companies and their business allies as "socialistic," with the government seen as a "malefactor when it intrudes its activities beyond its legitimate field," as one 1899 *Los Angeles Times* article declared, municipal ownership advocates were unsuccessful not only in their efforts at public ownership but in the arguments to establish public-planning criteria in the development of the transit systems. In an ironic postscript to the dismantling of the interurbans and electric rail service, the remnants of both the Big Red and Yellow Car lines were finally sold in 1958 to a newly constituted public entity, which then proceeded to dismantle the beleaguered and compromised interurban electric rail systems. Over the next twenty years, transit systems around the country, 96 percent of which were still privately owned in 1959, gradually began to be taken over by pubic entities. Public ownership had finally arrived but at a point where the system itself had collapsed and the private companies were ready to bail.[14]

This transfer of transit company ownership in Los Angeles in the 1950s from private to public was repeated in nearly every city in the country. What was regrettable about such a sale, as interurban streetcar historians Hilton and Due point out, was that it had not occurred earlier while the systems were still viable. As long as the Pacific Electric system stayed in private hands, Hilton and Due argued, "it was virtually impossible to keep its unprofitable service going, regardless of how much they [the interurbans] contributed to the solution of the traffic problem." Even the modest losses that the system eventually generated in the 1930s and then in the late 1940s could have easily been justified by a public entity engaged in overall transportation planning, given "the importance of continued use of the rail facilities in the over-all solution to the transportation problem in the area." Thirty years after the takeover and dismantling of the Yellow Car system and fifteen years after the last of the Big Red cars had abandoned service, the idea of a publicly funded and operated (but far more expensive and heavily subsidized) rail system was in fact reintroduced for Los Angeles, following the introduction of similar publicly operated systems in cities like San Francisco and Washington, D.C. A new debate emerged as a consequence—bus versus rail—and that debate also had class, race, and ethnicity dimensions. Meanwhile, L.A. became fully dependent on the automobile and the freeway, while the interurbans and all forms of interurban rail service became historical icons that were remembered for their nostalgic associations of an earlier Los Angeles dominated by rail rather than car. Over the years, the idea of a conspiracy gained more traction and even came to be incorporated as the plot line for *Who Framed Roger Rabbit?*, Hollywood's version of the demise of the electric streetcar.[15]

FROM PARKWAY TO FREEWAY

Before there were freeways, there were parkways. These dedicated roadways were designed to allow cars to travel long distances, and they had exits and entry ramps that did not interrupt the traffic flow or mar the visual experience associated with the roadway. The association of park and roadway was critical since the roadway was meant to pass through parklike or landscaped surroundings to provide a comfortable and pleasurable ride. The alignment was typically one of gentle curves designed for speeds that allowed a visual connection to the surrounding landscape.

This roadway concept was first identified in an 1866 report that Frederick Law Olmsted and Calvert Vaux provided to the board of commissioners of Prospect Park in Brooklyn, New York, recommending the addition of a parkway to the plans for the park. Olmsted and Vaux viewed parkways as carriageways, surrounded and contained by the park and designed for pleasure riding. With the advent of the automobile, the parkway became a specialized roadway designed to provide leisure driving and recreational opportunities as well as efficient transport. The first of these new automobile parkways, the Bronx River Parkway in Westchester County, New York, was completed in 1923. The most famous of the parkway designers was New York

City's Robert Moses, who oversaw the construction of several parkways and who ultimately linked parkways as a form of urban reconstruction in reconfiguring whole neighborhoods to make way for his new roadways. "You can draw any kind of picture you like on a clean slate," Moses said about the roads he planned in the dense neighborhoods they went through, "but when you're operating in an overbuilt metropolis you have to hack your way with a meat axe."[16]

In the 1930s, the modern parkway movement began to expand out of New York with the construction of several federal parkways, such as the Skyline Drive in Virginia, the Blue Ridge Parkway in North Carolina and Tennessee, and the Merritt Parkway in Connecticut. During the same decade, Los Angeles planners, aware of the parkway construction in eastern cities, started to envision the creation of "greenbelts across the city"—parkways that would address the region's increasing traffic and that could at the same time encourage "highway recreation" and "outdoor sightseeing." The focus on landscape was also explored through the concepts of the "townless highway" and the "highwayless town," as championed by advocates within the regional-planning movement, such as Lewis Mumford and Benton MacKaye, who saw such roadways as a connection for city residents with natural landscapes. But the parkway concept was also subject to competing claims and purposes. Despite some important innovations in landscaping and road design, such as Tennessee's Blue Ridge Parkway, parkways within urban boundaries (such as Robert Moses's neighborhood-busting speedways) were already by the late 1930s beginning to be identified by their engineer advocates not as pleasure roads but as a new, efficient transportation mode that emphasized speed rather than any connection to the places they passed through.[17]

Further compounding the debate over the purpose and design of dedicated roadways were the continuing debates about an overall transportation system as well as the impact of new roadways on neighborhoods, particularly within cities and between cities and suburbs. When the 1920 debate about a no-parking ordinance in Los Angeles was resolved in favor of the automobile, concerns intensified about congestion in the downtown district, the need for street widening to accommodate the growing number of cars, and ultimately the development of dedicated roadways. A 1924 Major Traffic Street Plan commissioned by the Major Highways Committee of the Traffic Commission was produced by the Olmsted-Bartholomew firm and Charles Chaney. It called for the "the development of an orderly and well-balanced system of thoroughfares of such width and arrangement as will facilitate direct and uninterrupted movement from center to center and, incidentally, facilitate movement within centers." The plan identified a north-south and east-west axis of boulevards as well as future parkways, arguing that a number of existing proposals for street widening would serve as the initial thrust of a comprehensive, auto-based transportation plan for the city and its connected suburbs. Key business leaders, such as Harry Chandler (who served on both the Major Highways Committee and the board of the influential Automobile Club) pushed hard for the city to adopt and implement the plan to enable Los Angeles, as the *Los Angeles Times* put it, "to tackle and solve the traffic problem, which must be solved soon and solved correctly if Los Angeles is to continue to grow and prosper."[18]

Though the Major Traffic Street Plan was approved in concept by voters and funding for implementation was provided, resistance by local residents to several street-widening initiatives caused them to be blocked or postponed, while an "orderly and well-balanced system of thoroughfares" remained more idea than implemented strategy for transportation in the city. The 1930 Olmsted-Bartholomew Plan commissioned by the L.A. Chamber of Commerce could be considered an extension of the 1924 Major Traffic Street Plan with its proposed parkway development. But several of the Chamber's leaders who had been at the forefront of the promotion of the automobile largely abandoned the 1930 plan, and some raised objections even before the plan was submitted. While funding remained one obstacle, divisions also arose over aspects of an approach that directly linked park development and access to beaches and playgrounds with the creation of parkways and land-based flood-control strategies. A stalemate around the development of a more comprehensive roadway program ensued, with opportunities for new dedicated roadways shifting to the intervention of the federal and state governments.[19]

During the late 1930s, as federal funding became available to build an Arroyo Seco Parkway, designed in part to follow the path of the Arroyo Seco stream between Pasadena on its way towards downtown Los Angeles, the transportation debates reached a critical point. On the one hand, parkways were seen as providing uninterrupted movement of vehicles and transportation efficiency as well as the overarching parkway experience of visually experiencing and connecting to the surrounding landscape. But even as the Arroyo Seco Parkway was being constructed and funding began to be sought for a second dedicated roadway that would pass through the Cahuenga Pass on through Hollywood towards downtown Los Angeles (a road that was subsequently known as the Hollywood Freeway), an important shift was also beginning to take place with respect to roadway design, funding mechanisms, and land-use and transportation planning. Transportation efficiency *and* aesthetic delight had been considered inseparable goals of parkway design, which in the early twentieth century was described as bioengineering—a marriage of architecture, landscaping, and civil engineering in a three-dimensional design. But by the 1940s, the goal of efficiency was already beginning to overshadow that of aesthetic delight. The focus of the multilane dedicated roadways, increasingly called *freeways* rather than *parkways* by California's transportation engineers and other automobile advocates, would be to move people and goods at higher speeds, with these roadways superimposed over the land with little or no attention to aesthetics, scenic pleasure, community values, environmental impacts, or land-use considerations.[20]

The transportation debates during the early 1940s—regarding the role of automobiles, interurban streetcars, and buses with respect to passengers and the role of rail and trucks with respect to freight—also extended to the question of the design of these parkways and freeways. A number of transportation planners assumed that the first of these parkways and freeways would accommodate cars, interurbans, and buses. Debates took place over whether to allow buses on the Arroyo Seco Parkway and an interurban streetcar rail line within the center strip of the Hollywood Freeway. As late as 1946, even as the interurban system began to be dismantled, traffic surveys pointed out that nearly half of those traveling within the central districts of the city were still using public transit. Based on such evidence, a joint report of the Board of Public Utilities and the City Engineering Bureau urged adoption of a policy "to use the Hollywood Freeway to the maximum extent for public transportation as well as for private automobiles." A year later, various transit advocates created a Rapid Transportation Action Group that subsequently issued a plan that included the creation of a countywide rapid transit authority, a nine-mile subway beneath the downtown business district, and new interurban streetcar routes in the center of each of the projected freeways, including the Hollywood Freeway, which was nearing completion. But a proposed rapid-transit bond measure was ultimately defeated by a coalition of suburban cities and developers as well as the Auto Club, which felt that the measure would delay freeway construction plans. It was also opposed by the Southern Pacific and Pacific Electric companies on the grounds that the bond measure "opened the door for a public agency to operate a competitive service."[21]

The question that needed to be addressed, according to transportation analyst Martin Webster in a 1949 article in *Engineering and Science,* was: "What role do we want mass transportation to fulfill in this area? Do we simply want to supplement the automobile, or do we want a self-sufficient integrated transportation system?" Despite the desire by some to pursue a more integrated transportation system, a key divide between auto and transit policies had emerged by the late 1940s in which transit fared increasingly poorly. On the one hand, policies were designed to accommodate the growing number of private automobiles through subsidized road construction and other initiatives that assumed that auto use was a private choice. Meanwhile, transit, which resembled more of a regulated private enterprise, was nevertheless viewed as a public type of enterprise that was subject to the decisions of private interests that ultimately failed to sustain an effective urban transit system that was becoming increasingly paralyzed. That divide ultimately extended to decisions about whether to link transit systems to the emerging new roadways that had become the central focus for transportation planning and funding.[22]

The shift from parkway to freeway in design and purpose and the parallel decline of the interurban streetcars indicated that the focus of transportation planning would be on accommodating the automobile. The concept of a rail line in the freeway center strip was not pursued with the Hollywood Freeway, and the various options for downtown Los Angeles, such as the subway link and even a proposed monorail system, were either not considered or not implemented. Instead, the focus in downtown Los Angeles continued to be "the exasperating Downtown parking problem," as the *Los Angeles Times* put it. As funding shifted from the city and the region to the state and several freeways began to uproot downtown streets, pressure from state policy makers focused on expanding parking opportunities, with one legislative committee calling for ten thousand new parking spots as a condition for obtaining state funds for further freeway construction.[23]

The shift in funding to the state for freeway design and construction in the 1940s was also critical in how an automobile- and freeway-centered transportation system influenced critical land-use, housing, park development, and other urban-planning issues. Until the 1940s, most transportation plans in such cities as San Francisco, Detroit, Chicago, and Boston as well as Los Angeles assumed a multimodal system, including a type of joint-use concept that allowed cars, rail, and buses to have access to the parkways, freeways, and surface streets. Key officials in Los Angeles (such as Mayor Fletcher Bowron and Charles Bean, the chief engineer of the Bureau of Public Utilities) continued to argue for a comprehensive transportation approach that included rail, bus, and automobile linkages. "Given the widespread interest in traffic calming, multi-modal planning, boulevard revitalization, exclusive roadway freight facilities, and investing in transportation facilities as part of redevelopment efforts, many of these plans appear surprisingly relevant today," transportation planner Brian Taylor wrote in 2000 about these earlier approaches.[24]

At the same time, it was also clear by the 1940s that parkways were quickly becoming freeways—that is, single-purpose systems whose primary goals were speed and efficiency. In Los Angeles, key figures in transportation decision making, led by city engineer Lloyd Aldrich and E. E. East of the Automobile Club, were also interested in creating a more expansive system that could cover the large urbanized area that extended beyond the city limits. A freeway plan pulled together by Aldrich, introduced by the Transportation Engineering Board in 1939, and largely adopted the next year by a Los Angeles City Planning Department report called for as many as twenty-one freeways to extend over 612 freeway miles that would be constructed over a fifteen-year period to blanket the region. The 1939 plan focused on the Hollywood Freeway as a crucial first step in developing such a system. It would be a potential model, have a toll feature that could help pay for its construction and maintenance, and be entirely within the city of Los Angeles, "thereby making for speed of decision, construction, and use." The 1939 plan, with input from Bean, who remained the city's foremost advocate of transit, also called for a "joint development" framework, including rail lines along parkway medians and other features that took into account the link between transit use and the new role of the automobile. The 1939 plan estimated that the improved transit service that would result would account for half of the benefits to be derived from the parkway plan.[25]

The question remained of how best to construct the parkway system. To build such a system, including completion of the Hollywood Freeway, required new sources of funding that neither the city nor the county could provide by itself. Most federal and state funding for highways at the time was rural-based and focused on farm-to-market needs as well as travel between cities rather than within cities. While the idea of a toll road was still being explored, it became apparent that the expressway plan required the city to look elsewhere for funding. That source became the State of California, with the funds based in part on highway user fees to be controlled and dispersed by the state highway department. Cities like Los Angeles, without their own adequate funding, agreed to this new arrangement, although decision making on route, design, linkages (or the absence of linkages) with public transit, location (and housing impacts), and land-use implications were now left in the hands of the state transportation engineers.

State legislation that passed in 1947 further elaborated this approach by incorporating parts of L.A.'s 1939 expressway plans into the state highway system by raising the gas tax to provide additional funding and by establishing a ten-year master plan that called for the development of a 14,000-mile state highway system. Most important, the state Division of Highways took charge of all freeway-development decisions within urban areas and in doing so essentially put an end to the parkway and multimodal transportation concepts. By 1949, when transportation analyst Martin Webster posed the question about whether a self-sufficient integrated transportation system was still possible, the outcome appeared to have already been set. Los Angeles—with its ambitious freeway plans, its change from a transit-rich to a transit-poor metropolis, and its high number of vehicles per resident (one car for every 2.5 people in 1949, compared to Chicago with one car for every 5.1 residents or New York City with one car for every 8.7 residents)—represented the new model for urban transportation. The highway engineers had triumphed, the interurban streetcars everywhere were beginning their rapid and ultimately fatal decline, and the debate at the federal and state levels, looking towards Los Angeles, was no longer about what type of transportation system to construct but how to make the freeway the defining feature of the American landscape and in the process transform the cities as well.[26]

Freeway City, Freeway Nation

By the 1950s, cars and highways had become central to the American experience. "America lives on wheels," U.S. Secretary of the Treasury George Humphrey remarked in 1955, "and we have to provide the highways to keep America living on wheels and keep the kind and form of life that we want." That same year, former General Motors president Charles Wilson, in his confirmation hearings for secretary of defense before the U.S. Senate, made his startling and still famous comment that he could not conceive of a situation where his former position could constitute a conflict "because for years I thought what was good for the country was good for General Motors and vice versa." First formed in 1942, a powerful highway lobby had emerged and continued to hold weekly roundtable gatherings. It was led by a self-constituted "Road Gang" of auto, rubber, oil, construction, engineering, and trucking interests as well as highway engineers and administrators. Cars were becoming cultural trend setters, thanks in part to features of the 1955 models, such as Cadillac's new fins, the Pontiac Star Chief's Strato-Streak V-8 engine and two-tone colors like "avalon yellow" and "raven black," and Chevrolet's "Hot Ones." In the process, car sales skyrocketed, jumping an astonishing 38 percent between 1954 and 1955 and creating in turn a type of "religion of the motorcar," as Lewis Mumford characterized the growing American obsession with the car. Also in 1955, Disneyland opened. With its vast parking lot (twice the size of the park and the largest in the country at the time), its Autopia car ride in Tomorrowland, and its connection to the ever-expanding Santa Ana Freeway, this fantasy theme park was accessible only via car and freeway, an exclusivity that was reinforced in part by the fact that the interurban streetcars had stopped running to Santa Ana a few years earlier.[27]

In 1955, California also took the lead in seeking to earmark as much as $310 million for a massive expansion of its state freeways. But the states, including California, were most focused on the debate then taking place in Congress about how to ensure that the "limitless needs" of new highway construction could be met. Highways were now seen as an inevitable and dominant feature of the country's mode of transportation. The issues that still remained to be resolved in the debate in Congress were not whether to expand the highway system but where the highways would be located, how they would be funded, and how their future development could be safeguarded from the problems related to annual funding requests and divided constituencies. It took another year before those issues could be resolved, but the legislation that resulted—the Federal-Aid Highway Act of 1956 (originally known as the National System of Interstate and Defense Highways)—

was both transformative by its reach and revolutionary in its impact on urban life and on future transportation systems. "Within the next fifteen years," Lewis Mumford wrote shortly after the legislation's passage, "it will be too late to correct all the damage to our cities and our countryside . . . that this ill-conceived and preposterously unbalanced program will have wrought." Indeed, about fifteen years later, Daniel Moynihan wrote that the 1956 legislation had "more influence on the shape and development of American cities [and] the distribution of population within metropolitan areas and across the country as a whole . . . than any initiative of the middle third of the twentieth century."[28]

The Federal-Aid Highway Act of 1956 was an enormous piece of legislation. In his memoirs, President Eisenhower characterized it as the "most gigantic federal undertaking" ever that "would change the face of America." It had a $333 billion price tag (costs that continued to escalate over time), included provisions for the construction of 41,500 highway miles, provided an extraordinarily high match of federal to state dollars (90 percent to 10 percent), had self-generating funding mechanisms (including the highway user or gas tax and dedicated Highway Trust Fund), and established rapid completion targets (with an original 1972 termination date that also changed over time). Among other key provisions, the legislation codified the central role of building freeways in the city by making available enormous new resources for urban highway construction and by identifying urban highway routes to be covered by the legislation through a document entitled the "General Location of National System of Interstate Highways" (also known as the "Yellow Book"). While a handful of cities, notably Los Angeles, Chicago, and New York, had embarked on major urban highway construction prior to the 1956 legislation, most cities had only limited or no highway programs in place or even in the planning stage. Even L.A.'s ambitious expressway plans first laid out in 1939 were still far from complete, while overall freeway construction in the state, including for Los Angeles, lagged behind a timetable established from the 1947 legislation, due in part to only modest increases in available funding through the state gas tax.[29]

Earlier federal legislation passed in 1944 and federal agreements established with the states in 1947 had provided for special federal funding for urban freeways. However, the criteria for funding of urban projects emphasized projects *between* cities. By 1956, only 480 freeway miles had been completed or were under construction in the country's twenty-five largest cities, compared to the 2,200 freeway miles earmarked for construction within cities that were identified in the Yellow Book. The Yellow Book also influenced the development of an important constituency and lobbying force consisting of city officials who had become advocates for passage of the legislation. Urban freeways had become a necessity for city residents, Detroit Mayor Albert Cobo argued in lobbying for the legislation, characterizing the urban freeway as "a picture of beauty." This urban lobbying group was particularly attracted to the legislation's massive funding and federal-state funding formulas as well as the shift away from the more exclusive focus in previous legislation and program implementation that had favored rural and intercity (as opposed to intracity) projects. City officials decided that they wanted to have a piece of the action, but in doing so any connection of highway building to urban land-use planning or multimodal transportation approaches was essentially abandoned.[30]

The anticipation of city officials about the new highway construction and its impact on such urban needs as reducing traffic congestion ("we can lick traffic congestion," Robert Moses exclaimed in an article published at the time by the Ford Motor Company) was palpable. Cities and states lined up to access funds and revisit plans that had been partially developed or that were still on the drawing board. Immediately following passage of the legislation, the federal highway bureaucracy also kicked into high gear. Within two months, more than $3.7 billion had already been apportioned to the states to begin projects, while the first contracts were awarded for construction of 110 freeway miles, including the very first project awarded to the Missouri Highway Department. California also quickly jumped on the bandwagon, determined to become its own mini-freeway nation, with plans to construct as many as 12,000 freeway miles, while spending as much as one-fourth of its entire annual budget over a twenty-year period. The California system reached its peak in 1966 when 341 freeway miles were built, more than 10 percent of the 3,000 freeway miles built nationally.[31]

But almost as soon as construction began, problems also began to appear, including growing community opposition, particularly when plans for freeways went through established neighborhoods rather than existing transit corridors. By 1958, the first sustained opposition emerged regarding freeway plans in Reno, Nevada. Seven years later, the city of San Francisco became the first place to actually derail a project when the "rebellion against the bull-dozing, home-displacing, land-gobbling freeway monster," as *San Francisco Chronicle* editor Scott Newhall put it, finally caused the proposed Embarcadero Freeway along the city's waterfront to be shelved. Protests sprang up in New Orleans, in Florida, and, by 1970, in thirteen different cities where protests raged. In Los Angeles, a proposed 7.5-mile freeway that would have provided an extension of the Hollywood Freeway to the Route 405 freeway on the west side and that would have cut through parts of Hollywood and the City of Beverly Hills, became a lightning rod for protests and was eventually abandoned in 1975. A major battle also erupted in Los Angeles over the proposed Century Freeway that was designed to go through several low-income neighborhoods to provide a link between two other freeway systems. The highway engineers were particularly interested in locations for their freeways that either took away park land (where rights of way were cheaper and easier to acquire) or forced the dislocation of residents in low-income neighborhoods, since, it was assumed, both the cost of displacement and the level of opposition were more manageable. Even when opposition developed in low-income communities (such as the Boyle Heights Anti-Golden State Freeway Committee, which fought the proposed splitting of its neighborhood through a freeway interchange, the elimination of the area's major greenspace, and the displacement of 1,400 families), the highway builders continued their march through the neighborhoods. The Highway Trust Fund, moreover, had no provision for housing-displacement costs, thus providing additional incentive to the states and the cities to target low-income residents to reduce their own costs. This was urban land use at its most inequitable and socially and environmentally destructive, without any consideration other than the straightest and quickest way to take people from one freeway to the next and eventually to their final destinations. Even those considerations seemed suspect: one freeway extension between the 405 freeway and a type of no-man's land east of the Marina Del Rey area by the Pacific Ocean (initially named the Nixon Freeway by the California Legislature in the 1960s until after Nixon's 1974 resignation, when it was renamed the Marina Freeway) was dubbed by local residents as the freeway that went nowhere and seemed to have no purpose other than its propensity to incur accidents.[32]

By the early 1970s, at a point when the interstate highway system was to have been completed, freeways had become as negative a symbol as suburbs. Congestion during rush hours increasingly extended to longer stretches of the day and was as visible a problem as ever, even though cars had full and near exclusive claims not just of freeways but of surface streets as well. Air pollution had become one of the preeminent environmental issues, with the more visible pollutants in cities (like nitrogen oxide and carbon dioxide) linked to cars and freeways. The 1970 Clean Air Act Amendments even included one component of the legislation, never effectively pursued, that called for demand-management strategies to help reduce automobile use. Books and magazine articles spoke of "superhighways" as "superhoaxes" or "Autokind vs. Mankind" struggles. But the die had also been cast. "It is just about impossible to get a major highway program approved in most large American cities," Daniel Moynihan wrote in his 1970 article in *The Public Interest*, although Moynihan also noted that such opposition was already too late, given that most systems had already been built.[33]

As opposition to freeways increased in the 1970s, new interest in mass transit emerged. Some of the interest represented a type of nostalgia, while also suggesting a kind of desperate hope that the presumably inexorable trend of the abandonment of transit in the face of an all-pervasive use of the automobile and the freeway could be reversed. This nostalgia was also pronounced in Los Angeles, where its mayor (Tom Bradley) and several other leading public officials (such as L.A. county supervisors Baxter Ward and Kenneth Hahn) hoped to reconstruct a new version of the Big Red Cars. With the Century Freeway in trouble, Hahn, for example, called for the development of a rail line in the middle of the proposed freeway to kick-start L.A.'s new interest in rail while providing another rationale for building a freeway that had already condemned a number of African American working-class homes in Hahn's district.[34]

But the rail revival was compounded by the enormous new costs of constructing such a system, particularly with the loss of its historic rights of way, the lack of the type of dedicated funding stream available for freeways, and the problem of attracting riders in cities like Los Angeles that had become car and freeway dependent. Los Angeles, similar to most other cities and regions, had by the 1970s experienced a shift in ownership of transit systems from private to public. In L.A., the two major private transit systems were sold in 1958 to a public entity, the Los Angeles Metropolitan Transit Authority, for a purchase price of $33.3 million, with the funds allocated by the state. This entity shut down the last of the interurban lines by 1963, but Bradley, Ward, and other public officials now turned to its replacement, the Southern California Rapid Transit District (SCRTD), in their quest to revive a rail-based system.[35]

Through the 1970s and 1980s, the mission and programs of the new public transportation agencies in Los Angeles and most other urban areas were split between the desire of cities to reestablish a transit system (with a major focus on a rail revival) and the continuing implementation of projects designed to address the expansion and maintenance of the massive highway systems that dwarfed whatever transit systems were in place or under construction. Transportation policy in this period was at best piecemeal and often disjointed (expensive plans for rail revival, focus on freeway expansion to address congestion, and often deteriorating bus service that increasingly served low-income residents) and still largely divorced from any broader land-use and community-development approach. Among agencies, the California Department of Transportation (Caltrans), staffed primarily by its engineers and highway builders, continued to maintain overall authority over the state's freeway system. This included its connection to the funds flowing through the federal highway purse strings that had been set in place with the 1956 interstate highway legislation.[36]

In 1991, the periodic six-year renewal of that highway legislation (the "highway bill," as it was commonly known) was recast in some significant ways. Efforts to refocus transportation policy in cities, including provisions in the highway bill, had been developing through the 1970s and 1980s, including the campaigns to stop highway construction through urban neighborhoods, struggles around inadequate bus service that served low-income residents, and a developing environmental constituency that focused on the enforcement of the Clean Air Act and its policies regarding the failure to attain specific standards that had been set in the 1970 legislation. Many metropolitan regions, including the Los Angeles region, which maintained the lead as the most polluted region in the country, were in nonattainment and could conceivably receive significant if not crippling penalties if attainment levels were not reached (though target dates kept on getting pushed back). This social-justice movement to improve transit service and prevent further freeway construction in low-income neighborhoods combined with an environmental lobby focused on implementation of the Clean Air Act, which led to a series of informal meetings to challenge the entrenched highway interests and shift the focus of the highway bill. Enlisting the support of New York Senator Patrick Moynihan, who maintained his long-standing interest in restructuring the highway bill, this new coalition, organized as the Surface Transportation Policy Project, helped bring about a change in the highway bill that was named, in the 1991 reauthorization, the Intermodal Surface Transportation Efficiency Act (ISTEA). The key concepts of "intermodal" and "efficiency," combined with a new emphasis on transportation planning, created new opportunities—as well as funding streams—that could establish some alternative space in future transportation approaches.[37]

The passage of ISTEA in 1991 and subsequent reauthorizations of highway legislation in 1998 and 2005 did provide new incentives for various alternatives and mechanisms that continued the emphasis on transportation planning at the metropolitan level. But while the 1998 and 2005 bills maintained opportunities for alternative approaches that had been developed in the original ISTEA legislation, they also included classic pork-barrel features that emphasized particular highway projects, a process that had long characterized the implementation of the highway bills since the passage of the interstate system in 1956. The freeway remained king, and after fifty years of highway spending since the passage of the Federal-Aid Highway bill and after more than 46,000 miles of highways had been set in concrete, the shift to another transportation mode, at least at the federal level, seemed remote. A renewed focus on multimodal transportation policy, the goal of the social-justice and environmental advocates during the 1990s, had largely disappeared with the second Bush administration and a sharply divided Congress that primarily sought to capture a piece of the $286.5 billion in transportation dollars—mostly highway oriented—for local, regional, and state interests. For these advocates, the limited nature of the federal programs for alternatives shifted the focus to the state and local levels. In places like Los Angeles, the potential to develop a different way of designing transportation remained available even as the dominance of the car and the freeway remained an overwhelming part of daily life in the city.[38]

Cruising for Parking

In the early 1990s, while teaching at the UCLA urban planning department, I was struck by the title of a course called "Cruising for Parking." The professor, Don Shoup, had his students join him to drive around (or to go by bike, since the bike riders cruised at about the same speed as the car drivers) at various locations in Westwood Village until they could, like other drivers, find a space to park. Westwood was the neighborhood adjacent to the UCLA campus and included a number of shops, restaurants, and other stores and commercial buildings. Parking was a premium in Westwood (as it was for the UCLA campus), and it included both free as well as one- or two-hour metered parking along the streets. There were several parking lots in Westwood as well, though the cost for parking in those lots was considerably higher. Traffic was often congested along the surface streets within the Village, including in the evening when there was no longer a charge for the street parking. As Shoup and his students calculated the time and distance it took for drivers in Westwood to find a free or moderately priced metered parking space on the street, they discovered how enormous an environmental—and economic as well as traffic-congestion—problem cruising for parking had become.[39]

At the time, I didn't connect parking as a core urban-planning (and social- and environmental-justice) issue and was bemused by what appeared to be an idiosyncratic course title. However, I've subsequently come to appreciate Don Shoup's objective in his teaching and research—to document how "free parking," as University of Pennsylvania transportation analyst Victor Vuchic put it, was possibly "the most important factor in the encouragement of car commuting, . . . [representing] the dominant obstacle to diversion of trips from cars to transit or to any other mode." When people drive, they expect to park for free and have an accessible parking spot almost as a privilege. The National Personal Transportation Survey, in fact, estimates that drivers have been able to park free in 99 percent of the trips they take.[40]

Parking, like freeways, generates an extraordinary array of land-use, economic, and environmental impacts. Huge amounts of land in the city are dedicated to parking. The amount of black top required for off-street parking spots or open-air parking lots has become, along with freeways and surface streets, a major contributor to polluted stormwater runoff. Land-consuming parking spaces contribute to the urban heat-island effect. The increased costs and amount of land required by parking regulations for housing and commercial development are enormous. Streets are widened to accommodate parking spaces, and sidewalks are narrowed and in some cases eliminated to provide passageways for cars and places to park them. In downtown areas, the ratio of parking area to total land area can be substantial, as high as 81 percent in Los Angeles, which is higher than any other downtown area in the world. Parking in urban areas has become so ubiquitous that some cities, like Tulsa, Oklahoma (which, along with Oklahoma City manufactured the first generation of parking meters), have earned the label "parking-lot city" for their large numbers of treeless parking lots and spaces available for off-street parking.[41]

Parking issues plagued cities as early as the 1910s and early 1920s, when increased sales of the automobile began to have an impact on traffic flow and commercial activity. The short-lived downtown parking ban in Los Angeles in 1920 was repeated in other cities, but pressure from automobile advocates and various commercial interests sought to accommodate this new form of transport, and the bans were lifted. Instead of restrictions, cities began to develop new types of codes and rules as well as devices that could address the increasing role played by cars in the city. Parking requirements for apartment houses were first instituted in Columbus, Ohio, in 1923, while requirements for buildings were initially developed in Fresno, California, in 1939 for hotels and hospitals. The parking meter was first invented in 1933 to encourage a more rapid turnover of cars in the downtown commercial district and thereby increase the stores' customer base.[42]

Since those first rules and strategies about parking were developed in the 1920s and 1930s, parking has become extensive and invasive in the city. Its "big footprint," as Robert Dunphy of the Urban Land Institute put it, its requirements, and its presence have become accepted and routinized. Conservative estimates identify an average of four parking spots per vehicle. Office buildings also typically require four parking spaces per thousand square feet of office space, often called the "magic number" or "golden rule" that city planners use when establishing parking requirements for new buildings. That big-foot requirement is even more pronounced for other types of commercial buildings, such as stores and restaurants, where parking becomes *the* critical factor in location decisions. Restaurants, for example, may need as many as ten or more spaces per thousand square feet.[43]

The costs for such parking requirements are largely hidden. For example, an estimate of the cost of deck parking for commercial establishments in Akron, Ohio, has been calculated at $14,000 to $17,000 per spot, while a surface parking spot (accounting for lighting, landscaping, and other costs) reached as high as $5,000 per spot. Parking costs for housing, based on standardized requirements, almost invariably fail to distinguish the nature of the housing, the availability of transit alternatives, the potential for a transit-oriented housing development, the density of the area, or the affordability of the housing. Nonprofit developers engaged in affordable-housing development have estimated that parking requirements add 20 percent to the cost of each unit. "Parking requirements are a huge obstacle to new affordable-housing and transit-oriented development in San Francisco," noted Amit Ghosh, the head of that city's comprehensive planning unit. Ghosh also argued that such requirements had forced developers "to build parking that people cannot afford." "We're letting parking drive not only our transportation policies, but jeopardize our housing policies too," Ghosh warned.[44]

Most of those costs are not incorporated into the costs of driving, skewing the calculation of the relative expense of cars in comparison to other forms of transportation for drivers and public officials alike. Parking costs—whether for building owner, restaurant, or housing developer—are not priced separately but are instead "bundled" into other costs. These bundled costs range as high as 96 to 99 percent, disguising the actual cost associated with parking (and car use). When efforts are made to establish a price for parking, even at a subsidized rate, there is often an enormous outcry since bundled costs and the long history of favorable parking policies have created an assumption that parking is free and that free and accessible parking represents its own perverse type of a right to the city and its varied places. "The cost of all parking spaces in the U.S. exceeds the value of all cars and may even exceed the value of all roads," Shoup argues, estimating the range of parking subsidies from $127 billion to as high as $347 billion annually.[45]

Parking issues are further exacerbated by the failure to distinguish between urban and suburban land uses and density characteristics when parking requirements are established. The publication *Parking Generation*, published by the Institute of Transportation Engineers, for example, calculated parking and trip-generation rates on the basis of suburban sites with their ample free parking, absence of public transit, limited or no pedestrian amenities, and any kind of transportation demand-management approach. These differences were subsequently acknowledged though not effectively revised in a later edition.

Supermarkets or big-box stores also define their parking approaches on the basis of the large suburban lots where supermarkets and big-box stores have been located in the last several decades. Suburban land is cheaper than inner-city sites (and thus parking lots also become more affordable), while locations are generally linked to their proximity to freeway exits. These stores assume that people will come by car to do their shopping and therefore have little or no provision for those arriving by foot, bicycle, or public transit. Supermarkets, moreover, are also required to provide a set amount of parking spots per square foot, no matter where the location, even in urban areas that are far denser and more transit dependent (e.g., where more than 20 percent of the population who live within a half mile of a store do not have cars). This, in turn, is similar to the situation regarding affordable housing and acts as a major disincentive for supermarket development in inner-city areas, a situation that has become more critical as supermarkets have increasingly abandoned low-income urban communities for more attractive suburban sites.[46]

Unlike the fight against new freeway construction, parking critics generally used to be limited to a small number of transportation analysts. However, in recent years, several alternative approaches have begun to be explored. One recent change has been a developing public role in providing parking in commercial districts. Several cities have developed new arrangements with developers where developers pay a fee that helps pay for the public parking structure in lieu of having to provide a set amount of parking spots, often calculated on the basis of standard parking regulations that seek to maximize the number of available spots. Making parking a public responsibility opens up the possibility of developing transportation planning approaches that would seek to reduce the parking demand to minimize the amount of parking to be made available.[47]

The key to reduced parking is reducing the number of cars on the road, and making parking less attractive (by increasing its cost or making it less accessible) can also influence the decision of whether to drive. The relationship of parking to car use also has a spillover effect on the attractiveness of the noncar or reduced-car-use alternatives. Preferred parking for car pools has become an easy and obvious approach. More ambitious strategies include the concept of "ecopasses," which employers purchase at a discounted rate from transit agencies and then provide to employees, who give up their parking spot but are able to ride free on local transit lines. Employer, employee, and transit agency all benefit: the employer reduces the number of parking spots (and the significant costs in providing such spots), the employee rides free on transit lines and reduces or eliminates the cost (and stress) associated with the use of a car for commuting, and the transit agency increases the number of transit users, an important goal since nearly all urban transit systems have been underutilized since the decline of transit in the 1940s.[48]

Complementing such cost-related incentives designed to reduce car use and eliminate parking spots are strategies for increasing the cost of parking and parking cash-out programs. In addition, the increasing number of car-share programs that have emerged in recent years in several cities, including Los Angeles, seek to address the convenience factor within the car culture associated with the notion of having access to cars and to parking at all times and in all circumstances. These approaches have economic as well as cultural dimensions insofar as they begin to address the privileged status of driving and parking as a core aspect of daily life in the city.[49]

Perhaps more than any other issue, transportation planners have strongly argued that *free* parking continues to remain the central factor in making driving economically attractive and in establishing the "parking right to the city" culture. Increasing the cost of parking in downtown districts begins to make the true cost of parking visible to drivers, who now consider driving to be "cheaper" than taking transit. But increased fees for parking are often controversial, given the long-standing subsidies that have been embedded in parking policies. Some social-justice advocates have even argued that increased parking fees represent a regressive tax since car use is widespread. The parking cash-out concept utilized by some employers answers those concerns in part by providing a financial incentive rather than disincentive to give up parking and as a consequence has become popular among those most concerned about the equity issues. Car-share programs, in turn, focus on both cost and convenience incentives. Initiated in Switzerland in 1987, car-sharing companies began to operate in U.S. cities in the 1990s and by December 2005 had spread to as many as thirty-six cities, ranging in size from Los Angeles and New York to Aspen, Colorado. Car-share members often work at institutions (for example, at colleges and universities) that also provide incentives for reduced car and parking use and include families with at least one car who decide to not pay for an additional car. By having access to a car-share program, someone could address the convenience factor, particularly in a spread-out region like the Los Angeles area, where transit alternatives may not allow people to reach their desired destination.[50]

The issue of convenience has also been a critical factor regarding grocery shopping, which represents the second most frequent destination related to car use (commuting to the job represents the most frequent). Most supermarkets and big-box stores, with their vast parking lots typically based on standardized parking requirements, have failed to establish any type of transportation policy to accommodate those without cars or to help reduce car use and parking requirements. In transit-dependent communities, shopping-cart theft has emerged as a major expense for these stores because people without cars use carts to carry home their groceries. As a result, a few inner-city stores began to successfully experiment with a van service for shoppers who purchased a certain volume of groceries (for example, more than $30 worth). In many of these stores, the volume of business increased while the loss of shopping carts declined. In addition, markets that established joint-use programs with community-development organizations (such as a flagship program involving a Pathmark store in Newark, New Jersey) created van service programs as part of the agreement. The success of these programs was due in part to the increased sales that more than offset the cost of the van service; the Pathmark store, for example, became the most profitable in the chain. These programs, however, have not yet created an explicit link between a transportation alternative like a van service and reduced parking requirements. Such an arrangement could reduce expensive land costs for the store and make available land for other uses in dense inner-city communities where the need for land for parks, schools, and housing remains critical.[51]

Strategies such as grocery van services, car sharing, and parking cash-out and ecopass programs for employees, students, and commuters who travel to their place of work or study offer an integrated approach to reducing parking and car use. Success, however, requires programs and policies that can make more attractive other ways of getting around in the city besides the car. To be able to walk, to bike, to take a bus, and to ride a subway or streetcar often requires relating to the city in new ways and establishing policies and infrastructure that make it possible to do so. The right to park and drive a car in the city then becomes reoriented to include and prioritize the rights of the pedestrian, the bike rider, the bus rider, and the subway or rail user. Such a reorientation can then help transform the ways that the city itself is experienced.

Getting There without a Car

Pedestrians

In 1947, a young writer named Ray Bradbury went for an after-dinner stroll with a friend along Wilshire Boulevard and was stopped by a police car. The two strollers were asked what they were doing walking on the street. Bradbury, who was already a critic of L.A.'s increasing embrace of car culture, later turned the incident into a play that he called *The Pedestrian: A Fantasy in One Act*. "The Pedestrian" introduces two characters, Mead and Stockwell. Mead is a "frustrated but passionately curious man" who announces, as he prepares to take a walk: "There's the city waiting for us." Mead seeks out his neighbor Stockwell to join him on his walk. Unlike Mead, Stockwell doesn't venture out much but instead watches his TV, often "slouched in his chair, slumped like a statue that has not even tried to move, though it probably has the gift." Mead convinces Stockwell to join him, but as soon as they begin their stroll, they're stopped by a patrol car. Inside the car a voice calls out, "What are you doing out?" Mead answers, "Walking." "Walking where?", the voice from the car inquires. Mead replies, "Nowhere." "[There is] no such destination as nowhere," the car voice bellows. To which Mead replies, "Nowhere is a very *fine* destination." The radio voice then asserts, "Unacceptable data, Mead." When Mead looks inside the car, he discovers that no one is inside and that the voice he heard was caused by a mechanical or robotlike simulation. "Wouldn't that be a joke," Mead surmises, "if they died years ago, and the car goes on, all by itself, enforcing the law?"[52]

When a *New York Times* critic described the 1947 incident involving Bradbury nearly six decades later in 2006, he wryly noted that "even then, it seems, nobody walks in Los Angeles." As much as L.A. is associated with car culture, it is also seen as hostile to pedestrians. English critic Deyan Sudjic, in his discussion of what he called the "hundred-mile city," dismissively characterizes the type of pedestrian that is found on the streets in L.A. or in California as "mad, bad, or from out of town." But Los Angeles has not been alone in creating an inhospitable environment for pedestrians. As car culture and land dedicated to freeways, surface streets, and parking became more pervasive in cities around the country, barriers to walking in turn became more and more pronounced. As a 2003 article in *Planning* noted, the presence of "wide arterials, surface parking, and lack of a fine-grained street grid [had made] walking a challenge" in most urban areas during the last several decades.[53]

The challenge for those who want to walk also includes the problem of pedestrian safety. For walkers, crossing streets continues to be a dangerous exercise. In 2003 in the United States, 4,749 pedestrians were killed by cars, constituting about 11 percent of all car-related fatalities, while more than 127,000 pedestrians were injured. Of those pedestrian fatalities, about a fourth were sixteen years old or younger. Traffic speed and traffic volume, endemic to a car-oriented built environment, have been identified as most responsible for pedestrian injury and death. In areas where more people walk, such as New York City, pedestrian fatalities have been even greater, as much as 48 percent of all traffic-related deaths. Higher accident rates for children who walk have been identified with particular types of neighborhoods that lack places (such as back yards or open spaces) for children to play. The large number of pedestrian accidents may ultimately reflect how streets have been transformed into passageways, long and straight, with limited areas for walking compared to parking or flowing traffic—where pedestrians, as Henri Lefebvre put it, become the "hunted" and cars become "privileged." In contrast, more pedestrian-friendly streets that also include street trees have been found to reduce rather than increase the number of accidents, contrary to the widespread assumptions of traditional transportation engineering approaches, which favor the barren streetscape.[54]

Yet the research, resources, programs, and policies that address pedestrian safety issues do not correspond to the extent of the problem. The focus remains the car and car occupants, not pedestrians. "Despite the size of the pedestrian injury problem," three researchers pointed out in a 2002 article in the *British Medical Journal*, "research to reduce traffic-related injuries has concentrated almost exclusively on increasing the survival rate for vehicle occupants." That bias also extends to overall funding levels as well. In New York, as a study in 2000 documented, funding levels were as low as 5 percent for traffic safety programs compared to the nearly 95 percent available for car-related traffic safety initiatives. Similarly, Los Angeles has a high number of deaths of pedestrians in traffic-related accidents (about 30 percent of all fatalities), and it dedicates far fewer funds to pedestrian-safety programs. This disproportionate response has been particularly acute throughout the state of California, which has ranked last among the fifty states in spending for pedestrians. "While an average of $40 per person in federal transportation funds was spent on highway projects statewide, an average of just 4 cents per person was spent on improving conditions for pedestrians," a study on pedestrian safety and public spending in California by the Surface Transportation Policy Project pointed out. The STPP study also indicated that a high percentage of pedestrian victims were low-income and transit-dependent residents, including a high percentage of Latinos. A study of pedestrian accidents in Los

Angeles from 1994 to 2001 also noted that the "hot spots" for pedestrian accidents occurred in places like Hollywood, South Central, Pico Union, East Los Angeles, and Chinatown, all locations with large low-income or minority populations.[55]

In Los Angeles, the pedestrian rather than the car has been identified as the problem. In one 2006 episode that captured the attention of the media, Mayvis Coyle, an eighty-two-year-old pedestrian, received a $114 jaywalking ticket for failing to cross a long and wide intersection in Sunland, one of Los Angeles's automobile suburbs. At that intersection, which is four lanes wide and seventy-five feet across, the light remains green for ten seconds, and pedestrians have less than thirty seconds in total to cross the street before the light turns red. One resident told *Los Angeles Times* columnist Steve Lopez that he had lived in Sunland for ten years and "you can never make it across the intersection in time." After receiving the ticket, Mayvis Coyle threatened to take the police department to court. Lopez cheered Coyle on, arguing that instead of ticketing people crossing intersections, "anybody on foot or bike in this traffic-choked city ought to get a citation."[56]

Bicyclists

In the same *Los Angeles Times* column, Steve Lopez reported that bicycle riders were also getting ticketed for riding their bikes through intersections rather than walking them. Bike riders, like pedestrians, are today seen as an anomaly in a car-centered environment, despite the favorable climate. "There is no part of the world where cycling is in greater favor than in Southern California, and nowhere on the American continent are conditions so favorable the year round for wheeling," one 1897 newspaper article enthusiastically commented about the thirty thousand bike riders then in the Los Angeles region in that preautomobile era. More than a hundred years later, the proportion of bike riders to population has become much smaller, and the conditions for bike riders have become far more precarious.[57]

As with pedestrians, a similar pattern can be found by comparing bike-rider fatalities and injuries to programs that address bike-safety issues. The numbers of bike-rider fatalities in the nation have been relatively small, 622 in 2003, in comparison to the 42,643 people killed that year in car crashes, although as many as 59,000 bicycle riders were injured that year. Nevertheless, a New York study noted that as many as thirty-five bike riders were killed in New York State in 1999 and that there were literally no funds allocated for bike safety that year. There are, in fact, funds potentially available to states and localities for bike programs, including some through the federal highway funding stream. While funding is at the discretion of local metropolitan planning organizations (for example, the MTA in Los Angeles), the amounts actually utilized have typically been far less than the potential funding stream available, including for bike-safety programs that could be linked to safer bike routes to schools, for recreation, or even for job commuting. Some safety improvements have been noted in places where bike helmet laws have been instituted. However, the focus on bike safety has been nearly entirely on the bike rider rather than the car driver, and a general climate of fear has been created that identifies bike riding in the city as hazardous. Bike riders, like pedestrians, are not welcomed in environments where cars dominate. Bike paths are few in number, and facilities to accommodate bike commuters or shoppers at workplaces or stores are the exception. Few cities provide bike accommodation requirements for workplaces or commercial buildings that compare to their enormous and costly parking requirements. Even New York City, where bike riding clearly represents, at least in Manhattan, an efficient and convenient mode of transport in heavily congested street traffic, bike riders are seen as interlopers, and militant bike-riding organizations such as Critical Mass have even been subject to spying, infiltration, and provocative actions by law-enforcement agencies.[58]

Bus Riders

Bus riders have also had to contend with an inhospitable environment that has become bleaker as bus service continues to decline in Los Angeles and other cities. People continue to ride the bus in Los Angeles as elsewhere, though those who ride the bus and in what numbers have also changed. In

1982, a bus strike in Los Angeles forced more than 1.2 million bus riders to search for alternative transportation and ultimately created some horrific traffic jams on the freeways. The media characterized the bus riders as a representative cross-sample of Angelenos—professional and skilled workers as well as a handful of low-income residents. Bus riders were not distinguished by class or ethnicity. An editorial in the *Los Angeles Times* after the strike also indicated the importance of bus service as part of the city's transportation options, despite the clear ascendance of cars and freeways. The transit strike in 2003, however, failed to have much of an impact on the freeways, since most of the remaining 350,000 bus riders, many of them Latino and immigrants, did not own cars. Their search for alternatives was both more desperate and limited. Articles in the press spoke of the bus riders as a marginalized group distinct from the vast majority of car drivers and freeway users. Even the 100,000 commuters who used L.A.'s new rail lines were viewed differently. The assumption (not entirely accurate) was that those who took the new Gold Line or Red Line had alternatives—as middle-class commuters who also owned cars but who used the rail lines as an option rather than a necessity.[59]

The decline of bus service and the marginalization of bus riders have come about because of a number of factors, including a changing revenue stream, transportation planning choices, a widening economic divide, design factors, and the social and cultural factors that stigmatized bus use and created a climate of vulnerability and fear for bus riders. During the 1980s and 1990s, the focus of the transit agencies in Los Angeles, as in other metropolitan areas, began to shift towards the development of new rail systems, particularly commuter-oriented systems from suburban to downtown areas. The failure to maintain the tracks and rights-of-way of an earlier generation of rail lines made the shift to rail far more expensive and subsidy-driven than the modest subsidies that would have been required to keep the earlier systems intact. Though new funding was becoming available through federal funding sources, particularly after 1991 and the passage of ISTEA, the nature of that funding often ended up creating a zero-sum game regarding the availability of funds for bus and rail systems, while highway funding, which continued to grow, was earmarked separately. Many of the metropolitan agencies, including in Los Angeles, responding in part to a type of middle-class nostalgia for rail and the agencies' desire to reestablish a middle-class constituency for transit alternatives, began to shift their funding from bus to rail. They accomplished this by cutting back the number of buses on the road, increasing fares, and deferring maintenance, which created a new set of burdens and harsh realities for the bus rider who waited longer, sat in overcrowded buses, and had longer trip times.[60]

The shift in transportation approaches mirrored the changing economic conditions and demographics of the region. A pronounced income divide between low-income, increasingly immigrant people and wealthier residents was already becoming apparent by the late 1980s and early 1990s. Los Angeles especially felt the full weight of the deindustrialization that had been occurring since the 1980s with the loss of part of its manufacturing base and federal policies that lengthened the widening economic divide between residents. During this period, the demographics of the bus riders changed significantly, with increased fares, longer wait times at bus stops, and overcrowded buses becoming the norm. The reduction in resources also translated into an increasing number of bus stops that had little or no lighting and few shelters, while various studies noted that the lack of even the simplest safety design features made bus stops magnets for crime. In contrast, an increasingly larger share of transit resources began to go to the development of the incipient (renewed) rail system in Los Angeles, which included both the suburban-oriented Metrolink system that brought commuters from the eastern edge of the county in the bedroom suburbs of the Pomona and San Gabriel Valleys to downtown Los Angeles and the new Metro lines that crossed low-income communities, commercial centers, and middle-class enclaves as they also headed toward Union Station in downtown Los Angeles.[61]

The decline in bus service in Los Angeles stimulated a major environmental justice lawsuit that relied on the antidiscrimination provisions of Title VI of the Civil Rights Act of 1964. The lawsuit contended that the change in funding for buses in favor of the new rail systems constituted a violation of that act. The suit was initiated by a community-based organization, the Bus Riders' Union, and brought together activists, civil rights and public-interest attorneys, and transportation experts to highlight the consequences of this shift in transportation spending and the inequities that resulted in the orientation towards funding rail rather than bus systems. The suit, which resulted in a victory for the bus advocates and a court order requiring a specific shift in spending toward more buses and improved service, became a milestone in the transportation history of Los Angeles—and of the country itself. A transportation equity movement emerged in both rural and urban communities, extending the environmental-justice argument to a consideration of access (to jobs, shopping, and other needs) as a key dimension—and burden for some—of daily life.[62]

The court order also had major implications for transportation planning itself in Los Angeles. The continuing battles over implementation of the court order resulted in greater scrutiny about transportation choices. It also pushed the transportation agencies to explore some potentially innovative strategies that could begin to enhance the status of both bus and rail transit in the region. Such strategies included the development of rapid bus technology (the MTA's Metro Rapid Buses or BRTs) that were seen as a far less costly alternative to rail and could potentially establish a type of express-oriented system to quickly navigate major surface street routes. It also included the purchase of less polluting alternative-fuel buses that provided an important health improvement for riders, bus drivers, and local residents. The emphasis on development of new routes for rail lines continued to dominate transportation planning and the ever-present search for new funding sources, but as the rail system expanded, it began to include a broader cross-section of riders and more options to reach a greater range of locations. Interest in transit-oriented development (locating housing and commercial development near transit stops) became more accepted as a goal by the transportation agencies, although the importance of equity considerations (for example, linking affordable housing to transit stops) tended to take a back seat to the continuing search for a middle-class constituency for transit. Even the highway agencies began to champion high-occupancy (car pool) lanes, particularly when they were linked to expanding the funding pie for roads.

By the new century, the transportation picture had become more complex. The growing visibility of both bus and rail and the modest development of new constituencies of bike riders and walkers allowed Los Angeles to conceive, for the first time in more than fifty years, a transportation future that no longer remained the exclusive province of the car and the freeway. But while the desire for change in the mode of transportation had increased, the realities on the ground still meant that the struggle to overcome the dominance of the car and the freeway remained formidable—despite the experience of congestion, long-distance commuting, the automobile's higher costs and even greater hidden costs, and the car and the freeway's impacts on land use, the ambient environment, and the myriad of other global, local, and environmental and social issues that affect daily life in the city.

6

The Magic of ArroyoFest

> It is interesting, if not useful, to consider where one would go in Los Angeles to have an effective revolution of the Latin American sort. Presumably that place would be in the heart of the city. If one took over some public square, some urban open space in Los Angeles, who would know? A march on City Hall would be inconclusive. The heart of the city would have to be sought elsewhere. The only hope would seem to be to take over the freeways.
>
> —*Charles Moore*[1]

A Freeway Fiesta

Saturday, May 29, 2004, was another hot day on the freeway. My colleague Anastasia Loukaitou-Sideris and I had recently completed a report for the University of California Transportation Center on L.A.'s first parkway—the Pasadena Freeway, formerly known as the Arroyo Seco Parkway, the major transportation artery of the Arroyo Seco corridor and northeast Los Angeles. Anastasia was with her family driving south on that same freeway. Passing over the channelized Arroyo Seco stream, curving around and through South Pasadena on the way into the Highland Park neighborhood of northeast Los

Angeles, Anastasia was thinking about the freeway festival that had taken place the previous year in about the same spot and that concluded at Sycamore Grove Park. Near the hole in the fence that led up to the park, Anastasia's thoughts were suddenly interrupted. Flashing red lights and sirens startled her family and other motorists, interrupting the monotony of the freeway landscape. The police were chasing some bank robbers on the freeway, which was soon shut down, stranding motorists in their cars and transforming the highway into a vast parking lot in a matter of minutes.[2]

The traffic jam was a classic Sig Alert. But while the experience of such a massive delay was not entirely atypical, particularly on this stretch of the freeway with its winding curves and abrupt exit and entrance ramps, the scene that unfolded, as Anastasia recounted it to me, was not a typical outcome. A few minutes after the cars had been stopped to give way to the police investigation that ensued, a mariachi band with guitars and drums suddenly appeared on the freeway pavement. The musicians had been trapped in their car and had decided to come out and offer free entertainment to this captive, parking lot audience. In a matter of minutes, the freeway landscape of stalled cars and unhappy motorists was transformed as hundreds of people got out of their cars and began to rhythmically swing and clap to the sounds of the musicians. Within minutes, ice cream and water vendors realized that this parking lot scene of stalled cars and captive (and thirsty) passengers created an opportunity, and they made their way through the hole in the fence from Sycamore Grove Park and the neighborhoods bordering the freeway. Bike riders also appeared, eager to weave their way through the parked cars up and down the freeway and to celebrate this unusual freedom to occupy a space they could not have otherwise explored.

This massive traffic jam marked the second time that people rather than cars had gained the center stage on the Pasadena Freeway/Arroyo Seco Parkway. A little less than a year earlier, on Father's Day, June 15, 2003, thousands of bike riders and walkers had descended on the freeway to participate in an event we called "ArroyoFest". For those of us who had planned the event, this planned occupation of the freeway was designed to create a special

L.A. moment. But ArroyoFest was also linked to a broader agenda to connect the diverse communities along the Arroyo Seco corridor. Among other goals, we wanted to stimulate a rethinking of the roles played by the oldest freeway in the West as well as by the Arroyo Seco stream that followed the same path as the freeway. The bikers and walkers who rode and strolled on the freeway for four hours on that Father's Day and then attended the community festival at Sycamore Grove Park participated in a moment that captured the power of the imagination to conceive what might be possible and how difficult yet extraordinary it would be to get there.

A Different Kind of Corridor

Bounded by the first freeway of the West and the stream that stretches from the San Gabriel Mountains through Pasadena and several of the northeast neighborhoods of the City of Los Angeles before it joins the Los Angeles River north of downtown, the Arroyo Seco has a storied place in the history of the region. The Arroyo's banks were originally the home of native Tongva villages, Spanish ranchos, and the original settlement near the confluence of the Arroyo Seco and Los Angeles River where the City of Los Angeles was founded. During the late nineteenth and early twentieth centuries, the Arroyo area emerged as the most prominent of the early suburbs of central Los Angeles, a favorite location for the region's burgeoning movie industry, and the center of the arts and crafts movement in California. Within its sycamore-shaded canyon, poets, painters, and photographers gathered to interpret life in the "Southland." Charles Lummis, the best known of the Arroyo's residents, built El Alisal, his home at the Arroyo's edge and founded Los Angeles' first museum, the Southwest Museum, overlooking the Arroyo's scenic canyon. A number of historic bridges also span the Arroyo, including the Colorado Street Bridge, built in 1913, which became one of Pasadena's distinct landmarks and a common image on early twentieth-century postcards.[3]

The sense of place associated with the Arroyo included the notion of an Arroyo transportation corridor following the path of the Arroyo Seco stream bed. A sandy trailway connecting the villages of Tongva Indians, the Arroyo had been recognized by the first Spanish settlers as the most direct route from the administrative center of the Los Angeles pueblo to the most important church in the region, Mission San Gabriel. A logical and direct pedestrian route, this ancient roadway was to be adapted to speedier means of transportation—first horse wagons, then bicycles and interurban streetcars, and finally automobiles.

A route connecting downtown Los Angeles to Pasadena and even farther—linking the mountains to the sea—had been talked about since 1895. In 1900, the first vehicle traffic plan catering to the "mechanical marvel of the day: the bicycle" opened through the Arroyo. The plan for this "cycleway" was to connect, through "a great transit artery," the city of Los Angeles with Pasadena and other Arroyo corridor suburbs. It was conceived and sponsored by a local millionaire named Horace Dobbins, who used his inheritance to promote the idea and launch the project. Designed to establish a path on an elevated, multilane, wooden structure that provided grade separation, this scenic, transit-oriented proposed bike route would eventually be regarded as a precursor of the parkway. "The country passed through by the cycleway is the loveliest in Southern California, the route having been chosen with an eye to scenic beauty as well as to practical needs," one commentator noted as the bike route was being built. The route also passed along the east bank of the Arroyo Seco, "giving a fine view of this wooded stream, and skirting the foot of the neighboring oak-covered hills," another commentator declared. The bike route was never completed because of lack of funds, however, and Dobbins eventually sold the cycleway's right of way to the Southern Pacific Railroad.[4]

Visions of a scenic transportation corridor reappeared throughout the early part of the twentieth century. Discussion and debate about a roadway continued well into the 1910s and 1920s, influenced by the national impetus for the building of parkways. These discussions often crystallized into specific proposals and plans. A 1911 drawing by Laurie Davidson Cox depicted the existing Los Angeles parks linked by a series of new or enhanced roadways, including one connecting the northeast corner of Elysian Park to the southeast reach of an Arroyo Seco Parkway. The 1913 proposal for an Arroyo Seco Parkway by the Los Angeles Parks Commission envisioned a metropolitan parkway that passed through the cities of Los Angeles, South Pasadena, and Pasadena to the mountains of the National Forest Reserve and that also connected a series of parklands in the Arroyo to other parts of the city. "If the parkway should end at an intermediate point, not connecting with the rest of the city park system, its value to the city, as well as to sister communities, would be greatly decreased," the Commission argued. In 1915, J. B. Lippincott, then a member of the Board of Consulting Engineers for Los Angeles and a key figure in the development of the L.A. Aqueduct, linked the initial development of an L.A. River and Arroyo Seco flood-control program (proposed by the newly established Los Angeles County Flood Control Commission) with a roadway between Pasadena and Los Angeles that could take advantage of the construction of a flood channel along the same pathway. Perhaps most significantly, the Pasadena to Los Angeles roadway concept was elaborated by the 1924 Major Traffic Street Plan for Los Angeles, which proposed the first grade-separated parkway following the Arroyo Seco from Pasadena to Los Angeles. The 1924 traffic plan, moreover, strongly evoked the design concepts of the suburban parkways of New York that emphasized the connection of roadway to parklands and natural landscapes. And a 1930 document, the Olmsted-Bartholomew Plan for the Los Angeles Region, also proposed a comprehensive and coordinated system of large parks and a connecting Arroyo Seco Parkway representing a "North-South chain from the mountains to the sea through the heart of the city."[5]

The Arroyo's rich cultural history and attractive landscapes therefore represented an ideal location to implement parkway goals. By the 1920s and early 1930s, as the plans for an Arroyo Seco Parkway continued to be explored, its development had already begun to evolve into a kind of hybrid roadway. It would efficiently move large numbers of cars at a continuous speed, and it would allow pleasure driving along a serpentine roadway of variable widths that was well adjusted to topography and offered views and vistas of nearby and distant landscapes. In urban areas, considerable grading and planting would need to be undertaken to achieve a park effect, a design process that would be accomplished through a type of restorative landscaping to frame views and reconstruct natural appearances. Essential to the success of the parkway would be the land-sensitive location of the route and its ability to connect visually with adjacent parklands. Adjacent landscapes and parklands would be protected by parkway design. For example, the city of South Pasadena insisted that in laying out the parkway, the road needed to curve around that city's own Arroyo Seco Park to cause as little damage to the park site as possible. Eventually, approximately four thousands plants of various varieties would be planted so that "a brilliant showing of color would be maintained throughout the year." To further enhance the pleasure of the ride, engineers also sought to adjust the road's contours to fit the landscape and to install "rustic" rails on rubble parapet walls and wooden railings along on- and off-ramps. By achieving these core parkway design goals, it was assumed that the physical and historic landscape of the region could be displayed—and experienced—through the windshield.[6]

Parkway design also incorporated safety, utility, and efficiency objectives. Limited access to the road ensured an uninterrupted flow of traffic. Grade separation was provided when the parkway crossed other major arteries. Parkways were to be divided by wide median strips, and their pavements had generous widths. During the 1920s and 1930s, parkways were designed for passenger cars with speeds ranging from twenty-five to forty-five miles per hour. Speed was as much a function of efficiency as the car's ability to accelerate. As mentioned in the dedication ceremonies of the Arroyo Seco Parkway, motorists could save time "not upon traffic flowing unduly at high speeds, but on [the car's] ability to flow continuously at reasonable speeds without delays caused by cross-traffic and left hand-turns."[7]

Funding had been the biggest obstacle in developing the parkway, particularly during the Depression-era reductions in local funding sources. Construction began when federal funds became available through the jobs-related funding that had also been designated for the development of the flood-control channels. In addition, state funds were also tapped that turned the Arroyo Seco Parkway into a state highway under the jurisdiction of the California Highway Commission. Additional funding was obtained through gas tax revenues from the cities of Los Angeles and South Pasadena. A final hurdle was overcome by the change of a state law that guaranteed property owners living next to public highways access to roads built with public funds.[8]

But even as the parkway features were being developed, the speed and efficiency objectives associated with the transition to a freeway approach were becoming paramount. With the Arroyo Seco Parkway's funding in place, the shift in focus from parkway to freeway intensified. Parkways, considered products of a bygone era, had by the 1940s already lost favor among the traffic engineers and even regional planners who were affecting the final designs of other high-speed dedicated roadways like the Hollywood Freeway. However, the adjustment of the completed parkways, such as the Arroyo Seco Parkway, to the new freeway era turned out to be problematic at best, since they had been designed for different capacities (much fewer cars), different speeds (slower vehicles), and additional objectives (visual connection and recreational driving) that were not always compatible with the speed and efficiency objectives of the freeway.[9]

The Arroyo Seco Parkway eventually became the leading example of this anomalous situation, although initial goals such as speed, cost, and safety seemed to be met. Built in three major stages from 1938 to 1953, the first segment of the parkway, completed in 1939, cost less than $1 million per mile. This included the initial construction of the Arroyo Seco flood-control channel, as well as all the bridge structures, railroad relocations, utility reconstruction, and landscaping. For the construction of the parkway embankments, engineers utilized hundreds of thousands of cubic yards of material excavated from the Arroyo Seco Channel by the Works Progress Administration and from the Los Angeles River by the U.S. Army Corps of Engineers. This cut down considerably the cost of the parkway and was found to be "exceptionally low for a freeway of its character." Furthermore, the parkway was touted as a major economic benefit for motorists for its efficiency, with the Automobile Club of Southern California hailing the West's first freeway as saving each motorist six cents per trip from Pasadena to downtown Los Angeles.[10]

Traffic safety also remained of paramount importance. To reduce the possibility of head-on collisions, a six-foot median strip was designed. The shrubbery planted in the median was intended to shield drivers from the headlight glare of oncoming traffic. Fences were erected to separate traffic from children and animals from nearby properties, including the public park lands that crossed the parkway along much of its route. The lanes were eleven feet wide, and a shoulder of ten feet was originally planned for each side of the roadway. Different-colored types of concrete were used for different lanes to encourage drivers to stay in their lane. Other safety features included special safety lighting at all inlets and outlets of the parkway/freeway, warning and directional signals, and red reflectors and amber-colored flashers installed in curbs. A 1945 study pointed to these safety features as an explanation for the remarkably low ratio of traffic accidents that the parkway enjoyed compared to other major highways with similar traffic volumes. This appeared to be a true parkway-turned-into-freeway success story.[11]

However, even though the Arroyo Seco Parkway had been initially viewed as a model for roadway design, it soon began to lose some of its visual character and suffer from its hybrid design. Decked overpasses supplanted the decorated stone bridges. Wooden rails and sculpted roadside surfaces gave way to concrete sound walls, and the gently winding roadway lanes came to be replaced by flat and curveless ribbons. Even at the time of its dedication, the parkway's crucial goal of aesthetic appeal and connection to a natural landscape began to be redefined by the new language of freeway identified as "route" rather than "connection to place." The program that was prepared for the dedication ceremony emphasized that the parkway would become "the first completed unit of the proposed system of modern express highways which is absolutely essential in this, the fastest growing and most congested metropolitan area in the West, to provide for the safe and expeditious handling of traffic." The terms *parkway* and *freeway* had become interchangeable, and the goal would be to turn this "first freeway in the West" into "the promise of many more freeways to come," as California Governor Culbert Olson proclaimed at the dedication ceremony.[12]

Over time, the freeway functions that had been superimposed on a parkway design came back to haunt its engineer managers, and the roadway eventually came to represent a kind of freeway outlier. Originally built to carry 27,000 automobiles per day at 45 mph, those numbers had already increased to 70,000 cars per day by 1960. By 2003, the parkway carried an average daily traffic of over 130,000 cars (at its southern end) at speeds often exceeding the official limit of 55 mph (when not congested). The 1939 parkway plan by the Transportation Engineering Board had in fact argued that "a road speed of 60 miles per hour would be an inefficient and unsafe use of an expansive highway." At the same time, congestion continued to build throughout the day and even into the evening hours. According to Caltrans data from 2002, traffic heading south increased to a peak of 8,000 cars per hour in the middle of the parkway and about 14,000 cars per hour where it intersected with a connecting freeway, the Golden State Freeway (Interstate 5). Originally built for a leisurely drive, the parkway/freeway had only three (and only two at certain stretches) rather narrow lanes on its northbound and southbound sides and no effective shoulder areas for disabled cars. Given the greater volume of vehicles, higher speeds, and high accident rates (perhaps its best-known feature), bottlenecks became a daily occurrence for this main thoroughfare connecting Pasadena to downtown Los Angeles.[13]

When the California Highway Commission changed the name of the Arroyo Seco Parkway to the Pasadena Freeway in 1954, many of these problems were already becoming apparent. Drivers on its winding path had to navigate exit ramps at a 5 mph speed limit or merge onto the freeway from a stop sign into lanes where cars zoomed by at 70 miles per hour. "You wait at a dead stop for an opening in traffic," one motorist said of the experience of merging onto the freeway, "and then, with teeth clenched, you push the accelerator all the way to the floor and see if anything was gaining on you." The Pasadena Freeway drive had become "more of a chore and less of a pleasure," a California Department of Parks and Recreation 1967 report put it.

Given its increasing accident rates, its lack of emergency areas, and its narrow lanes, by 1971 the parkway-turned-freeway had been able to earn the designation as "the world's longest parking lot."[14]

More than anything else, accidents were a key to the problem of congestion. The parkway that became a freeway was simply not designed for the type of high-speed driving where motorists often ignore posted speed limits. As a result, fast driving along its tight curves and narrow lanes became one source of traffic collisions, while the on and off ramps presented the most serious safety issue due to their lack of merge lanes. The data indicated that the percentage of total accidents on the parkway increased as the distance from an on-ramp or off-ramp decreased. Aside from the entrance and exit accidents, the majority of the accidents on the parkway/freeway happened because of speeding (61.8 percent) or because of an improper turn (17.2 percent).[15]

Today, the Pasadena Freeway can be considered the least safe and most unpredictable route in the Los Angeles region. The historical features of a parkway have made this parkway/freeway hybrid stubbornly resistant to a full-scale conversion to a contemporary highway. On several occasions, Caltrans officials tried to explore such design changes as widening and straightening the lanes and lengthening or redesigning the exit and entrance ramps. But these efforts were blocked by the historic designation of the parkway, the high cost of reconstructing the route (such as redoing the concrete bridges), various property and right-of-way barriers not anticipated when the parkway route was first conceived, and legislation from the 1970s that prevented the taking of park lands for freeway expansion. Though no longer a parkway, the Pasadena Freeway remained constrained by its history, an uncomfortable position for its freeway managers, who were increasingly subject to the anger and concerns of nearby residents who used the roadway and were affected by the high number of accidents and near misses. The question, as parkway historian Tim Davis put it, was whether this "bad parkway" ("bad" due to the pressures at its origins to serve freeway goals) could also become a type of freeway of the future ("an interesting proto-freeway," as Davis put it). To make that transition from parkway to freeway and then perhaps to a new type of hybrid based on its residual parkway features could help redefine the very purpose of a roadway, its connection to other forms of transport, and its relationship to the places it passed through.[16]

The Stream That Became a Concrete Flood Channel

When it was being constructed in the late 1930s, the Arroyo Seco Parkway was considered a "wondrous feat" that could enable drivers to be "launched like a speedboat in a calm, spacious, divided channel." But the stream that resided below the parkway was also in the process of becoming an unobstructed channel. Before this transformation, the Arroyo Seco stream, a 46.6-square-mile watershed tributary to the Los Angeles River, had long been considered one of Southern California's greatest natural resources. From high in the San Gabriel Mountains north of Pasadena, even today the Arroyo Seco's stream flows freely until it leaves the mountains and the Angeles National Forest and enters the urbanized segment of the Arroyo corridor near NASA's Jet Propulsion Lab at the California Institute of Technology in Pasadena. In this urban stretch where the stream is now channelized, it meets the freeway for the last six miles of its journey before emptying onto the broad concrete plain of the Los Angeles River, next to the one-time rail yards and overpasses just north of downtown Los Angeles.

Before its urban stretch was channelized, the Arroyo stream flowed freely and was filled with trout. Willow and sycamore trees grew along its banks, providing habitats for aquatic life and birds. Generations of Tongva Indians and European settlers were inspired by its beauty and lived partly off its bounty. Arroyo Seco, however, means "dry gulch," and the stream developed a reputation as both a "spacious water course" (as Father Crespi noted in his journals) and also a channel that would sometimes turn dry after it reached the area below the mountains on its path towards the L.A. River. As development increased during the early years of the twentieth century, the river bed was without water for most of the year, except when periodic storms filled the stream's banks and flooded adjacent lands. A number of proposals emerged during this period to address the problem of flooding, including the development of park lands along the Arroyo to absorb those occasional flood waters and to enhance the Arroyo's visual attractions. In 1917, Pasadena's Arroyo Park Committee, chaired by one of L.A.'s most celebrated architects, Myron Hunt, put forth a plan (in conjunction with landscape architect Emil T. Mische) for the lower segment of the Arroyo, the area within Pasadena that was most vulnerable to problems of flooding. The plan recommended that the Lower Arroyo be restricted to walking and bridal paths and trails that would be planted only with native plants. In 1924, Pasadena's city manager, A. G. Koiner, introduced the idea of establishing the Arroyo as a wildflower preserve.[17]

But developments also began to take their toll. After the major floods in 1914 cut through the banks of the Arroyo in several sections, walls were built to contain future flood waters in one area of the Arroyo, with costs reduced in part by the use of prison labor. In 1922, Pasadena's best-known institution, the Rose Bowl, was built within the Arroyo Seco watershed, a decision that was also controversial. One critic, Pasadena city manager Stuart French, argued in a 1924 article that "the location of the Stadium in the Arroyo is most unfortunate for the landscape development and improvement, as it now stands in the center of the area like a huge ant hill." "Even though its rocky slopes may be made to bloom, it will forever intercept vistas and will prevent a 'natural flow of landscape,'" French complained, arguing that "this pile of concrete must be made to set *into* the landscape, not *on* it."[18]

Given the changing nature of the urbanized stretch of the Arroyo, a "natural landscape" approach that was linked to strategies for flood management appeared increasingly problematic as developments edged closer to the stream and the major storms of 1914, 1934, and 1938 shook the region. In earlier periods, the stream swelled tremendously during winter storms and raging floods occurred periodically, but these maters were absorbed by the agricultural lands or open spaces that abutted the stream. The increased development along the Arroyo increased a sense of vulnerability, particularly adjacent to the Lower Arroyo, reaching a crisis point during the 1930s as significant numbers of homes and businesses in that part of the Arroyo were swept away by storms. Similar to the floods that furthered the efforts to channelize the Los Angeles River, the floods in the Arroyo ultimately facilitated the construction of the flood-control channel that followed the parkway's path up to where the stream joined the river north of downtown Los Angeles.[19]

Establishing a link between the flow of the stream and the Arroyo transportation corridor was proposed prior to the construction of the flood channel and the parkway. In 1907, City Beautiful advocate Dana Bartlett argued that the Arroyo stream, with its "natural growth of trees and shrubs, live oaks, and sycamores," could be made "one of the most charming drives that any city could desire." The 1924 Major Traffic Street Plan raised the possibility of creating a five-lane highway on each bank of the Arroyo Seco. The 1930 Olmsted-Bartholomew plan for the Chamber of Commerce focused on the stream and the development of parkways, arguing that establishing park lands near the Arroyo could serve as a flood-management strategy and open the way to the development of a scenic parkway transportation corridor. But the pressure to lay concrete for both roadway and stream became irresistible by the 1940s, and, in the process, the flood-control and freeway-oriented engineer-managers ultimately took charge.[20]

As these changes began to occur, the Arroyo Seco stream lost much of its appeal and, for a time, some of its constituency. Tributaries became encapsulated or diverted into buried pipes. These changes, combined with the dams of the Upper Arroyo, altered the hydrology and ecosystems of the Arroyo watershed. Like the Los Angeles River itself, the stream became a mass of concrete jutting. The parkway/freeway was built above the stream through diverse neighborhoods and park lands, and Arroyo residents began to lose the connection to a sense of place that the Arroyo had once provided.

Rebirth the Stream, Recreate the Parkway

By the early 1990s, community concerns about congestion, high accident rates, and the deteriorating visual quality of the Pasadena Freeway had peaked. At the same time, a renewed interest in park and open-space development within the Arroyo corridor, linked in part to interest in an Arroyo Seco watershed approach, also began to emerge. The attention to the stream beneath the Arroyo Seco flood-control channel as a place for restoration and reinvention was initially restimulated in 1985 through an exhibit entitled Arroyo Seco Release: A Serpentine for Pasadena by the artists Newton Harrison and Helen Meyer Harrison at the Baxter Gallery at Caltech. The Harrisons, who spoke of their art as a type of environmental "conversational drift," created a series of imaginative works on the theme of what to do about the Arroyo Seco flood channel. The exhibit included a multimedia installation of collages, maps, photography, and poetic hand-written texts that envisioned transforming the channel into an urban preserve and even covering and planting a new landscape over it. A second art showing in 1987 on these themes called String of Pearls at the Armory Center of the Arts was followed the next year by a visioning exercise in the form of a master plan for the Lower Arroyo by the environmentally oriented landscape architecture program at California State Polytechnic University at Pomona.[21]

These artistic and planning initiatives began to be complemented by new environment-oriented advocacy regarding the Arroyo. In 1989, as plans for the twentieth anniversary of Earth Day took shape around the country, a volunteer with the Tree People organization approached the city about hosting an Arroyo Seco Earth Festival. The Arroyo Seco requires care, the festival organizers told city officials, suggesting that a festival could help build constituency support for the kind of envisioning and advocacy work that had been stimulated by the earlier art exhibitions and student landscaping projects. Four sites within the Arroyo were selected to provide environmental awareness as well as to bring attention to the Arroyo. Over 35,000 participants attended the festival, and although subsequent festival events declined in the number of participants and media interest, the 1990 Arroyo Seco Earth Festival helped generate a new level of advocacy for Arroyo Seco restoration and new park development.[22]

Following the Earth Festival, city officials initiated a planning process that was designed to identify programs and policies that could help reenvision Pasadena and begin to rethink what could be done with the Arroyo. A new park, the Hahamongna Watershed Park, was established in 1992 as a recreational area and for flood protection and water-storage purposes for the Upper Arroyo near the site of the Arroyo's first flood-control facility, Devil's Gate Dam. The new park's name was derived from the name of a chief of the Tongva tribes and is translated as "flowing waters, fruitful valley." During the development of the park, advocates for new soccer fields struggled with those who wanted to create open spaces and natural habitats. The open-space advocates ultimately prevailed, with one advocate suggesting, in a fit of hyperbole, that even though "it doesn't look like much now [the park] has the potential to be Pasadena's version of Yosemite."[23]

The idea of Arroyo stream restoration was given an additional boost in the 1990s through an unusual arrangement that involved environmental tradeoffs in the development of what could be considered a type of reinvented nature in the Arroyo. In the early 1990s, Browning Ferris, the waste-management firm that managed a landfill at Sunshine Canyon that received L.A.'s garbage, was granted permission to expand the landfill to accommodate an additional 16.9 million tons of waste. To do so, Browning Ferris was required to establish a separate riparian habitat as a condition of the landfill's expansion. The site selected was a stretch of the Arroyo Seco flood channel below a small dam site under the Colorado Street Bridge. With more than $5.5 million in funding from Browning Ferris, a project was launched to divert water from above the small dam to create a separate restored stream that also turned parts of the area into marshland. Some of the exotics were then removed, and reseeding was undertaken to bring back Arroyo willows, cat tails, mulefat, alder, and other woodlands and riparian habitat. New wildlife returned to the area, including different species of birds, such as hawks that helped "patrol" the area by reducing some of the gopher population. "If you bring the water, they will come," Pasadena biologist Michael Long said of the stream diversion and the new life that developed around it.[24]

Although the restored stream project was greeted initially with enthusiasm, Arroyo advocates also began to fear that the project could conceivably become a poor model for the broader agenda of stream and open-space transformation. In the spring of 2000, nearly three years after its initial development, one of the sessions of the Re-envisioning the Los Angeles River program brought together a number of scientists and other experts to evaluate the efforts to reinvent and restore. The session identified the success of the project in providing a restored stream running parallel to the flood channel to bring back some nature in the city as well as the barriers that still remained in maintaining and extending the initiative. While providing a kind of self-contained restoration initiative, some of the Arroyo advocates questioned decisions that significantly inflated the project's price tag by including such efforts as underground water tanks, an expensive irrigation system, and an approach that sought to guard against an extended drought. The $5.5 million funding for a self-contained project particularly rankled. "Hell, with $5 million we could have probably restored the entire Arroyo," the Arroyo's leading advocate, Tim Brick, complained to me several years later. "While it's wonderful to have a restored stream," Brick argued, "by spending those kinds of funds and pursuing some of the approaches that were adopted, it undercut the symbol of restoration and what it could mean to the Arroyo."[25]

In the wake of these various initiatives, a new organization, the Arroyo Seco Council (subsequently the Arroyo Seco Foundation), was organized to help coordinate advocacy, research, and new initiatives around the Arroyo watershed. In 1999, the group partnered with the Northeast Trees organization to undertake a major watershed study to identify areas throughout the Arroyo that could become available for new watershed-management approaches, open-space development, and stream restoration. The two groups hosted a number of community meetings to discuss possible stream-restoration or related urban-greening strategies, brought in a range of researchers and consultants, and sought to increase the visibility around the Arroyo as part of the study's objectives. The study also sought to identify what it considered feasible strategies for controlling floods that related to the Arroyo's natural flow. These opportunities included providing pervious surfaces for water to percolate into the ground rather than run off into the channels, daylighting stream sites that were buried under the channel, and diverting areas along the stream into constructed wetlands and retention areas. Although removing concrete would not be practical in all spots along the Arroyo, especially where structures were close to the stream, some areas, the study noted, could be naturalized, helping the Arroyo experience at least a partial rebirth as a more natural canyon and stream.[26]

Released in 2001, the Arroyo watershed study increased the focus on stream restoration and had perhaps its biggest impact on its primary audience—key public officials, including the Arroyo's engineer managers. That same year, the California Resources Agency declared the Arroyo to be one of ten model watersheds for the state of California. The next year, the Army Corps of Engineers undertook a reconnaissance study that sought to answer whether the Arroyo in fact constituted a possible opportunity for restoration. And by 2003, as the plans for ArroyoFest were also coming together, the idea of renewal of the stream no longer seemed just a product of an artist's imagination or the landscape designer's visioning exercise.[27]

While the Arroyo stream advocates elevated the idea that a new approach to the flood channel was not only possible but feasible, Arroyo corridor alternative-transportation advocates also began to envision a different approach to the parkway that had become a freeway. Community concerns about freeway issues—including the high volume of accidents, the lack of maintenance, and the absence of sound barriers—intensified in the early 1990s. In response, a forty-person task force appointed by California State Assembly member (later state senator) Richard Polanco sought to explore strategies to reduce accident rates, establish connections to the area parks, develop bike paths, and enhance the visual quality of the roadway. The task force, which included parkway advocates, community group representatives, and a key ally within Caltrans, sought to focus the attention of highway planners on two core strategies—achieving historic parkway status, which could lead to landscape changes consistent with the original parkway concept, and developing traffic-calming approaches (including a reduction of the speed limit to its original 45 mph) to help reduce accidents and ultimately relieve congestion. Prior to the development of the task force, Assemblyman Polanco had successfully introduced legislation that designated the freeway as a State Historic Parkway, part of California's Scenic Highway system, which also offered the opportunity to seek a reduced speed limit and parkway-rehabilitation initiatives that had become part of the task force's deliberations.[28]

Though the original task-force recommendations were not directly pursued and the California Historic Parkway status became more symbolic than substantive, the Caltrans freeway managers welcomed the efforts, in part to shift the focus from community anger to an exploration of how obtaining historic parkway status could provide new funding and visibility for the roadway's historic significance. An internal Caltrans task force also had been convened prior to Polanco's task force to explore possible options for the freeway. These included posting signs to indicate that the freeway was in fact an historic parkway and that the roadway had not been built for modern freeway conditions, indicating that the parkway had to be driven in a different manner (i.e., more slowly). The Caltrans group also explored freeway redesign to reduce the high accident rates, the most visible and contentious issue it faced. This included changing the exit and entrance ramps, widening lanes, and even taking park lands for redesign purposes, but these proposals encountered continuing and still insurmountable barriers regarding redesign initiatives. The group also considered briefly reducing the speed limit but concluded that such changes were limited by legal and political constraints. "There's always a reluctance to reduce speed; if anything the pressures are just the opposite," Caltrans participant Chuck O'Connell recalled about the agency discussions at the time. Caltrans staff also raised the problem concerning the "85th percentile speed rule" whereby traffic engineers conduct a speed study on a particular freeway, plot the speeds on the roadway, and then set the speed limit at the 85th percentile, with 55 mph the statutory limit. "Obviously most drivers are going faster than 55 mph," Caltrans staff member Larry Loudon recalled the discussions. "We worried that if we lowered the speed limit we would be taken to court," Loudon said. The agency decided not to pursue lowered speed limits or related traffic-calming measures. Moreover, the speed-limit issue had not been pursued further by Polanco and other policy makers even though they potentially had an opportunity to do so through a legislative route. Thus, the shift from parkway to freeway remained intact even as it had become increasingly untenable—that is, the roadway originally built for lower speeds and a "pleasure ride" experience became the freeway dedicated to speed, though such an objective had become increasingly difficult to achieve when inappropriate speeds led to accidents that in turn created major congestion for a roadway not designed to accommodate those higher speeds.[29]

The focus for the Caltrans engineers as well as the parkway advocates eventually shifted from traffic-calming issues to the historic parkway designation and related opportunities at the federal level. In 1999, the American Society of Civil Engineers designated the parkway/freeway a National Engineering Landmark. In 2000, a report was issued on the parkway's history and changing status as part of the Historic American Engineering Record. That same year Caltrans, along with several community partners and the Santa Monica Mountains Conservancy, was awarded a federal grant to develop a Corridor Management Plan that would serve two purposes—to enable the parkway/freeway to achieve National Scenic Byway status and to identify opportunities for making landscape changes along the parkway/freeway route, for enhancing the parkway features, and for highlighting its historic status. The grant to develop a Corridor Management Plan in turn brought together many of the same players who had participated in the community task force five years after it had been disbanded and also linked the initiative to the national groups that were most focused on parkway and historic-preservation issues. With the roadway's increased visibility, these various advocates were successful in getting the U.S. secretary of transportation to designate in 2002 the Arroyo Seco Parkway as a National Scenic Byway, potentially freeing up future resources to pursue the objectives being defined in the Corridor Management Plan.[30]

By 2003, the parkway advocates, several of whom were participating in the Corridor Management Plan, had begun to ally with a range of other community players. These included transit advocates who were focused on the soon to be opened Gold Line Metrorail service from Union Station to Pasadena, bike advocates who dreamed of recreating Horace Dobbins's cycleway or other bike paths along the Arroyo, neighborhood activists who focused on the need for open space, affordable housing, and other core community needs, immigrant-rights groups focused on the changing demographics of the Arroyo corridor, and the river and stream restorationists who saw opportunities for linking parkway changes to changes to the flood channel. New kinds of coalitions were available, and the focus on the parkway that had become the freeway seemed to offer the most immediate and visible symbol of the need for community and environmental change. With several studies about the parkway/freeway history and its present conditions in the works, a Caltrans bureaucracy caught in the bind that the highway engineer managers had themselves created, and the desire to highlight the possibility for change among disparate players needing some way to link agendas in a visible way, an imaginative yet improbable plan for action began to take shape.

Biking the Freeway?

Soon after the last of the Re-Envisioning the Los Angeles River events occurred, I met for lunch at the Señor Fish restaurant in Eagle Rock near the Occidental College campus with Tim Brick, who was still several years away from his election as board chair of the Metropolitan Water District of Southern California. Before coming to Occidental, I had met with Tim from time to time to talk water politics, but I had not followed as closely his work around the Arroyo. Now that I was at Occidental and a neighbor of the Arroyo, I wanted to find out what opportunities for change were available in the Arroyo corridor. Tim suggested three possible initiatives for the Urban and Environmental Policy Institute—researching opportunities and highlighting strategies to green the Rose Bowl, undertaking research as well as organizing around the Arroyo Seco stream restoration and related open-space development, and becoming involved with the parkway/freeway issues. The parkway issues most intrigued our institute staff, particularly an idea that had floated around for a couple of years about a bike ride on the freeway. This idea seemed improbable, and yet it would be extraordinary if it were to occur, given Los Angeles's reputation as freeway-centric.

Following that discussion, UEPI and the Arroyo Seco Foundation convened a number of community, transportation, and environmental groups to explore the development of an Arroyo Corridor collaborative that could integrate the agendas of the different groups. The meeting included Diane Kane, an architectural historian with Caltrans who had become a passionate advocate for restoring as many of the parkway's historic features as possible and for connecting the parkway with other greening and historical-preservation initiatives along the Arroyo. Kane had participated in Richard Polanco's community task force, had been the lead Caltrans figure in pushing for historic signage along the parkway/freeway, and had emerged as a liaison with the community and environmental groups. She had also become a visible participant among the national historic-parkway networks and had convinced Caltrans to cosponsor a national conference on parkway issues with the National Trust for Historic Preservation in 1998. A force to be reckoned with, she insisted that we needed to use the word *parkway* instead of *freeway* when discussing the Arroyo Seco Parkway, despite or perhaps because of its name change to the Pasadena Freeway and the cultural and transportation management associations with those words. Every time during the Arroyo Seco Corridor collaborative meetings that I or anyone else would use the words "Pasadena Freeway," Kane would say, "Arroyo Seco Parkway." It reminded me of Lewis MacAdams's battles with the flood-control engineers when he insisted on calling the Los Angeles River a river rather than a flood-control channel: language matters.[31]

The Arroyo Seco collaborative meetings identified a wide array of projects, neighborhood activities, policy initiatives, and research and educational programs that were taking place throughout the Arroyo corridor, although many of the groups were unaware of each other's programs. The Arroyo corridor was split between the three cities of Pasadena, South Pasadena, and Los Angeles and ran through several economically and ethnically varied neighborhoods within the city of Los Angeles, such as Highland Park, Mount Washington, and Eagle Rock. It suffered from a lack of attention from the media and policymakers. The distinctive history of the Arroyo—the native Tongva societies, the early pueblo, the arts and crafts culture, the architecture and the museums, the stream and the parkway—provided one type of linkage. Yet a more contemporary history of demographic change and immigrant communities—the Philippine Center in Eagle Rock, the Latino neighborhoods in Highland Park and Lincoln Park, the African American neighborhoods of Pasadena—provided an additional context, and possible challenges and opportunities regarding the advocacy around social and environmental change.[32]

As the Arroyo Seco collaborative meetings sought to identify common ground as well as ways to highlight existing initiatives and agendas, the idea of the bike ride on the parkway/freeway kept surfacing. An event coordinator had been solicited to explore the possibility, but the barriers seemed overwhelming. The advocates for establishing a Pasadena to Los Angeles cycleway also were enthusiastic about the idea but failed to interest other bike advocates in the Los Angeles region, who felt that smaller, incremental steps for creating localized bike lanes and bike paths had priority. The Caltrans staff, whose agency would be the first but not the only entity to review whether to allow any freeway closure to occur for whatever purpose, remained skeptical and viewed the previous discussions about the concept as far-fetched and not likely to advance beyond the idea stage.[33]

The staff at UEPI, however, loved the idea and immediately began to brainstorm and envision what such an activity might entail. Closing the freeway for whatever public purpose represented an important objective in its own right, a symbolic yet highly suggestive illustration of residents' right to the city. A bike ride on the parkway/freeway could highlight the ideas that bike use could be linked to commuting to work and that an unrestricted bike ride on the 8.2 mile parkway/freeway might ultimately be faster than commuting on a roadway subject to accident-related congestion. We also identified other potential activities. For example, a pedestrian aspect to the event could allow for a different kind of visual experience of the corridor. A significant constituency of equestrians also used some of the park land in the Arroyo. Perhaps an event involving a freeway closure could be inaugurated with horseback riders trotting onto the freeway to lead the procession of walkers and bike riders. What particularly appealed to us was the challenge of linking multiple constituencies and agendas through an event whose scale and unusual nature captured the attention of the public about those various agendas. Similar to the advocacy around the Los Angeles River, a freeway closure for a bike ride, a pedestrian stroll, a horseback ride, and countless other activities for participants who might bring their skateboards, rollerblades, or baby strollers could offer the opportunity to begin to envision a different kind of Los Angeles. And as Diane Kane continued to remind us, a bike ride and walk on the freeway focused attention on the fact that the so-called Pasadena Freeway, as she called it, had once been a parkway. One of the key agenda items for an event would be to explore how the Pasadena Freeway could reestablish itself, at least in part, as a reconstituted Arroyo Seco Parkway or at least as a new type of hybrid freeway in a region where the car and freeway culture was deeply entrenched.

To explore some of these opportunities further and to better establish a link between research and action, a three-way academic course and research collaboration between UCLA, Occidental College, and the California Institute of Technology was developed in 2001 and 2002. I joined with several other faculty and students from the three schools and a number of community partners to explore the range of Arroyo Corridor issues, with the concept of the bike ride and walk on the freeway providing a backdrop to the collaboration. Participants included Professors Anastasia Loukaitou-Sideris and Richard Weinstein from UCLA, Marcus Renner and Beth Braker from Occidental, and Bill Deverell from Caltech. Courses included a UCLA design studio, policy analysis and environmental research from Occidental, and a Caltech historical exploration of the changing Arroyo corridor. Loukaitou-Sideris and I separately undertook an analysis of the parkway's roots and the consequences of its shift to a freeway operation, including such factors as accident rates, while Renner and Braker explored the historical ecology of the Arroyo in the context of initiatives for developing new habitat and open space along the path of the Arroyo stream. These collaborations—along with the substantial research being pursued through the Arroyo Seco Watershed Study, various parkway studies including the Corridor Management Plan, and the analysis of the range of community issues such as housing needs, K–12 issues (including an Arroyo-based teacher training and curriculum initially developed by two area activists), and community and economic development questions—provided a baseline of data and analysis about opportunities for reenvisioning the Arroyo.[34]

While this research helped situate the issues and inform any future action, there was no direct guidance on how to close down the freeway on a Sunday, make it possible for people to gather on it, and link such an event to the projects and initiatives to transform the Arroyo. As we began to discuss the freeway event, we knew we had to operate at several levels. This included mobilizing community participation as well as pressuring various agencies to allow such an event. To do so, we needed a broad coalition of groups that could develop ownership of the event and see its connection to their own agendas. We needed support and facilitation by local elected representatives and various administrators and agency personnel to establish enough legitimacy and persuasion to move the transportation bureaucracies to the point where various permits could be obtained. We needed the media to provide coverage of the event and to identify the political and environmental contexts in which it was taking place, a task not often assumed by the media. We needed strong word-of-mouth and cyberspace interest in a period when the kinds of communication outlets such as Myspace or the blogosphere were just beginning to take form. And we required developing the resources to pull it off. Could this actually be done? Would it be a case of smoke and mirrors, a kind of Wizard of Oz exercise? After all, we told ourselves, people don't ordinarily shut down a freeway in Los Angeles (except when making big-budget movies), although trying to do so seemed as though it could be almost as instructive as making it happen.

Film fans await the sun to set for a Hollywood Forever Cemetery film screening.
Photo: Hollywood Forever Cemetery Web site

Strolling along the new rubberized jogging path at Evergreen Cemetery in Boyle Heights, East Los Angeles. *Photo:* James Rojas

The logo of the Evergreen Jogging Path Coalition. *Photo:* James Rojas

Boyle Heights mural. Courtesy of the East L.A. Community Corporation

Hanging out in Boyle Heights. Courtesy of the East L.A. Community Corporation

Barbed wire guards the entrance to the L.A. River. Courtesy of the Urban and Environmental Policy Institute, Occidental College

La Gran Liempieza: Down by the L.A. River. Courtesy of the Urban and Environmental Policy Institute, Occidental College

Tim Brick, Arroyo Seco advocate and chairman-elect of the Metropolitan Water District of Southern California speaking at the opening of Hahamongna Park in Pasadena. Courtesy of Tim Brick and Arroyo Seco Foundation

Chinatown school children explore the Cornfield brownfield site in a Re-Envisioning the Los Angeles River program. *Photo:* Andrea Misako Azuma

Strolling past the Not a Cornfield site, facing south toward downtown Los Angeles. *Photo:* Andrea Misako Azuma

Los Angeles Mayor Antonio Villaraigosa at the groundbreaking event for the Los Angeles State Historical Park (the Cornfield). The author, with the white cap, is seated in the back row to the mayor's right. *Photo:* Roy Stearns, California State Parks Department

Poet and Friends of the L.A. River founder Lewis MacAdams speaking to Occidental College students down by the River. *Photo:* Andrea Misako Azuma

Playing along the diverted Arroyo Seco stream. Courtesy of the Arroyo Seco Foundation

Biking on the path by the channelized part of the Arroyo Seco. Courtesy of the Arroyo Seco Foundation

Bike riders enter the Arroyo Seco Parkway, more widely known as the Pasadena Freeway, on a misty morning on Father's Day, June 15, 2003. Courtesy of the Urban and Environmental Policy Institute, Occidental College

Bike riders and rollerbladers on the Pasadena Freeway: ArroyoFest, June 15, 2003. *Photo:* Virginia Renner

Cart wheeler, rollerblader, and walkers on the Pasadena Freeway for ArroyoFest, June 15, 2003. *Photo:* Jerilyn Mendoza

Walking on the Pasadena Freeway: ArroyoFest, June 15, 2003. *Photo:* Tony Lin

Traffic jam on the 710 freeway, as the trucks make their way to and from the Port of Long Beach. *Photo:* City of Long Beach, California, Web site

Immigrant Rights demonstration in Los Angeles, May 1, 2006. *Photo:* Elizabeth Medrano

Environmental Justice organizer Penny Newman. Courtesy of Center for Community Action and Environmental Justice

Port activist Jesse Marquez looking out at the Port of Los Angeles. *Photo:* Andrea Hricko

Community activist Angelo Logan talks to a City of Commerce resident about pollution from the rail yards and the freeway. *Photo:* Andrea Hricko

Entering the South Central Farm. *Photo:* James Rojas

Gardeners and their families gather at the South Central Farm. *Photo:* James Rojas

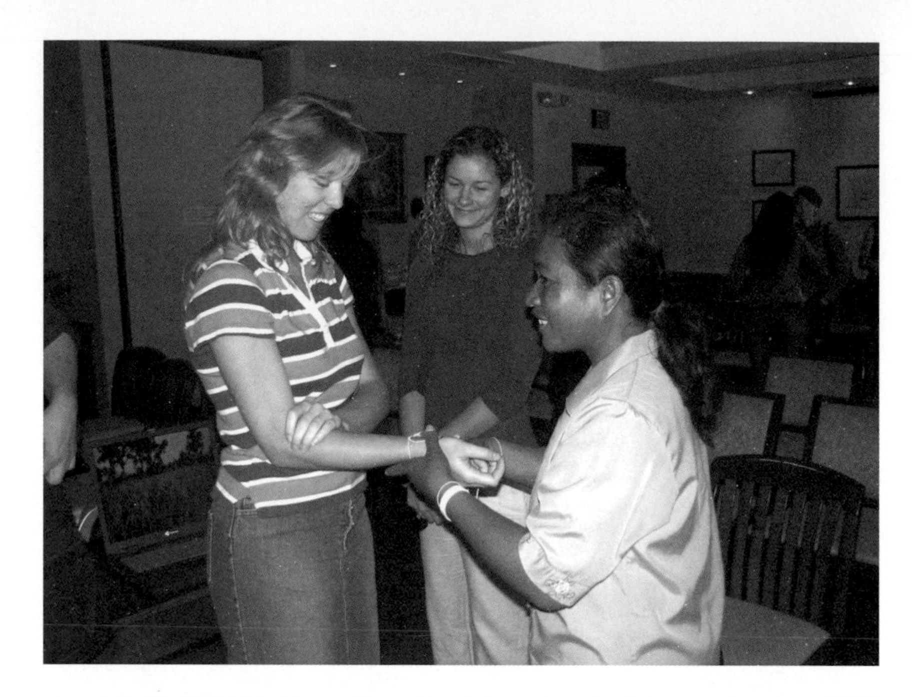

Organic Jasmine Rice Thai farmer Kanya Onsri places a bracelet for hope and peace on UEPI staff member Amanda Shaffer's wrist while UEPI staffer Jessica Gudmundson looks on.
Photo: Ellen Roggemann

Working the System to Shut Down a Freeway

To put on this event required strategies to pressure the transportation agencies that issued the permits, had final approval, and could dictate the terms for any freeway closure. But what also quickly became apparent was the need not only to coordinate the management of the event but also to exercise the political skills that would ensure its community character and its part in an ongoing agenda for change. UEPI had experience in mobilizing and coalition building. As we planned the freeway closure event, our Center for Food and Justice was playing a key role with other groups in getting the Los Angeles Unified School District to ban the sale of sodas in its schools. What we didn't have was direct experience in event management, which would eventually require some kind of consultant arrangement. The UEPI staff person given the task of exploring the feasibility of the event was Marcus Renner, a community and environmental educator who had been engaged in research on the ecological history of the Arroyo. Meticulous and detail oriented as well as strongly committed to the idea of community ownership of the process, Renner soon became totally absorbed by the intricacies of planning the event.

At one of the Arroyo Seco Corridor collaborative meetings, Diane Kane asked two of the Caltrans staff who would have initial responsibility in decisions about permits whether this event might be feasible. The concept at this point was still limited to the idea of a bike ride on the freeway, and we were able to offer few details on how, when, and where such an event might take place. To our surprise, the Caltrans staff did not dismiss the idea out of hand, though it was also clear that they did not assume that it would get much past the idea stage. We took their rather bemused response as a signal that we could proceed to the next stage in our planning, as if a closure of the freeway for such an event was in fact feasible. Our assumption that the absence of a direct veto of the event enabled us to move forward was critical, since the key to making it happen was overcoming the sense that a freeway was totally off limits and that any right to the city stopped at the freeway entrance ramp.[35]

With the collaborative of Arroyo groups and participants committed to ArroyoFest, the name given to the bike ride on the freeway event, community meetings were held to explore its possible dimensions. To broaden its participation, a "walk on the freeway" component was added to allow those who wanted to walk rather than ride a bike to participate. A community festival at the terminal point of the event was also added to include music, booths for sponsors and community groups, a local dance ensemble, and short presentations to touch on the event's key themes, which included community economic development, the environment, historic preservation, watershed management, and stream restoration. Local equestrian groups, neighborhood councils, various bike organizations, transit advocates, planning organizations, environmental-justice and mainstream environmental groups, parent and teacher groups, and several K–12 schools all became engaged, identifying possible actions and event activities both prior to, during, and after the event. Discussions were pursued with Sweat X, a local garment cooperative consisting of former sweatshop workers, about an ArroyoFest teeshirt that would publicize the event and also create visibility and funding support for the garment cooperative. An October 2002 date was initially selected, and the organizing and event-management planning (including efforts to obtain the permits to allow it to happen) was launched.[36]

The organizing aspects were exciting though challenging. When groups and residents learned that a walk and bike ride on the freeway might happen, they were a bit dubious about whether it could actually take place but also excited by the possibility. Typically, at any organizing meeting for the event, the first question was always, "Do you really think Caltrans is going to allow this to happen?" Moreover, bringing together such a wide array of projects and organizations in the Arroyo Corridor as well as the metropolitan region through an event that included a linked agenda seemed problematic, particularly where potential conflicts were possible. Tensions between groups and agendas needed to be overcome, such as the subtle and not so subtle differences between those advocating for affordable housing and those championing historic preservation, those with parkway goals and those with bike commuting goals, promoters of stream restoration and open-space development and backers of more recreation places like soccer fields, people promoting job development and people pushing for environmental change. There were conflicting views on what to do with the homeless encampments along the banks of the Arroyo. There were also challenges associated with the class and ethnic differences and personality disputes that erupted within and among community organizations, particularly the neighborhood councils that had been established in the city of Los Angeles. These organizations, established in 1997 as advisory bodies through an election that brought about a change in the city's charter, were initially designed to head off threats of secession from parts of the city such as the San Fernando Valley and also to provide a sense of greater access to neighborhood decision making, particularly with respect to land-use and planning decisions. In several areas of the city, the neighborhood councils tended to be dominated by homeowner groups, several of whom opposed aspects of key social-justice goals, such as affordable housing or homeless services. Immigrants who did not have legal status, moreover, were often not able to participate in the councils, a particular concern in the Arroyo corridor with its large Latino and Asian immigrant populations. We also encountered sharp personality conflicts and boundary disputes within and between some of the councils, compounding the difficulties in mobilizing for an event that was also seeking to strengthen and extend key community goals that not everybody necessarily shared or assumed was important.[37]

Once the process of organizing the event began in earnest and its actual realization seemed more realistic, some tensions between groups and over issues subsided or receded to the background. Planning for the event provided its own important organizing opportunities, including developing an Arroyo policy agenda linking the four themes, designing K–12 teacher-training workshops to provide opportunities for schools to engage in the themes of the event (such as the history of the Arroyo, stream restoration, and alternative transportation ideas), developing new organizing and coalition initiatives from the watershed-advocacy networks, and renewing attention toward parkway/freeway issues that were once again under consideration. To help provide visibility to the organizing and alert the permitting agencies that support for the event extended beyond a handful of groups and individuals, a luncheon presentation was arranged in January 2002. Those who gathered at the faculty club at Caltech included elected officials from the three cities involved (Los Angeles, Pasadena, and South Pasadena), key state figures, such as Secretary of Resources Mary Nichols and Assembly Member Jackie Goldberg, and several Caltrans representatives.[38]

Despite the preevent organizing, the event remained a complicated if not improbable goal. Discussions with Caltrans officials were initiated, and several potentially event-breaking obstacles were identified. One was the need to obtain expensive liability insurance. Another was the need to overcome problems regarding plans to divert traffic from the freeway during the hours of the event; the plans also required approval from various city transportation agencies. We knew that obtaining permits and approvals quickly from the three cities also required formal resolutions by their city councils. And we also needed to obtain approval from the California Highway Patrol, which would be responsible for addressing vehicle, pedestrian, and bike rider safety. As Renner recalled, a kind of Catch-22 emerged: "We needed the permit to get the funding; we needed insurance to get the permit; we needed to work out the details of the event to get the insurance; but all the details wouldn't be settled until we got the permit." Such details included the route and hours of the event, which in turn were connected to the plan for diverting traffic, compounded by the overriding desire of Caltrans (and the city transportation agencies) to minimize any disruptions to the flow of freeway or surface traffic. And the city agencies wouldn't provide their permit until Caltrans issued its permit and vice versa.[39]

As these issues began to be addressed, the difficulties for meeting the deadlines for an October event loomed large. Funding was also a huge concern, particularly since the focus for the event was on building community and establishing an Arroyo identity rather than on obtaining a commercial sponsorship and hefty fees from participants. Occidental College's role was another issue that emerged, given UEPI's leading and visible role in developing the event and negotiating with the various permitting agencies. While a steering committee of community groups had been established that made decisions governing the event, the committee was an ad hoc entity that didn't have the authority to sign the paperwork for the permits and the liability insurance or to solicit funding from nonprofit public entities or foundations (those exempt from federal income taxation under section 501(c)(3) of the Internal Revenue Code) for obtaining the funds and was not seen as authoritative enough, given its ad hoc nature, for the permitting agencies to assume that the planning for the event involved what they would consider to be a legitimate and substantive organization. UEPI through Occidental had been able to address some of those obstacles and thus emerged as the organization of record, but the college's financial and legal officers worried about the college's own risks and liabilities. Though the college had been supportive of UEPI's community orientation and hybrid role as academic entity and community-based organization, ArroyoFest represented a major shift in the college's role in and responsibility for such an unusual event.[40]

Our goals—to persuade the permitting agencies to agree to a plan that could divert traffic and close down the freeway, raise enough funds to meet the event's myriad expenses, reassure Occidental College about its risk exposure, keep the community groups together and focused on ArroyoFest's capacity to extend their own agendas, and still meet our own deadline of an October date (chosen in part to occur prior to the beginning of L.A.'s limited but sometimes extensive rainfall season)—seemed overwhelming, despite the excitement and anticipation about the event. In May 2002, an event planner with experience planning bike events on surface streets was finally hired after an anonymous donation of $25,000 was received. But just three months later, on July 17, the event planner returned from a meeting with transportation agency officials to tell us that he had had "a Jesus moment" and had come to the steering committee meeting that night to pull the plug. The event was just not feasible in the remaining time available, he argued, throwing it back to the committee to decide what to do.[41]

The choices seemed bleak. To cut our losses and decide that an event of this magnitude was too fraught with problems and cut against the grain of not only the transportation agencies but the freeway culture itself seemed to acknowledge that the freeway remained off limits. To postpone the event to a later date would simply delay having to overcome the obstacles, with the additional problem of loss of momentum. Some of the community groups were especially upset with the drain on their own limited resources that had already occurred and worried that ArroyoFest was detracting from their own agendas. At UEPI, we shared some of these concerns and particularly worried that as the only group providing staff resources we were severely understaffed for such a large event. But Renner, whose own learning curve about overseeing an event of this magnitude had significantly increased, came up with new timelines, new fundraising targets, and possible funding opportunities. A far more focused steering committee was reestablished, and a plan of action that included seeking out a high-profile event planner was developed. In discussions with the college administrators, a decision was also made to establish a separate nonprofit organization (under IRS Code section 501(c)(3)) that could separate the risk and liability questions from the issue of UEPI's own involvement. Though discouraged, the collaborative of groups, with a few exceptions, decided to push forward, still inspired by the vision of ArroyoFest as a community-building event and as an imaginative way to reenvision Los Angeles. The new date would be June 15, 2003 (Father's Day and after the rainy season).[42]

At first, things appeared to move slowly, with Caltrans and the other transportation agencies still willing to talk but requiring responses to issues like traffic diversion that appeared difficult to answer. A breakthrough came when the new event-management consultant (the group that managed and owned the rights to the L.A. Marathon) changed the tenor of the discussions with the transportation agencies. Experienced in traffic diversion issues on surface streets, with a high profile and well connected politically, the new event consultant proved to be an important asset in increasing the credibility of the event, even though the framework for closing the freeway and diverting traffic to other freeway loops had already been developed prior to the L.A. Marathon group being hired. The plan to limit the event to 6 to 10 a.m. on a Sunday, block off the entire route of the Pasadena Freeway/Arroyo Seco Parkway from the downtown Los Angeles interchange to Pasadena, and divert traffic to other freeways that circled around the route rather than through surface-street diversions persuaded the transportation agencies to give their assent.

By spring 2003, the possibility of pulling off the event became more likely, even as the shadow-dancing between transportation agency and community group, supporters and skeptics, continued. New media coverage prior to the event, including the Arroyo Seco Parkway research study undertaken by myself and UCLA professor Anastasia Loukaitou-Sideris, increased the credibility factor. Funding started to become available. This included a few small grants from foundations, from public agencies (including from the Metropolitan Water District through its community funding initiative, recognizing the watershed dimension of the event), and from various individuals (including a private donation from the president of Occidental College). The community-policy document issued by UEPI provided a framework for linking ArroyoFest's key themes of environmental change, community economic development, river restoration, historic preservation, and community identity. Two years of organizing had established a strong network of supporters and event activists, enormously facilitating the new outreach for a June 15 event, even as Caltrans still deliberated whether—and when—to issue the final permit. Nevertheless, word throughout the Arroyo Corridor began to spread that a bike ride and stroll on the freeway was this time really going to happen, and the growing enthusiasm of the various community groups at one point even led Caltrans staff to request that we somehow stop the flood of calls urging the agency to grant permission to close the freeway.[43]

Despite the increasing profile of the event and the ability to overcome each of the obstacles, the feeling that we were still engaged in a type of smoke and mirrors exercise remained with us literally up to the several days prior to June 15. Just ten days before the event, at a meeting of all the principals among the permitting agencies (including Caltrans, the transportation agencies from the three cities, and the California Highway Patrol), the Caltrans staff announced that they had issued their permit to close the freeway, to the astonishment of the staff from the other agencies. While these agencies had also granted their permission, thanks in part to the buy-in from elected officials who had become responsive to the community interest in the event, they had assumed that the final approval would simply not be forthcoming, since, in Los Angeles, freeways ordinarily are only closed when there are Sig Alerts. "You mean this is really going to happen?" one of the meeting participants asked after the Caltrans staff acknowledged that all the permits were now in order.[44]

Flowers and the Sounds of Silence on the Freeway

A heavy fog settled over the Arroyo Corridor in the early hours of June 15, 2003. More than three thousand bike riders began arriving from every direction to line up in front of the entrance to the freeway/parkway. Caltrans had placed one of the signs that Diane Kane had been promoting—Historic Arroyo Seco Parkway—at the entrance way. During the next hour, thousands more would also converge at four different locations through the Corridor, from north to south, eager to walk on top of this historic parkway. The destination point for the walkers and bikers was Sycamore Grove Park in the Los Angeles neighborhood of Highland Park about midway between the two points where the parkway had been closed. Nearly a hundred murals were draped over the historic bridges and overpasses that intersected the parkway, produced by participating K–12 schools to provide color and imagination for the event. More than seventy booths were set up by community and advocacy groups, ethnic food vendors were visible throughout the park, and local bands assembled for the festival as bike riders and walkers found their way through the hole in the fence that led from the parkway to the park grounds.[45]

The bike riders took off just as the fog began to lift, while those who walked the parkway joined soon after. The stories about the event that we received through e-mails, word of mouth, and Web postings in the days that followed captured what many characterized as a magical moment for Los Angeles. Since it was Father's Day, a number of families came out, and most of the bike riders were there for the sheer pleasure of riding a freeway rather than to demonstrate speed and prowess as some riders at bike events liked to demonstrate. People who entered the parkway by bike or foot were exuberant, and while the horseback riders were never able to secure their liability insurance, they were able to ride their horses along the concrete channel of the stream below the parkway.

For bike riders, the experience revealed how a bike ride not only provided pleasure but could potentially serve as an alternative form of transportation. One South Pasadena mother e-mailed us that after "having received some information two days prior to the event, I managed to get my father involved for the bike ride. . . . Best of all, both of my children (daughter age 10 and son age 13) had the opportunity to cycle on the freeway with a significant sense of purpose! Such a reward and life lesson that my children and I will never forget! Everyday that we drive the 110 [Pasadena] Freeway, both children are still amazed that we cycled so far in such a short time!"

The bike riders, walk participants, and residents adjacent to the freeway also noted how uniquely silent it had become that morning and how much they appreciated their chance to connect to the green space and natural surroundings of the Arroyo. One participant noted that while he knew that parks lined the parkway, "Seeing and experiencing them as I went by was magical. I could feel the cool air coming out of the tree-covered parks. I always knew the parkway was built to be beautiful, but seeing it at the appropriate speed clarified my vision." One of the speakers at the Community Festival who lived close to the freeway in Highland Park spoke of how disorienting—and liberating—it was to "open my window in the morning and hear birds and the wind and breathe the air in a way I had never experienced before."

Part of the focus of ArroyoFest was exploring and promoting alternative transportation options to the car and the freeway. The Gold Line, the new light-rail system that runs from downtown Los Angeles to Pasadena and whose route paralleled the parkway and the stream, was close to completion. Part of the message of ArroyoFest was identifying car and freeway alternatives—such as the Gold Line, expanded bus service, new rapid buses that sought to mimic rail service, commuter and recreational bikeways, and pedestrian pathways. Each contributed to the possibility of connecting rather than separating from the communities they passed through. Several ArroyoFest participants noted that the organizing for the event brought increased attention to alternative transportation options, based in part on the distribution of the thousands of fliers and posters showing bus and bike routes between the Festival at Sycamore Grove Park and the four starting points for the walk and bike ride. One ArroyoFest participant described how she had sought information on how to get back from the park to her home, about a mile and a half away in the Hermon neighborhood. One of the ArroyoFest speakers who lived nearby in Highland Park and was a frequent bus rider provided information on the 81 bus that stopped by the park and continued north to a point just a few blocks from the participant's home in Hermon. "You know, I've lived seventeen years in Hermon and I never knew about the 81 bus, let alone how to get to various places such as the park. And it actually stops close to my work in downtown L.A.," she exclaimed, recognizing an opportunity she had never seized before.

The experience of nature in the city was also a theme present throughout the day. One tour took participants to nearby Debs Park, where the Audubon Society had established its first major inner-city park and nature education program. During this event, several participants (including those who lived nearby) acknowledged that they had not previously been aware of the park. The Audubon Society tour leader later noted what a noticeable difference there had been in the sounds in the park. "I could hear birds sing and not just parrots, and I saw red-tailed hawks, bullock's orioles, and red-shouldered hawks nesting near the Parkway," he subsequently e-mailed us. One of the local leaders of a major national environmental organization (Environmental Defense) participated in one of the walks with several friends, including the twelve-year-old daughter of a family that lived in a nearby neighborhood. "There was a quite extraordinary moment that symbolized to me the power of ArroyoFest," the environmental leader told ArroyoFest organizers. "My friend's daughter was walking with us and at one point let out a shriek. 'That's a passion flower,' she cried out, pointing to a delicate flower growing along the edge of the freeway. 'I know it, because I studied it, but I never thought I'd actually see one!'"

Aftermath: Bike Rides on the Freeway Every Week?

ArroyoFest identified both the strengths and the weaknesses of the new transportation, open space, watershed, social and environmental justice, and community-development movements in the Arroyo Corridor and the broader goal of trying to reinvent a city and a region that had long been associated with the car and the freeway and their transportation, land-use, and environmental impacts. The excitement following the event was palpable. James Rojas, founder and chief organizer of the Latino Urban Forum, sent out a missive to his group exclaiming, "ArroyoFest went beyond our expectations about creating a linear temporary plaza where the community could come together. It struck a chord in L.A., where people from all walks of life were able to experience a peaceful and silent freeway. Elderly women with parasols, Latino families, hipsters, and just regular folks were there. ArroyoFest suggested that even a car-oriented city like L.A. can change its ideas about freeways."[46]

Efforts to rethink class, race, and ethnicity differences, sometimes experienced in relation to the kind of diversity that could be found among and between different neighborhoods and cities along the Arroyo corridor, also emerged as part of the ambience of ArroyoFest. A neighborhood leader involved with the Mt. Washington Association, one of the middle-class neighborhood organizations that had a booth at the festival, told an ArroyoFest organizer that the event had been enormously revealing and transformative for herself and her organization. "For some time we have been disconnected and sometimes in conflict with the groups from [adjacent, low-income, and largely Latino neighborhoods] Highland Park and Cypress Park. But working together on ArroyoFest and having a chance for our organizers to talk with each other at the booths was eye-opening. We've now had an opportunity both for collaboration and communication and quite possibly a sense of partnership for the future." "The power and energy [of what we've done] is in (re-)creating a 'place-based' identity in the middle of the city, which cuts completely against the grain in greater Los Angeles region," Marcus Renner wrote in a memo to UEPI staff two weeks after the event. ArroyoFest, Marcus argued, allowed one to see how an event of that kind showed "the power of place to bring diverse communities together."[47]

"Let's do it again and do it more often" also became a post-ArroyoFest rallying cry. "Why not shut the freeway down every year?" several of the organizers e-mailed each other, and one person began to tout the idea that the freeway ought to host a bike ride and a stroll *every week*. "They do it in Central Park in New York and Golden Gate Park in San Francisco," several e-mail correspondents argued. Others noted the successful bike ride that had caused the Lake Shore Drive in Chicago to shut down, an event that took place on the same day as ArroyoFest. Even Mary Nichols, ordinarily cautious and focused on how to accomplish incremental change and arguably, at the time, the most important and powerful environmental official in the state, was caught up in the post-ArroyoFest enthusiasm and e-mailed me that a bike ride each Sunday morning was an idea that ought to be explored.

But the obstacles for recreating ArroyoFest and ensuring its future remained substantial. Several weeks after ArroyoFest, a meeting was arranged between myself, Marcus Renner, the event manager from the L.A. Marathon group, and officials from Caltrans and the California Highway Patrol. The Caltrans and Highway Patrol staff expressed their overall satisfaction that no major problem had occurred but that any future event had to be more limited, particularly in duration. Following the meeting, we discussed with the L.A. Marathon consultants whether and how ArroyoFest could be institutionalized. The L.A. Marathon consultants argued strongly about the need to better commercialize the event to make it financially and logistically viable. "You need to brand it," they insisted, and by doing so, we could attract sponsors like Toyota and other companies who wanted their own name associated with the event. We worried, however, that while "Toyota Presents ArroyoFest" would be the kind of approach that might turn ArroyoFest into a successful commercial venture, such an approach directly undermined the power of ArroyoFest as a community event and as a symbolic counterpoint to the dominant car and freeway culture.

Discouraged about the immediate choices available for a future event but inspired by the moment itself when the freeway went silent and the walkers and bikers turned it into a public space, I thought about Charles Moore's sardonic comment about the freeway as the heart of the city and its takeover as Los Angeles' version of a revolutionary moment. But the neighborhoods within the Arroyo Corridor and Los Angeles itself were not in the midst of or about to experience a revolutionary moment. Changes were happening, though. New types of coalitions had emerged involving connections between different neighborhood groups as well as new citywide and regional networks. The exploration of a different kind of approach—to view the Arroyo Seco or the Los Angeles River as a watershed rather than a concrete flood channel or to develop an integrated transportation policy rather than a freeway-centric approach—was no longer simply a romantic idea but was becoming central to the debates about the direction of policy. Caltrans or the water agencies had not been transformed, but some subtle and important shifts in the culture of the agencies had begun to occur. New groups, such as the Alliance for a Livable Los Angeles and the Los Angeles Alliance for a New Economy, were linking housing, food, transportation, jobs, and community development as part of an integrated agenda and were influencing the political discourse. Tim Brick's election as MWD chair suggested how much water politics in Los Angeles had changed. New bike groups in the Arroyo Corridor and the region were also formed—such as the Bicycle Kitchen, Northeast Los Angeles (NELA) Bikes!, and Cyclists Inviting Change thru Live Exchange (CICLE)—and undertook a number of community bike events. This included an international Bike Summer event in June 2005 with a bike ride down Figueroa Boulevard that culminated at Sycamore Grove Park as well as a "car-free" evening of music and celebration for bike riders at the

Cornfield park. Nature and community in the city—the underlying themes of ArroyoFest—had become part of the search for that place-based identity for Los Angeles, recasting the city that had been the very symbol of the loss of nature and community and the rise of a dominant freeway culture.[48]

Los Angeles was changing in numerous ways. The city had a new mayor who spoke of his own story of transformation and of his vision of greening the city and making it the symbol of community based on difference and diversity. It had a changing population that reflected those differences and influenced the patterns of development and the creation of new kinds of place-based identities. Nearly three years after ArroyoFest and nearly two years after a police chase had turned the Arroyo Seco Parkway into an open-air mariachi performance, another invasion of the freeways took place, and it was led by high school students who insisted on the rights of L.A.'s newest residents. Although not Charles Moore's "effective revolution of the Latin American sort," the freeway once again emerged as a symbolic place in the heart of Los Angeles, and while seizing the freeway did not indicate that a revolution had occurred, it symbolized the place where change needed to occur, whether occupied or silent.

7

City of Migrations

> The modern city can turn people outward, not inward; rather than wholeness, the city can give them experiences of otherness. The power of the city to reorient people in this way lies in its diversity; in the presence of difference, people have at least the possibility to step outside themselves.
>
> —*Richard Sennett*[1]

> The urban can be defined as a place where differences know one another, and, through their mutual recognition, test one another, and in this way are strengthened or weakened. Attacks against the urban coldly and lightheartedly anticipate the disappearance of differences.
>
> —*Henri Lefebvre*[2]

Into the Streets

It was an extraordinary day in Los Angeles. In the weeks leading up to that Saturday in March 2006, students spoke to each other through a Myspace.com Internet link, while priests in the Latino parishes were handing our *gran marcha* flyers and urging their parishioners to show up on Saturday for the demonstration. Spanish-language DJs spoke in unison about the need to come out and show a community united: "We are all immigrants. We are all hard-working contributors to the common good." The DJs told their listeners, "Wear a white tee shirt. Wear it as a symbol of peace. Make sure there is no litter on the ground, and come by bus, by foot, by Metrorail."

By 7 a.m. on March 25, the buses from Boyle Heights, Pico Union, Koreatown, and Wilshire Boulevard were already jammed. People started out by foot five miles north at Echo Park, across the bridges that crossed the L.A. River from East L.A. Farmworkers, down from Bakersfield, and from even farther north in the Central Valley, occupied Broadway before the first of the buses arrived. The word also spread in the high schools and middle schools to walk out to underline the message of the demonstration. While a few students had seen the HBO film *Walkout* that week (it describes the 1968 high school walkouts in Los Angeles that protested poor conditions in the Latino schools and asserted a "Chicano" identity), the March 2006 walkouts were spontaneous, boisterous, and all about core values and rights. "We are all the children of immigrants" became the common refrain. Leaving their schools, some students even made their way up the freeway ramps. Freeways, after all, were the symbols of community uprooting and environmental degradation of the neighborhoods where the homes and the schools of these children of immigrants lived.

From where this March 25, 2006, demonstration began, it was impossible to see to its ending. The marchers—a half a million to a million strong—converged on City Hall, where they were welcomed by the new mayor, Antonio Villaraigosa; the Los Angeles cardinal, Roger Mahoney; and leading activists in labor and immigrant-rights groups. Villaraigosa, who had participated in the 1968 Chicano student walkout, applauded the marchers for their call for respect and dignity for all who lived and worked in L.A. and other regions around the country. He also wanted the students to return to school, but they chanted back, "Hell no, we won't go." This was a moment to define what was new and what was possible. "People were connected and chanting," Urban and Environmental Policy Institute staff member Elizabeth Medrano later told me about the collective spirit that animated all who were there. "For me, this was a day when everything seemed possible," she said.

While this was a special Los Angeles moment, it was not exclusive to this region. As other demonstrations began to unfold around the country—40,000 demonstrators in Salt Lake City, 500,000 in Chicago and Houston, 50,000 in Denver, 15,000 in Nashville, and 25,000 in Salinas, California, in a city that had a population of less than 40,000—something else began to happen. The demonstrations became a moment of change and community identity experienced throughout the country. The country was caught between hostile and inflammatory rhetoric and legislative restrictions and a new definition of community that crossed borders. Rooted in L.A.'s (and increasingly the country's) cultural, social, and demographic diversity, such a definition of community based on diversity and difference suggested Richard Sennett's argument about the power of the city and the possibility of reinvention.

The Culture of Immigration

> Immigration is not just a political issue. It's a cultural one, and at this moment the U.S. is witnessing a cultural revolution that all this music is a part of. What many don't know yet is just how much the U.S. is being enriched by it.
>
> —*Omar Valenzuela (music producer of Mexican regional music)*[3]

Los Angeles has always been a city of migrations, constantly drawing on new populations that have helped redefine and reinvent the city and the region. But Los Angeles has also spawned a culture of hostility towards immigrants. In doing so, policies of displacement and discrimination have been created, two-tier economies have predominated, and a culture and politics of fear and ethnic panic have continually resurfaced, based on what former California

Governor Pete Wilson called, in full campaign mode in 1994, "the flood of illegal immigrants."[4]

The focus of this ethnic panic has also been something of a shifting target in Los Angeles and elsewhere in the United States. The negative portrayal of the immigrant as the "other" has reflected changing demographics as well as changing patterns of migration and settlement. But what makes one type of migration similar to or different from others? Is there a single definition or a common frame of reference, or is the fear of the immigrant as the other a social construction with powerful class, race, and ethnic reference points?

The most basic definition of *immigrant* is someone who resides in one country and was born in another country. *Merriam-Webster* defines that person as someone who "comes to a country to take up permanent residence." The political and legal definition of *immigrant*, as applied by the United States Immigration and Naturalization Service, is that of an alien (any person who is not a citizen or national of the United States) who is admitted to the United States as a "lawful, permanent resident." Both types of definitions distinguish between those who cross borders within this country and those who cross from another country to this country.

There is no easily recognizable term to identify migration within a country. The most proximate term—*inmigrants* (defined by *Merriam-Webster* as those who "move into or come to live in a region or community especially as part of a large-scale and continuing movement of population")—is neither widely used nor an effective distinction between types of migrations. This is particularly true for Los Angeles, with its periodic migrations from other countries and from other regions within the United States.[5]

The concept of crossing national borders becomes problematic when the border represents, as historian Patricia Limerick has argued, a "social fiction" that nevertheless plays a presumed role in protecting national sovereignty and cultural integrity. Writing in 1987, Limerick said that the longstanding desire to conquer and control the U.S.-Mexico border "continued to exist only in the imagination" and that "when politicians in the 1980s bemoaned the fact that America lost control of its borders with Mexico, they dreamed up a lost age of mastery." "In fact," Limerick argued in her book *The Legacy of Conquest: The Unbroken Past of the American West*, "from the Gulf of Mexico to the Pacific Ocean, the Mexican border was a social fiction that neither nature nor people in search of opportunity observed. That proposition carried a pedigree of decades, if not centuries." Moreover, the remapping of the border in the nineteenth century through conquest was designed to establish a racial as well as territorial divide, since the goal of conquest was to "take as much land and as few Mexicans as possible," as Limerick put it, a goal that has remained elusive for more than 150 years.[6]

Migration ("to move from one country, place, or locality to another," according to *Merriam-Webster*) rather than *immigration* or *inmigration* better reflects the complex movement of people (and goods, capital, and production facilities) across borders and into and out of cities and regions, including Los Angeles. From the founding of the pueblo in the late eighteenth century, migration emerged as the dominant characteristic of Los Angeles and its changing population and economy. The region's continuing migrations, beginning in the late eighteenth century, led to the establishment of the missions and early ranchos and the development of some of the area's first industries, each characterized by the continuing conflicts and integration of native and migrant populations. After the founding of the pueblo, native populations became a source of cheap labor, performing "virtually all manual labor operations in the Mission and on the ranchos," as Carey McWilliams put it. This cheap labor supply allowed for the development of the large hacienda system in the early nineteenth century and foreshadowed what McWilliams called factories in the fields that came to be based on Mexican as well as Asian and American migrant labor a hundred years later. The pueblo remained under Mexican control after that country obtained independence from Spain in 1821. But the transition to American rule in Los Angeles—following the 1846 to 1848 "Manifest Destiny" war against the Republic of Mexico—set in motion a period of turbulence and racial hostility that would later mark the arrival of each new migrant group.[7]

In later years, some Los Angeles boosters presented an idyllic view of Los Angeles in the 1850s as a type of multicultural Garden of Eden after its occupation by the European and American immigrants who took over the land. "The population of the place may be described as of four equal classes, Americans, Europeans, Spanish Californian and Indians," Theodore Hittell wrote of the area, describing an idyllic land of "luscious fruits, of many species and unnumbered varieties," "a water [rippling] musically through the zanjas," and "delicious odors from all the most fragrant flowers of the temperate zone [that] permeated the atmosphere." But this was also the period that witnessed the "unpronounced sentence of death" of the native culture and populations, as historian John Weaver put it. It was also a period when a virulent and deeply racist campaign took place against 80 percent of the population—the Spanish-speaking people who still resided in Los Angeles in the "homes for the defeated," as the Americans called the houses in places like Sonoratown near the heart of the old pueblo where the Los Angeles River crossed on its path southward. "Blood flows in the streets—justice weeps. All is anarchy," one Eastern journalist wrote in his diary in 1857 about Los Angeles, describing the violence perpetrated by the new American immigrants against long-time Mexican residents. With the rapid takeover of the land by the Americans and the imposition of their new and harsh order, Los Angeles had quickly become a place that "belonged now to the victors," as historian Bill Deverell put it.[8]

It wasn't just the Mexican and Native populations that became subject to the abuses of the victors. By the 1860s, migrant Chinese laborers, fresh from their work on the intercontinental railroads, also formed new settlements in Los Angeles. Similar to the Native American population, the Chinese migrants provided another source of cheap labor for the local economy. But nearly as soon as the "heathen Chinee" (as they were frequently characterized after Bret Harte's 1870 poem "Plain Language from Truthful James" popularized the term) established their roots and helped shape some of the earliest and most productive economic developments in the early Anglo period, such as the citrus and fishing industries, they were subject to discriminatory measures undermining their activities, while a racially motivated culture of hostility regarding the "Mongrel aliens" took root. During the last decades of the nineteen century, the Chinese were subject to a series of restrictive immigration measures (including a statewide measure modeled after efforts to bar blacks from entering the state) and became subject to a form of terror that contributed to the uprooting of their communities. The 1871 massacre that led to the murder of nineteenth Chinese men and boys, one of the worst race riots in the history of the United States, was symptomatic of the use of force to drive the Chinese from the fields, their camps, and their homes and became the basis for restrictive immigration legislation. "It is impossible for Caucasians, with their higher Civilization, to maintain the struggle for existence alongside the Chinamen," one commentator wrote in 1891, arguing that the Chinese "must either give up their civilization or—go to the wall." This reign of terror lasted for several decades until the Immigration Act of 1924 restricted future Asian and Southern and Eastern European immigration.[9]

The fear of immigrants, in a region based on continual migrations, also took on a strong nativist and class dimension. The large influx of Anglos into Los Angeles, fostered by the railroads in the late 1880s and continuing up through the 1930s, for example, led to appeals from business and political leaders to attract "the right kind of person" or the "better class who are industrious, intelligent, and progressive," as Harrison Gray Otis put it. To Otis and his newspaper, the *Los Angeles Times*, the wrong kinds of people were especially the "vicious, ignorant, and unfit foreigners," who at one time or another included Chinese laborers, Japanese farmers, Mexican farmworkers, Croatian dockworkers, and various other immigrant groups from Europe. Policies were needed to "protect American citizenship from invasion by large numbers of vicious and ignorant foreigners," the *Times* insisted, urging that Los Angeles not be made a dumping ground for "the scum of Europe."[10]

The linked issues of race, class, and ethnicity bound the debates regarding migrants and the cultural and political attitudes and actions that followed. The hostility towards the Chinese, Southern and Eastern Europeans, Mexicans, Filipinos, and Japanese that dominated the political and cultural discourse through the 1920s resulted in a series of draconian immigration policies that eliminated the influx of Asian immigrants into the United States and redefined acceptable immigration in racial terms, spurred in part by vigilantism and by theoretical justifications of racialist arguments (such as eugenics).

Although the federal 1924 immigration legislation did not exclude Mexicans on the basis of a racial category, Californian fears about Mexican immigration became more pronounced during the 1920s. The number of Mexicans had nearly tripled in the census count between 1910 and 1920, and those numbers continued to increase throughout much of the 1920s. Calls for immigration restriction against Mexicans during this period also had a strong class as well as racist component, with distinctions made between the displaced elite of rancheros (considered "whiter" due to a presumed purely Spanish ancestry), who had become part of a small and diminishing Mexican American middle class, and the much larger landless agricultural workers and laborers, many of whom traveled back and forth across the border and who were seen as dark-skinned and more Indian than white. And while an earlier generation had spoken of ethnic "alien indigestion" when referring to Southern and Eastern European immigrants, whose large numbers were seen as creating barriers for successful assimilation into the American melting pot, as early as World War I Mexicans (and Asians) were said to constitute a type of "racial indigestion." By establishing a racial argument, the anti-immigrants effectively "rejected the idea of the melting pot altogether," according to immigration historian Mae Ngai.[11]

During the 1920s, this strong nativist sentiment led to administrative actions that limited the number of permitted entries of Mexicans, increased the number of deportations, and also began to blur the distinction between legal and illegal regarding those who had already settled in the United States. As a result, by 1930 and 1931, the number of legal Mexican immigrants entering the United States dropped sharply. A December 1931 editorial in the *Los Angeles Times* argued that the "ebb of migration" due to the reduced entries and increased deportations had essentially eliminated the problem of unwanted immigrants.

Yet just a few years later, a fierce new debate was launched regarding the "idle hordes" that were "invading the state." But these were not Mexicans, Chinese, or Japanese, who continued to experience hostile racial backlash and exclusionary policies. These "idle hordes" were the Anglo migrants of the Dust Bowl, and the hostility about their Depression-era search for work substituted class tensions for racial arguments. "Return the transients to their homes," migration opponents argued, utilizing arguments about the need to prevent "citizenship" status (that is, California residency) and to avoid unwanted burdens and expenses. Opponents also called for deporting the Dust Bowl migrants back to the states that they came from, and one L.A. County supervisor, William A. Smith of Whittier, even filed a petition on behalf of agricultural workers from the "Mexican colony of La Verne," complaining that the Dust Bowl migrants were taking away the jobs of the Mexican laborers who had been working in the citrus fields. "These dust bowl migrants, by agreeing to work at a lower wage, offer a threat to the present standard of living and the wage scale which has evolved in this country," Smith complained.[12]

While hostility toward migrants had, in different periods, dominated the public discourse in Los Angeles and California, a kind of studied ambiguity among key conservative L.A. business and political figures, including Otis's son-in-law, Harry Chandler, also emerged with respect to the Mexican border and Mexican immigration. Even as Mexican immigration began to decline in 1929 and 1930, prior to the Dust Bowl migrations of the 1930s, the chair of the committee on immigration of the U.S. Chamber of Commerce was arguing for the need for a new round of Mexican immigrant labor for agriculture. A January 1930 editor's note in the *Los Angeles Times* supported that position, insisting that Mexican laborers were doing the kind of agricultural labor that "white Americans refuse to do." The border also represented economic opportunity. Several California-based landowners, including the Chandlers, owned large tracts of land on both sides of the border, while Los Angeles businessmen took ownership of various enterprises in border towns such as Tijuana. Even after the U.S.-Mexico Border Patrol was created in 1924 as part of federal immigration legislation and a policy was instituted of forced repatriation (initially encouraged in the early 1920s by vigilante groups not dissimilar from the Minutemen vigilantes of 2005 and 2006), some L.A. business interests and agricultural growers sought to limit the role of the Border Patrol to lessen the number of repatriations of those who were now defined as illegal aliens. The Mexican migrant worker, Harry Chandler declared before a congressional committee, was "less quarrelsome and less troublemaking than any other labor that comes into America," underlining his and other business leaders' measured ambiguity.[13]

As tensions over the Dust Bowl migration subsided and the country prepared for war, the push for renewed use of Mexican labor increased, even as deportations of Mexicans continued to take place throughout the 1930s. The development in 1942 of the Mexican Farm Labor Supply Program (better known as the bracero program), instituted as part of a broad set of policy initiatives regarding U.S.-Mexico relations, provided yet another controlled supply of cheap labor. Administered by the federal government but largely controlled by agricultural interests in California and from other parts of the country, the bracero program enabled growers to keep farmworker wages low, undermined efforts at unionization, and reinforced a continuum of discrimination against Mexican American citizens, bracero transient laborers (also "legal" but temporary), and "wetbacks" (the Mexican migrants who crossed illegally into the United States and worked in the fields but without the sanction of the bracero program). Some braceros became wetbacks and vice versa, a process encouraged in part by the seasonal demands of the growers, and migrants drifted to the cities and employment in places like hotels and restaurants. After World War II, the bracero program (which was averaging about 200,000 Mexican workers a year in the United States, continued to be extended as growers argued that this foreign labor was "more dependable than our local workers" (according to one Santa Ana grower's testimony at a 1948 hearing). The need for this low-wage workforce remained a theme that repeated itself throughout the history of the program. It also "signaled the consolidation of industrial farm production as a low-wage enterprise beyond the reach of federal labor standards and workers' rights," as Mae Ngai described the impact of the intertwined system of bracero and undocumented farm labor.[14]

During the 1940s and 1950s, the bracero program continued to be extended well past its original war-related sunset date. It was finally terminated in 1964 after growers began to seek out labor-displacing machinery while relying on the continuing supply of undocumented, nonsanctioned migrants who crossed the border. But by then, the sharp cultural divide and continuing hostilities extended to the entire Mexican community, including both recent Mexican migrants and Mexican Americans born on the U.S. side of the border. The policies that had flowed from the 1924 federal immigration legislation through the bracero program significantly extended the concept of illegal immigration and also created a more extensive pattern of discrimination. Those Mexican migrants born outside the United States assumed the status of "unassimilable foreigners" while those born inside the United States became "alien citizens." These were "American citizens by virtue of their birth in the United States but who are presumed to be foreign by the mainstream of American culture and, at times, the state," as Ngai put it. The 1942 Sleepy Lagoon murder trial of twenty-three Mexican American young men and the 1943 zoot-suit riots where United States servicemen targeted Mexican American zoot-suiters reinforced cultural fears and ethnic panic that were not simply associated with illegal immigrant status.[15]

Nevertheless, the most vulnerable Mexicans continued to be the newest migrants, pejoratively named "wetbacks" for their act of crossing the waters of the Rio Grande to get into the United States. The wetback was a "commodity migrant," as Jorge Bustamante put it, who was subject to the whims of the agricultural industry's manipulation of border policy. Mexican migrants were also subject to continuing xenophobic campaigns, such as "Operation Wetback," instituted in 1954 by the Eisenhower administration and its new commissioner of immigration, who had been a West Point classmate of President Dwight D. Eisenhower. This campaign, designed to stop "history's greatest peacetime invasion," was deemed a success just one year after it was instituted. That claim, like the claims made in the late 1920s, was soon forgotten, as the repeat pattern of what Carey McWilliams called "entry, work, repatriation" continued to characterize the flow of labor across the border. The border itself became both a transit way, often dangerous, and an instrument of control. Commanding at once economic and political significance, the border was "like a cunningly designed filter that separates the economic utility of the Mexican illegal entrant from the rest of his cultural makeup," wrote union organizer and Mexican migrant-labor analyst Ernesto Galarza. "Never a participating member of the community or society, the wetback lives anthropologically in no-man's land," Galarza said of the migrant's status.[16]

This status of uncertainty, exploitation, and a cultural no-man's land was reflected in (and rebelled against) in the language, music, dress, and redefinition of community of the Mexican migrants and their children who lived in Los Angeles throughout the twentieth century's history of entry, work, repatriation—and settlement. George Sanchez writes of this pattern in his description of the 1926 recording of "El Lavaplatos," the first commercial recording of a new Mexican musical genre called the *corrido*, which provided a narrative that was familiar to and provided identity for the corrido audience of Mexican migrants. In this new music, the connection between the migrant's new home and place of origin remained powerful, even as the Mexican migrant sought to establish whatever roots were possible in a place that encouraged migration but pursued a politics and culture of displacement. To wear the zoot suit, the baggy pants, and loose-fitting outfits that became popular among young Mexican American and African American males in the late 1930s and 1940s also asserted an identity that historian Stuart Cosgrove called a "refusal: a subcultural gesture that refused to concede to the manner of subservience" for Mexican Americans in Los Angeles.[17]

During the mid- and late 1960s, the search for cultural identity and an oppositional politics led to the development of the Chicano movement in the Southwest and California that paralleled the rise of a race- and ethnic-based politics throughout the country that included the African American, Puerto Rican, and Native American communities. In Los Angeles, the African American community grew significantly during the 1940s and 1950s, due to migration from several states in the Southern United States, whose residents were lured in part by the availability of jobs in the expanding manufacturing sector, such as the aircraft industry. By 1965, when the Watts riots convulsed Los Angeles and the nation, the modest economic gains of the war and postwar years had given way to chronic problems of poverty and unemployment, lack of political power, and a dominant culture that ignored or undercut the cultural expressions of its various migrant and minority populations.

The protests of the 1960s and the search for a type of cross-border identity for Los Angeles's Mexican populations shifted for a time the nature of the debate about the politics and culture of migration away from the prevailing assumptions about acculturation, assimilation, and hidden messages about "racial indigestion" and "alien citizens." For the Mexican community in Los Angeles and elsewhere in the Southwest, the term *Chicano* symbolized a new identity that was based on the culture of a community that was part migrant, part Mexican, and part American. This new assertion of a community identity even swept up moderate and assimilated Mexican Americans, such as *Los Angeles Times* columnist and KMEX (Spanish-language) TV news director Ruben Salazar. In a *Los Angeles Times* column written shortly before he was killed by a tear-gas projectile while taking a break from covering the 1970 Chicano moratorium demonstration against the Vietnam War, Salazar spoke of Aztlán, a region that linked the five southwestern states in the United States with Northern Mexico and a concept that challenged the notion of the border as a cultural as well as national boundary. "Chicanos explain that they are indigenous to Aztlán and do not relate, at least intellectually and emotionally, to the Anglo United States," Salazar said of this new articulation of a community identity.[18]

Despite the assertion of a new Mexican American identity and a growing political influence and ethnic consciousness of Latinos in Los Angeles, New York, and other cities, the immigration issue continued to reappear as the central factor framing the political and cultural debates in Los Angeles and elsewhere. While a Latino "ethnic principle of organization," as Felix Padilla put it, strengthened the political and cultural links among groups as diverse as Mexicans, Central Americans, and Puerto Ricans, the overriding significance of *immigration*, (legal, quasi-legal, and illegal) influenced the status and issues associated with Mexicans and, increasingly by the 1980s, Central Americans and other migrants from Latin America and Asia as well.[19]

As the bracero program shut down, the immigration legislation enacted in 1965 and subsequent legislation in 1986 primarily influenced the patterns and the extent of the new migrations of the 1970s, 1980s, 1990s, and the first decade of the twenty-first century. Among those immigrating to the United States, the numbers from Mexico after 1965 were far greater than those from any other country. Yet the quota for the legal entry of anyone immigrating from Western Hemisphere nations was "restricted to a level far below actual and already documented numbers for Mexican migration," as Nicholas de Genova pointed out. Thus, the new "equitable" quota systems that allowed for the same number of legal immigrants from Europe and Latin America also intensified the level of illegal immigration from Mexico and other Latin American countries.[20]

The Latin American migrations combined with the huge influx of Asian immigration to turn Los Angeles into an increasingly different place where minorities were now majorities, the highest-rated radio station was in Spanish, and even some of the automobile suburbs and edge areas like San Bernardino witnessed rapid demographic change. At the same time, similar to the debates of the late nineteenth century, the 1920s, and the early 1950s, calls intensified to deport illegal residents and to deny them publicly funded health care and education and even the right to work. The push to militarize the country's Southern border, which intensified during the 1980s and again after President Clinton's Operation Gatekeeper initiative in 1994, became the foundation of nearly all immigration policy proposals during the contentious debates of 2005 and 2006. It further underlined what Leslie Marmon Silko characterized as the development of a "Border Patrol State" that was designed to repel the invaders from the south and reconstruct a demographic *status quo ante*.[21]

Language itself became contested terrain, as the English Only movement sought to link the American identity to assimilation and acculturation, despite the enormous demographic changes and globalization effects that were reshaping the country. During the immigration legislation debates in 2006, which included efforts to designate English as the country's national language, Tennessee Senator Lamar Alexander, characterizing the issue in ways reminiscent of the hostility a hundred years earlier toward Southern and Eastern European immigrants, worried about the large numbers of people that were being allowed into the country and the need to "still make sure they become Americans." "A lot of the uneasiness and emotion over this immigration debate is from Americans who are afraid we are going to change the character of our country," Alexander argued. But the "character of the country"—and particularly places like Los Angeles—had already changed. The evolving dynamics of place, cultural identity, political power, continuing and expanding migrations from Latin America and Asia, a global economy, and its urban extension the global city had thoroughly transformed Los Angeles—and had made it the harbinger of the changes that many cities throughout the country were also beginning to experience and to resist.[22]

The Global City

In 1960, census figures identified Los Angeles County, with its six million residents, as second only to the New York metropolitan area in population, and the vast majority of its residents were recently arrived to the region. That year, the *Los Angeles Times* ran a series of articles seeking to answer the question "What is L.A.?" and exploring how the population changes in the region defined the city's "personality." The articles suggested that the huge migrations to the region were made up of Anglo middle-class (or aspiring middle-class) men and women from other regions in the United States who did not have specific cultural ties to their places of origin. This factor—migrants without roots—provided the basis, according to the *Times*, for a type of uniform American identity for Los Angeles, a place that was "more American than the rest of the country, because we haven't assimilated as many other cultures." L.A.'s residents, in turn, exhibited a "kind of anti-cosmopolitanism, in which there is no experience socially with other than one's kind." This notion of Los Angeles as placeless—part of a great "homogeneous core" spread throughout L.A.'s vast suburban land mass of separate cities and communities, each sharing this uniform identity—was already becoming the dominant image of the city and the region. The image was scorned by its detractors but celebrated by its champions, who continued to preen with satisfaction as L.A.'s population continued to grow and its geographic edges continued to extend outward. No longer viewed as the end point of migrations from the Dust Bowl or the rural communities in the Midwest, the domestic migrations to Los Angeles were now seen as primarily interurban. People were moving from city to city, "so that people bring a uniform living pattern to a new city."[23]

Unanticipated by either its boosters or its critics, Los Angeles by the end of the 1960s reached a new tipping point in relation to its identity. The Watts riots and the assertion of new Chicano identities were the most visible manifestations that such uniformity no longer held, while new migration patterns were also changing the idea of a homogeneous core. In 1960, African Americans constituted the largest minority group in L.A., with residents of Japanese ancestry the next largest but significantly smaller minority. The count of Mexican Americans remained uncertain, a function of the confusion generated by census categories and by the continuing migrations of legal, illegal, temporary, and permanent residents.[24]

By the 1970 census, the population mix in Los Angeles had begun to change significantly, stimulated by new immigration legislation and a global demographic upheaval (and economic restructuring) that began to transform Los Angeles and numerous other cities and regions around the world. The Hart-Celler Act of 1965—federal immigration legislation designed primarily to increase European immigration and high-skilled labor, increase quotas, and yet keep immigration increases to a modest level—had two immediate impacts for Los Angeles. It set in motion enormous increases in its Asian immigrant populations, whose exclusions were now lifted, and it created a new pattern of predominantly illegal Latino immigration from Mexico and Central and South America (and from the Caribbean island countries to New York and other East Coast cities). The cross-border migrations to Los Angeles, a result of political upheavals and economic shifts in the immigrants' countries of origin, paralleled the vast migrations from rural areas to the cities of the developing world, Europe, and the United States. In just a few decades, the world's population shifted from two-thirds rural to nearly half urban, creating vast new megacities of the developing world that far outstripped in size their counterparts in the United States and Europe. Mexico City grew from a population of 8.8 million in 1970 to 18.1 million in 2000, while Mexico's border cities, like Tijuana and Matamoras grew even faster, fueled in part by a border industrialization program and the vast migrations to the north.[25]

These demographic shifts were influenced by the economic restructuring and reorganization of local, national, and global economies. The deindustrialization of the 1970s, 1980s, and 1990s that affected Los Angeles and numerous other regional economies in the United States was also linked to the rapid industrialization that was taking place in the developing countries and their megacities. A number of theorists and researchers began to describe a new type of urban form in the age of globalization, identified as "the global city," "the world city," or "global city regions," among other conceptual terms. One of the origins of the global city concept could be found in the Maoist notion of the location of power in a central place that was able to control vast territories but that would ultimately fall when the countryside overwhelmed the power located in the global city. But the global city that emerged at the end of the twentieth century did not represent the classic idea of an urban core that controlled the periphery. Instead, what evolved from the industrial city of the nineteenth century and the early and mid-twentieth century represented different constellations of power, different types of economies, different types of global relationships, and different kinds of demographic and spatial characteristics. In the developing world, megacities were rapidly industrializing cities that would ultimately emerge in the twenty-first century as "production motors not of national economies but of the global economy," as geographer Neil Smith put it. Analysts such as Saskia Sassen argued that key changes in such global cities as New York, Tokyo, and London included their role as "highly concentrated command points in the organization of the world economy" as well as urban economies characterized by a shift from manufacturing to finance and specialized service firms. These changes also reflected an increasingly stratified social structure where the gap between the wealthiest and poorest residents had increased enormously.[26]

Though often left out of some of the early "global city" literature, the changes in Los Angeles that emerged during the 1970s, 1980s, and 1990s reflected much of the same economic restructuring and income stratification that characterized other global cities. L.A., for example, had more millionaires than any other region in the United States, while it also had developed a huge low-wage sector based significantly on its large immigrant populations. As a result, Los Angeles and a number of other global cities in the United States and Europe became "regions of extreme polarization" where first-world and third-world conditions resided simultaneously and a middle-class sector continued to shrink. The concept of food insecurity, for example, was identified during the 1970s and 1980s in developing countries that had witnessed the rapid erosion of self-sustaining rural economies and food systems due in part to the pressures of globalization to establish export markets and the consequent huge migrations to the cities. But by the early 1990s, the term *food insecurity* had begun to be applied to domestic U.S. circumstances as well, where cities like Los Angeles witnessed increasing numbers of its residents, including those employed in low-wage jobs, become dependent on an emergency system of food shelters and food banks. While Los Angeles had emerged as home to the largest concentration of individual wealth, a growing number of its residents who worked in many of the expanding service industries experienced such conditions as food insecurity, homelessness, lack of affordable housing, lack of health insurance and appropriate medical care, and poverty-level wages.[27]

Moreover, Los Angeles added a critical new dimension to the global city phenomena—the extraordinary role of the border in the life of the city. Through the 1980s, 1990s, and the first decade of the twenty-first century, Mexican and other southern migrations swelled the Latino population in Los

Angeles, reaching majority status in the city and near majority status in L.A. County and becoming the largest ethnic group in California. The phenomenon of the blurring of the lines between the "illegal alien" and the "alien citizen"—which first emerged in the late 1920s and had been extended in the 1940s and 1950s during the life of the bracero program and the efforts to close the border—reemerged divisively in the 1980s and 1990s and even more sharply during the debates over immigration policy in 2005, 2006 and 2007. At the same time, a powerful new political and cultural immigrant identity reappeared first in massive demonstrations opposing Proposition 187, a 1994 California ballot initiative that sought to deny publicly funded social services to illegal immigrants and then during the 2006 immigration debates in Congress.

These political, cultural, and demographic shifts were also taking place during periods of major economic restructuring that redefined Los Angeles as a global city with a changing economy that was increasingly defined by its relationship to global economic investments in trade, finance, accounting, legal services, and other businesses servicing the global economy. While L.A.'s law firms, financial-service groups, and cultural industries became global in their reach and activities, a large number of non-U.S. companies also began to invest in the L.A. economy in commercial real estate, entertainment companies, and global trade-oriented operations. At the same time, the region's manufacturing base shifted from the heavy manufacturing sector (characterized by higher-paid unionized jobs in automobile, aerospace, and other durable-goods production) to low-wage, light-manufacturing sectors (such as food processing, toy manufacturing, garment, and furniture production). This shift was rapid and extensive. In just four years, from 1990 to 1994, for example, heavy manufacturing lost as much as a third of its jobs. By 2000, the fastest-growing sector in manufacturing continued to be in the light-manufacturing, low-wage sectors of the L.A. economy, which now represented nearly half of all the jobs in the region.[28]

This growth area of the L.A. economy also coincided with the rise of the sweatshop economy that relied on low-wage immigrant workers, many of them undocumented and subject to abuse by their employers. Some of the major scandals of this period were associated with sweatshop industries that used immigrants as a type of indentured labor. This situation was dramatically revealed by the 1995 exposure of a "virtual slave-labor camp" for seventy-two Thai immigrants, who were trapped by guards and barbed wire as they worked seventeen-hour days in one garment-industry operation in the suburban community of El Monte. Some service industries, such as janitorial services, intentionally sought out low-paid immigrant workers as part of a restructuring process that included displacement of a largely African American unionized workforce; reliance on toxic chemicals to speed up the cleaning process; and the subcontracting of arrangements between big commercial buildings (several of them owned by real estate companies with significant international investment) and their subcontractors, who provided the building owners with a low-wage, nonunion, immigrant workforce. Some of the largest of those subcontractors were in turn either owned or invested in by international firms.[29]

These economic shifts were also related to the expanding global trade and production activities that increased the size and the extent of global economic activity during the same period that the immigration patterns had also changed dramatically. Complex global production, trade, and capital flows are not new to the region or to the United States. When the oil industry took root in Los Angeles in the 1920s, helped establish an important manufacturing base, and influenced the development of several of L.A.'s suburban-industrial clusters, L.A.-based companies like Doheny, Richfield, Union Oil, and Occidental Petroleum were consolidated or expanded into new entities that became part of multinational networks that operated their extraction, production, refining, and distribution activities on a global scale. In the early 1950s, Los Angeles–based Cyprus Mines established a partnership with one of the companies that had built the Hoover Dam to participate in a Pacific Rim version of the old Atlantic triangle trade. The partnership, which was called Marcona, operated a massive iron operation in Peru, which provided the ore for the Japanese steel industry. Marcona built a fleet of convertible ore carriers to move the ore to Japan, and the carriers then sailed to Indonesia, where they picked up crude oil to cross the ocean and deliver the oil to West Coast refineries. These refineries served the huge oil-consuming markets in Los Angeles and California as well as other parts of the country. But even as some refineries abandoned their Los Angeles locations due to air-pollution regulations and some of the largest oil companies (such as ARCO) abandoned their Los Angeles headquarters, Los Angeles still remained pivotal to the infrastructure of services related to the global oil industry. At the same time, the Los Angeles region continued to be the home of several oil refineries that supplied more than half of the refining capacity for California and maintained a significant operable crude-oil distillation capacity, with L.A.'s twenty-two refineries able to produce nearly two million barrels of oil per day.[30]

The border's effect on the economic, social, and cultural life of Los Angeles and other U.S. regions also figured prominently in the development of the mid-1960s' Border Industrialization Program—the *maquiladoras*, as the border plants were known. The term *maquiladora* is derived from the Spanish word *maquila*, which references the "value-added" payment that millers received from peasants for grinding corn. The border region became the home of industrial parks for factories that assembled and processed products to be exported to the United States and other industrialized countries. The *maquiladoras* were eligible for duty-free imports of raw materials, machinery, parts, or tools that were used for the plants. The Mexican border plants were expected to be able to compete more effectively, given their location close to U.S. markets, with other "runaway production" that had drifted to the new low-wage plants in Singapore, Taiwan, the Philippines, and South Korea. At the same time, it was hoped that the development of new jobs on the Mexican side of the border, which coincided with the termination of the bracero program, would reduce the level of Mexican immigration into the United States by employing, among others, the one-time bracero workers who would otherwise become, border-plant advocates argued, a new generation of wetbacks.[31]

The *maquiladoras* provided another illustration of the "social fiction" of the border. These "export-processing plants," as they were initially known, were effectively controlled by non-Mexican companies, with most top management and investments coming from the United States, while other countries (such as Japan or several European nations) also assumed a stake in this new extension of globetrotting production. Mexican subcontractors provided the labor, primarily female workers drawn from the interior, where cheap wages (wage rates initially averaged about fifty cents an hour), hostility to unions, lack of enforcement of environmental regulations, and poor living conditions (such as substandard housing) prevailed. While employment, primarily in such sectors as garment and electronics, grew modestly at first, by the 1980s the border plants had expanded substantially in the number and types of operations and in numbers of Mexican workers employed. The population of the border cities also grew astronomically in this period due to employment in the border plants and cross-border activities, including the continuing flow of migrants. The border plants eventually came to be seen as contributing to rather than reducing the immigration flow. The *maquiladora* garment plants, for example, had over several decades transferred as many as 100,000 garment jobs from the United States to Mexico, many of them due to plant closures and job losses in Los Angeles. But those labor flows eventually began to include border crossings in both directions, as *maquiladora* garment workers began to migrate north to cities like Los Angeles and New York. This shift was a direct result of the discrepancy in wages between the two countries, even including the sweatshop wages in the United States plants. By 2006, Los Angeles–based union leader Christina Vasquez would comment that previously "migrants came north and they learned to work a sewing machine here. Now we see them coming from Mexico already knowing how to sew. Ten years ago, we never saw that." A study by Hunter College sociologist Margaret Chin also documented a labor-flow shift and found that as many as 75 percent of the garment workers in New York City were, as Chin put it, "pretrained" workers from the *maquiladora* plants.[32]

The social fiction of the border was perhaps most revealing in the 1994 North American Free Trade Agreement (NAFTA), which extended the *maquiladora* concept of a duty-free and open-ended flow of goods—and capital—to the entire North American continent. NAFTA did not shift production away from the border region. The most rapid expansion of the *maquiladoras* occurred during the 1990s as NAFTA was being negotiated and after it was implemented. But neither did NAFTA protect Mexico from the loss of jobs when other globalization forces interceded, including China's leap onto the world stage as the leading center for low-wage employment and its rapid transfer of export-oriented production facilities to China's own "border economy" in its southern provinces. While border-plant wages increased from an average of 50 cents an hour to $1.50 an hour during its first forty years (far lower than equivalent wage increases in the United States), wages in the Chinese plants averaged about 50 cents an hour, one-third the Mexican rate. When the U.S. economy went into recession in 2001 and China's low-wage and export-based economy rapidly expanded, the Mexican border plants declined significantly, with 800 plants closed and 200,000 jobs lost. Only after the U.S. economy began its recovery did the border plants, beginning in 2004, once again expand, with talk of Mexico's comparative geographic advantage of its proximity to the United States suggesting possible future expansion.[33]

By 2006, NAFTA and the border-plant economy had precipitated an even tighter U.S.-Mexico economic relationship. Mexican exports to the United States, for example, jumped from 76 percent in 1992 to 88 percent in 1998, while U.S. penetration into the Mexican economy also grew substantially, with NAFTA facilitating those shifts. Whether in relation to the flow of capital and goods or to the migrations and their patterns of work, deportation, reentry, and settlement, the border could no longer be considered either fully open and inviting or truly closed. The border continued to reflect the studied ambiguity that had characterized U.S.-Mexico relationships, with U.S. policy makers encouraging integration of the two economies while remaining fearful of the implications of the Latino presence in the U.S., a product of the long and tortured history of the regulated and unregulated migrations that had passed through the border. This history, including the connections of Los Angeles to its Mexican past and the rise of NAFTA and other globalizing forces, made more transparent the idea of the border as social fiction. The migrations of goods, capital, productions facilities, and people seemed destined to continue at its own measured pace—despite fences, National Guard troops, workplace raids, and nativist rhetoric.[34]

Movement of Goods

The increasing role played by Los Angeles in the global economy and that economy's consequences for both immigrant and nonimmigrant communities were perhaps best symbolized by the "goods movement." This awkward phrase has come to represent Los Angeles' central role in the shipment and distribution of goods flowing in and out of the Ports of Los Angeles and Long Beach, the largest port complex in the United States and fifth largest in the world behind Singapore, Hong Kong, Shanghai, and Shenzhen. Goods coming from places like China arrive at the ports and are transferred onto the trucks and railroads that take these goods through the heart of central and southeast Los Angeles toward their ultimate destinations, including outlets like Wal-Mart, which are spread throughout the country.[35]

By way of illustration, in an editorial for the journal *Environmental Health Perspectives*, the University of Southern California's Andrea Hricko portrayed the prototypical journey of a $9.97 doll. It was purchased at a big-box retailer in a suburban Chicago community and produced by low-wage workers in an Asian country, whose lower production and wage costs undercut their Mexican border and American toy industry counterparts, including toy-production facilities in both the Los Angeles and Chicago areas. Hricko recounts how such a doll, packed with ten thousand other dolls, would likely be placed in a container and loaded onto a ship that held four thousand other containers filled with other goods like shoes and electronics. Arriving at the Port of Los Angeles or Port of Long Beach—where 40 percent of all U.S. imports arrive—the doll's container would first sit idle (sometimes as long as a week), given the backlog of containers due to the expanding volume of goods coming into the ports. The doll's container would finally be transferred to a truck (that had also sat idle awaiting its cargo) that would take it to a rail yard close to nearby schools and homes, where the container would continue its journey east. Meanwhile, other containers, including those with other dolls, would be loaded onto trucks that would take the dolls for repackaging at a million-square-foot megawarehouse built on former dairy land in the Inland Empire at the urban edge of Southern California. Similar to the journey of its rail counterpart, the just-in-time packaged doll would in this scenario be shipped from the warehouse by truck to its ultimate retail outlet destination.[36]

During its 8,000-mile journey, the doll would become responsible for new and expanding sources of pollution—including hazardous emissions from the low bunker fuel used by the ships as well as other pollutant discharges, such as nitrogen oxides, sulfur oxides, and particulates. To illustrate the magnitude of these environmental and health impacts, 128 tons of nitrogen oxides are emitted daily at the ports—significantly more than the 101 tons of nitrogen oxide emitted daily from all six million cars in the Los Angeles region. The diesel trains, their rail yards, and the diesel trucks also generate a myriad of pollutants, particularly the diesel-related particulates that have emerged as the most serious toxic air-pollution health hazard in the Los Angeles region, especially in the low-income communities where the transfers, the freeways, the rail yards, and the warehouses have been located. These communities also suffer from the noise, the twenty-four-hour glare of lights, and the unsettling of daily life that the rail yards and other goods-movement activities have caused. And many of these same environmental and community burdens are also experienced at the Asian port cities where the goods have been produced and shipped.[37]

Yet even as these impacts are being felt, nearly all of Los Angeles and California's political and business leaders have proclaimed that goods movement has become the economic driver for the region, illustrating the value of globalization and Los Angeles' share of the benefits it brings. The two ports and the goods movement have in fact grown enormously in the past two decades. As of 2005, they accounted for a third of the goods coming in and out of U.S. ports and handled more than 14 million twenty-foot equivalent units (TEUs) (the measure related to the size of the containers carrying the goods), with a trade value of more than $200 billion. It is predicted that the amount of cargo coming through the ports is likely to double and possibly even triple by the year 2020. Three-quarters of the trade through the ports is produced or consumed outside L.A. The goods movement—whether its expansion can be accommodated, how this will happen, and whether its community and environmental impacts can be addressed—could reveal, these boosters suggest, the future direction of the city itself.[38]

While the two ports, less than ten miles apart, remain rivals for their share of traffic and also fall under the jurisdiction of two different port commissions from the two cities, they have nevertheless come to be considered a single entity in relation to the custom duties, regional impacts, infrastructure issues, and related community, environmental, and health impacts. The growth in port activity has paralleled some of the major changes in goods movement during the last several decades and the changing technologies of the ships and the ports that have facilitated those changes. This has included the introduction of containerization (packing the goods in large metal containers that are then more easily transported and transferred into and out of the ships). First introduced in the mid-1950s, containerization has influenced the development of deep bulk ships and ports and facilitated the development of the new production centers and export-oriented economies in places like Singapore and China.

When these changes were beginning to occur during the 1950s, the Port of Los Angeles, also known as "Worldport LA," had a long and storied history of labor militancy and community action. Although considered a major West Coast port, it was smaller than several other U.S. and overseas port complexes. Its counterpart in Long Beach was dubbed by Aristotle Onassis the "world's most modern port" in the 1950s due to its early experimentation with rail-transfer operations and conversion to containerization. Beginning in the 1950s, the two ports capitalized on their location as the favorable route from places like Hong Kong and Singapore (which emerged as the two largest port operations in the world) as well as Japan (where the rapid increase in automobile imports also helped stimulate port growth). By the 1970s and 1980s, Los Angeles had positioned itself as a "gateway city for the Pacific Rim," as L.A. mayor Tom Bradley put it. "We're simply on the right ocean," Long Beach Port spokesperson Yvonne Avila happily commented to the *Los Angeles Times* in 1995.[39]

In 1995, Long Beach broke ground for its new Pier A Container Terminal at a 170-acre site where in 1930 the Ford Motor Company built an assembly plant for its Model A cars. Adjacent crude-oil production facilities had housed sumps for the disposal of hazardous drilling fluids that were used in the nearby Wilmington oil fields. The site had previously been leased to a waste-disposal company that managed a commercial disposal operation for the waste oils, sludges, heavy metals, resins, alcohols, paints, and organic wastes from the oil drilling, refining, and other industrial activities. As a result, the area had achieved Superfund status as a hazardous waste site, and the development of the Pier A facility became controversial for the major environmental hazards that were encountered during the dredging and disposal of the soils. Weathering those criticisms in part by touting the economic benefits that were associated with port expansion, Pier A was able to be completed in just a couple of years.

Soon after, in 1998, ground was broken for construction of the Alameda Corridor, a long-awaited rail link from the ports through the heart of central Los Angeles. Designed to consolidate all port-generated rail freight traffic, plans for the Alameda Corridor had been proposed as early as 1984 by the Southern California Association of Governments. Completed in 2002 at a cost of $2.4 billion, it allowed upward of fifty freight trains a day to pass quickly in and out of the two ports by avoiding conflicts at two hundred at-grade crossings between downtown L.A. and the ports. With these infrastructure changes, the two ports became massive operations in global trade and were poised to benefit from a massive jump in imports from China, which had already become their leading importer and exporter. The ports' managers and governing bodies, deeply connected to the shipping, rail, and trucking industries, had assumed significant autonomy in their decision making and influence in the region. The ports operations and the flow of goods across oceans had effectively become integrated into what Carolyn Cartier called "a maritime world economy," a complement to the twenty-first-century global city.[40]

But while the ports expanded, their community, environmental, and health effects for residents, truckers, and dock workers also became increasingly apparent. Communities adjacent to the ports and the rail yards in places like Long Beach, Wilmington, San Pedro, and the City of Commerce were for the most part low-income and predominantly Latino, including many residents who were recent immigrants. Many of the independent truck drivers were Latino, including a number of undocumented immigrants. Both community residents and workers inhaled the pollutants that originated from the ships, delays at the port, and idling ships and trucks. Air pollutants were also generated in the rail yards, again due in part to idling diesel engines, and intensified with the massive number of big-rig diesel trucks that rambled onto the freeways on their way to various warehouses, rail yards, and retail destinations. Communities that bordered the major freeway route from the port, Interstate 710 (the Long Beach freeway), were most subject to continuous exposures from the stream of big-rig, diesel-polluting trucks that clogged the freeway and hemmed in cars brave enough to compete along this route. Major health risks have been identified from each leg of these journeys—from the particulate matter, nitrogen oxides, and other goods-movement-related pollutants that have produced increased risk of death, increased heart-related concerns, increased asthma, and stunted lung growth in children (capable of creating a lifetime of health problems). Other research has pointed to the continuous high noise levels (an increased factor in the case of the goods movement, as the ports shifted to twenty-four-hour operations) that have led to sleep disturbances and myocardial infractions. And near the ports and the freeways, seven-foot tall empty cargo containers imprinted with the names of shipping firms have been transforming communities like Wilmington into a globalization-induced type of "container land," as the *Los Angeles Times* put it. "We have become the sacrifice zones of the goods movement," one community activist angrily declared of the health, environment, and quality-of-life effects of the nearby rail yards and freeway traffic in his City of Commerce neighborhood.[41]

Perhaps more than any other issue, the constant passage of trucks on freeways, particularly Interstate 710, symbolized that notion of community-sacrifice zones. Originally called the "Los Angeles River Freeway" when plans were first drawn to establish a route along the river bed from the southeast communities to the port, the problems associated with the I-710 freeway also became a critical concern to goods-movement advocates who worried about the delays caused by freeway congestion but did not consider the community impacts that were associated with that congestion. Initial plans to expand the freeway to reduce congestion similarly ignored those impacts. At the same time, the congestion factor, more than any other goods-movement issue, had framed more than two decades of discussions regarding potential strategies for intervention. By 1985, as the volume of goods moving through the Port of Long Beach began to increase significantly due to its shift to containerization, the I-710 freeway was considered essential to an effective growth transition. That year the roadway, initially designated California Route 7, a state highway, had its status changed to its Interstate 710 highway designation. This was partly in response to the heavy truck traffic coming out of the port. Daily traffic counts were as high as 175,000 trucks and cars at the Santa Ana freeway interchange at the City of Commerce southeast of downtown Los Angeles, a major transit point for goods-movement traffic. Interstate highway status also meant that more federal highway funds would be available for maintenance and upgrades. Problems of congestion on I-710, however, increased significantly rather than decreased over the next two decades, and related delays at the ports made efficient and rapid movement a top priority for advocates. Even the opening of the Alameda Corridor, which facilitated the continuing expansion of goods traffic, did not effectively overcome the bottlenecks along the freeways and at dockside.[42]

The congestion problem was dramatically highlighted during a 2002 conflict between the Pacific Maritime Association and the International Longshore and Warehouse Union (ILWU) that resulted in a slowdown and ten-day lockout of the union workers until a settlement was reached. Trucks sat idle while some goods (particularly imports more sensitive to delays, such as perishable food products) lost their value. Fears of bottlenecks rather than concerns about community and health effects dominated the debates about what needed to be done. For example, just two years after the lockout and another congestion episode, a *Fortune* magazine article warned that "with the boom in imports from Asia, this place [the Port of Los Angeles] is out of control. . . . It's hard to overstate the ripple effects of the chaos. Just ask Toys "R" Us, which had to build ten extra days into its supply chain; or Sharp Electronics, which had to fly in television parts from China; or toymaker MGA Entertainment, which lost some $40 million in revenues when it couldn't deliver its bestselling Bratz dolls on time to big retailers." "Getting them out," the article said of the goods backlog, "is like driving in Calcutta during rush hour."[43]

With port operators, shipping and railroad companies, truckers, retailers, and other goods-movement promoters arguing that immediate action to relieve congestion was necessary, a series of additional infrastructure changes were proposed, including the expansion of I-710, new rail yard and transport operations, and twenty-four-hour port operations. California's new governor, Arnold Schwarznegger, also sought to fast-track a Goods Movement Action Plan that incorporated many of these proposals, while seeking to accommodate and ultimately expand trade opportunities with countries like China, which was also experiencing its own port bottlenecks. As one Action Plan press release put it, this "coast-to-border" system that was headquartered in Los Angeles was "critical to the national goods movement network." At the same time, an alternative NAFTA corridor was being proposed as a cheaper though longer route from the Asian ports to the Mexican port at Lázaro Cárdenas, where goods would then proceed along a rail or truck route through the border town of Laredo up through U.S. Interstate 35 to the key Midwestern transit way station for goods-movement traffic at Kansas City, Missouri. The port operations at Lázaro Cárdenas and the truck and rail traffic through the NAFTA corridor increased significantly after the North

American Free Trade Agreement took effect in January 1994, as had traffic through competing ports on the West Coast at Seattle-Tacoma and Oakland since the 2002 lockout. By 2006, however, the Los Angeles ports, with their coast-to-border system of facilities, still remained the major transit route for the flow of goods to and from Asia and the focus for action among goods-movement advocates.[44]

The strategies for expansion (such as the Goods Movement Action Plan), however, met with resistance from the community groups that focused on health and environmental impacts, the dock workers and truckers who raised concerns about occupational health and safety issues, and the environmental and health researchers who argued that the region was being subject to unacceptable risks. A new L.A. Harbor Commission, appointed in 2005 by L.A.'s new mayor Antonio Villaraigosa, quickly inserted itself into the discussions, arguing that expansion was not possible without first addressing the community, environmental, and health effects from the port, truck, and rail operations. The regional air-quality agency, the South Coast Air Quality Management District, whose own position had been influenced by researchers and community activists, explored the idea that port activities should be considered a single "stationary source," similar to how oil refineries and or other industrial facilities were regulated. This resistance highlighted the dilemma of goods movement: it could expand by crossing oceans and borders without economic or political barriers, and yet it was generating resistance that questioned the value and purpose of globalization itself. As the traffic of goods transformed regional, national, and global economies and intensified the development of global cities, the identity of the region was at stake.[45]

Transformed Region, Changing Country

In December 1979, an unusual meeting was held at the downtown offices of the Los Angeles Chamber of Commerce. Members of the United Neighborhood Organization (UNO), an East Los Angeles–based, predominantly Latino group, met with sixteen members of the Community Committee (formerly known as the Committee of 25), a group of L.A.'s major business figures that included department store head Edward Carter, Union Oil's Fred Hartley, and MCA chair Lew Wasserman. UNO, an affiliate of the Industrial Affairs Foundation, founded by Saul Alinsky, often confronted powerful bodies like the Community Committee to demand increased jobs, living wages, more funding and improved conditions for public schools, and affordable housing for the communities it sought to mobilize for action. The Community Committee members, who had long-standing reputations as behind-the-scene power brokers, were, it would soon become apparent, no longer the dominant players in the region. They had been eclipsed by other global-city forces who focused on a changing global economy that had less connection to or effect on local and regional issues.

When pressed at the meeting about conditions in East L.A., the Community Committee members tried to communicate interest and concern. "I guess I never focused on East Los Angeles," Carter commented at one point, noting that he had been vaguely aware that the Latino population in the region had increased significantly. "Do you have movie theaters in East L.A.?" Lew Wasserman inquired, while a phone company executive stated, "We've tried to be helpful." But tension in the room suddenly increased when Union Oil's Hartley wondered what kinds of jobs Mexicans held and, if there were any rich Mexicans, "Where would they live?" The oil executive then warned that an explosive "Quebec situation" was unfolding in Los Angeles, with two divergent languages and cultures. "This has to be stopped," Hartley argued.[46]

In some ways, the 1979 meeting symbolized two kinds of forces passing each other within a changing Los Angeles. In 1979, few Latinos were represented among elected officials or business figures of influence in the region, and none appeared in groups like the Community Committee or its predecessor, the Committee of 25. An earlier generation of Latino activists had focused on school and neighborhood issues and less on electoral politics, although the election of city councilman and later congressional representative Edward Roybal from East Los Angeles had galvanized Latino activists and provided them with their first taste of potential political power. During the 1960s, the rise of the United Farmworkers (UFW) union (led by César Chávez), the 1968 student walkouts at several predominantly Latino high schools in Los Angeles, and the 25,000 demonstrators at the 1970 Chicano moratorium who protested the Vietnam War and the war's high proportion of Latino and African American casualties, mobilized activists on community issues, cultural identity, and movement rather than on electoral politics.

The immigration issue continued to hover over the Latino community in Los Angeles and in other southwestern Latino communities and had a potential to marginalize all Latinos as outsiders. The response to immigration had nevertheless created a political divide among Latino groups and activists. On the one hand, groups like the UFW had opposed the bracero program and were concerned that undocumented labor had become a device by growers (or hotel and restaurant or janitorial service employers) to establish a low-wage workforce and undermine efforts at unionization. On the other hand, key figures like Bert Corona, a former union organizer and long-time community activist, identified the immigrant-rights issue and its role in ethnic marginalization as a crucial cause for Latinos and claimed that the demand for legalization and immigrant-rights protections needed to become the community's core civil rights issue.[47]

During the 1980s, new efforts to elect Latino candidates were more successful as particular enclaves established large Latino majority districts. Competing Latino political machines emerged during this period, and a potential divide between African American and Latino political interests reflected the changing demographics in areas like South Los Angeles, where once predominantly African American neighborhoods (and electoral districts) became partly (and in some cases, majority) Latino. This electoral dimension of Latino politics in some ways represented a type of classic ethnic route to political power rather than a demonstration of a new type of equity-based political agenda that cut across ethnic and racial lines. While generally liberal in orientation and still responsive to specific community concerns, some of the leading Latino politicians in this period saw themselves as establishing a political fiefdom—a more progressive *cacique* as opposed to a role as community organizer and mobilizer of new constituencies.

The 1970s and 1980s in Los Angeles also reflected a political "feel-good moment" that was associated with the 1973 election of Tom Bradley as mayor of Los Angeles (and his subsequent reelection during the next twenty years prior to the era of term limits). At the time of his election as the first black mayor of a city without a majority black population, Bradley created a cross-class and cross-ethnic majority predominantly of low-income blacks in South L.A. and middle-class Jews from the Westside. A dignified and stately figure, Bradley also became an important and welcome symbol of a city that was becoming increasingly diverse and multiethnic. The city's successful hosting of the summer Olympics in 1984 without any major embarrassments (such as massive traffic jams) represented the high point of this feel-good era. At the same time, Bradley encouraged the rise of L.A. as a global city. He welcomed new international investment, supported the unencumbered growth of the port and related transportation corridors, but was unable to prevent or even soften the deindustrialization forces that helped precipitate the low-wage, two-tier economy that emerged full blown at the end of the Bradley era in the early 1990s.[48]

The 1992 civil disorder in Los Angeles took many political and civic leaders by surprise and eventually reflected the collapse of the feel-good era and political alignments that had held for nearly twenty years. The 1993 election of Republican businessman Richard Riordan as mayor precipitated a further reconfiguring of the political landscape and also exposed the economic and cultural divides that had been building during much of the 1980s. Immigration once again emerged as a contentious and volatile issue, culminating in the passage of California Proposition 187 in 1994. This measure sought to prevent access for illegal immigrants to social services such as public schools, public health care, and food stamps. Proposition 187 became the trigger for a demonstration of nearly 100,000 Latino youth and adults—documented, citizen, and undocumented—who waved flags of the United States and Mexico. This event highlighted emerging and complex Latino politics related to immigration issues and the changing political landscape in Los Angeles.[49]

The second half of the 1990s and first years of the 21st century have witnessed the rise of new political alliances and social movements in Los Angeles, even as traditional business and political elites have become more diffuse and less potent and as cultural and demographic changes have reshaped the neighborhoods and various aspects of daily life in the region. In response to these changes, the labor movement in Los Angeles has undertaken what amounted to a seismic shift by focusing on the low-wage, largely immigrant workforce in the service sectors and even in parts of the construction trades and other industries. Led by such unions as the Hotel Employees and Restaurant Employees Union, the Service Employees International Union, and the United Food and Commercial Workers Union, by 2000 the Los Angeles County Federation of Labor had, for the first time in its history, become an active champion of the rights of undocumented immigrants, including their demands for legalization. This approach also facilitated a reversal in the position of the national AFL-CIO.

Two events that year identified how the change in the labor movement also pointed to the new kinds of coalitions beginning to emerge around the immigration issue. An April 2000 strike by the janitors' Local 1877 of the Service Employees International Union under the banner of the Justice for Janitors campaign, made visible and dramatic the realities of the low-wage sector. The events around the strike were capped by a boisterous march of the janitors from downtown Los Angeles, through the mid-Wilshire corridor, and into Beverly Hills and Century City that brought people from the high rises into the streets to flash victory signs and even hand dollar bills to the chanting though startled janitors and their supporters. For the janitors, the strike "was about respect" and "showed we're part of a bigger movement," as one striker, Jesús Pérez, told the *Los Angeles Times*. Two months later, a Town Hall meeting of more than twenty thousand people that included participants from the County Federation of Labor, the Catholic Church, and immigrant-rights group far exceeded expectations in the turnout and furthered the notion that a new kind of social and political alliance had emerged.[50]

Politically, this Latino-labor alliance, as it came to be called, and multiple economic-, social-, and environmental-justice groups, including the immigrant-rights activists, coalesced in 2001 around the candidacy of Antonio Villaraigosa for mayor of Los Angeles. A second-generation Latino from the eastside who was a former union organizer, student activist, and member of the State Assembly, Villaraigosa became the speaker of the State Assembly before leaving office due to term limits. In the 2001 mayoral election, Villaraigosa ran a spirited campaign that was part movement building and part a traditional electoral effort that relied significantly on labor's increasingly powerful political role in the region. Though he came in first in the primary, Villaraigosa lost the runoff against a moderate Democrat, James Hahn, who engaged in a successful negative campaign that characterized Villaraigosa as supportive of drug dealers. Hahn was also able to fashion an unusual coalition of conservative white suburbanites in the San Fernando Valley with African American voters who remembered fondly Hahn's father, a county supervisor who had represented South Los Angeles and had helped nurture a number of future black office holders.[51]

Despite the loss in 2001, the constellation of forces that had come together to support Villaraigosa continued to play a role in the reconfiguration of the L.A. political landscape. Even the events of September 11, 2001, and the immigration fears that it triggered in the country didn't directly undermine the political directions and cultural changes taking place in the Los Angeles region. In a 2005 rematch between Villaraigosa and Hahn, Villaraigosa this time emerged the victor by a landslide, helped by his ability to establish a broad coalition that crossed ethnic and racial lines, even as he also became a symbol of the emerging Latino majority in the city. Both during the campaign and after his election, Villaraigosa continued to emphasize the power and value of Los Angeles as a diverse city and a city of differences that needed to be appreciated rather than feared. At the same time, he would often appear in the Spanish-language press and media and sought to make visible the contributions of immigrants and the low-wage service jobs they filled. At events in hotels, he recognized the workers serving the meals and also, on many occasions, recognized those "who clean our toilets" and others who undertook work in the least attractive and lowest-paying jobs at the heart of the low-wage service sector. Villaraigosa also became a target for some talk radio personalities who urged their listeners to send toilet bowl brushes to City Hall, where, according to one Villaraigosa deputy, thousands of toilet brushes arrived in the mail addressed to the new mayor.[52]

The immigration issue had not surfaced during the 2005 mayoral election, itself an indication of how much Los Angeles had changed even as immigration rhetoric intensified around the country. In 2005 and 2006, complaints about the invasion of illegals from Mexico extended throughout the country to places where there were few immigrants and to states like Arkansas, Georgia, Colorado, Nebraska, and Kansas, where various industries like meatpacking and poultry operations employed low-wage immigrant workers. The increased militarization of the border, which had been earlier stimulated by the Clinton administration's "Operation Gatekeeper" program of fence building and heightened security, was adopted by vigilante groups like the Minutemen in 2005 and 2006. During this period, legislation was introduced in a number of communities, such as Costa Mesa in Orange County and the city of San Bernardino in the Inland Empire at the eastern edge of the Los Angeles region, that paralleled legislation passed by the U.S. House of Representatives. This had been introduced by Wisconsin Republican and advocate James Sensenbrenner.[53]

Just as California Proposition 187 did in 1994, the Sensenbrenner bill, which was passed by the House in December 2005, stimulated a rapid and passionate mobilization of residents and citizens in Los Angeles and throughout the country. The immigration demonstrations that took place between March and May 2006 drew families and the young and the old to the streets. Relying on the Internet, word-of-mouth communication, Spanish-language radio, and the informal Latino labor and social movement alliances that had developed in Los Angeles and had begun to spread to other regions in the country, the demonstrations suggested a potential paradigm shift regarding the country's political and cultural make-up, even as hostility to illegal immigrants reached new heights. Proclamations of the dignity and value of immigrants and their community building role sought to counter the inflammatory rhetoric, actions, and policies of the anti-immigrants. This conflict was pivotal in how community—and a sense of place—could be defined in the global city. Los Angeles was changing and the rest of the country was beginning to experience the same kinds of changes, conflicts, and tensions that had occupied much of the history of the region and its relationship to the problem of "racial indigestion" and the social fiction and cruel realities of the border.

8

Social Change across Borders

The Hispanic World did not come to the United States. The United States came to the Hispanic world. It is perhaps an act of poetic justice that now the Hispanic world should return.

—*Carlos Fuentes*[1]

Unlike past immigrant groups, Mexicans and other Latinos have not assimilated into mainstream U.S. culture, forming instead their own political and linguistic enclaves—from Los Angeles to Miami—and rejecting the Anglo-Protestant values that built the American dream.

—*Samuel P. Huntington*[2]

Celebrating Immigrant Farmers: A 9/11 Story

In September 2001, National Public Radio was preparing to run a story on American Airlines pilot John Ogonowski and his connection to immigrant Hmong farmers in Massachusetts. Ogonowski had become involved with the New Entry Sustainable Farming Project, a program designed to provide recent immigrants with the opportunity to use their skills to farm, when the local Farm Service Agency in Westford, Massachusetts, contacted him, looking for land to make available for the Southeast Asian families living in nearby Lowell. Ogonowski, who loved to garden himself, embraced the idea and made the land behind his home available to the farmers. The area was given the name White Gate Farm, and it became the immigrant farming project's first "mentor farm" and training site for new immigrant farmers.[3]

Through the White Gate Farm project, Ogonowski emerged as a key local-food and immigrant-farming advocate. He was a founder and active member of the Dracut land trust that helped to preserve a substantial amount of local farmland in the town from development. In 2001, the trust negotiated the purchase of about fifty acres of land about a mile from Ogonowski's house. With some of his own land not available due to a major gas pipeline installation, he also made the trust land available to the immigrant farming project, and about a dozen Cambodian families began to farm the land that same year.

The project was one of the first to identify an important new phenomenon in agriculture—the increasing numbers of immigrant farmers, primarily Latino and Asian, who had recently arrived in the United States. NPR was one of the first media outlets to pick up on the trend, and in August 2001 Ogonowski was interviewed by Susan Shepherd of NPR's *Living on Earth* program about the immigrant farming community and the work of the mentor farm. Ogonowski spoke of his own love of farming and how he had become involved through his desire to offer an opportunity to others who also loved agriculture. "I think once a person is a farmer, they're a farmer for life. They're hooked," Ogonowski told Shepherd. "I don't know if the children are going to be active in it, but they may be. The Cambodians bring their whole families out here. You'll see whole families weeding, so they may take a liking to it." Ogonowski praised the hard work of the Cambodian families and told how much it had meant to him to be able to offer them this opportunity. "I have three daughters, and they're good workers," Ogonowski said of his own family. "They pick blueberries and pumpkins, and hopefully they'll continue so I can retire."

Shortly before the interview was to air on NPR, John Ogonowski was piloting American Airlines Flight 11 from Boston to Los Angeles when it was hijacked and then crashed into the World Trade Center. NPR ran its interview with Ogonowski a few weeks later, but the mood after 9/11 did not lend itself to a story celebrating the work of immigrants, and it wasn't picked up by other media outlets. Instead, the airwaves and the public discourse were dominated by a suspicion of immigrants, who were lumped together as a potential menace to communities and to the nation.

In the years that followed, I often recounted John Ogonowski's story and his regard for the immigrant farmers, particularly in talks that contrasted the dominant, global-oriented food system with alternative pathways, including a regionally based community food approach. Our work, through the Center for Food and Justice at Occidental College's Urban and Environmental Policy Institute, had begun to identify an immigrant-farmer-to-immigrant-consumer pathway through farmers' markets, ethnic food markets, innovative programs like farm to school in immigrant neighborhoods, and an emerging fair-trade food pathway between immigrants and their countries of origin. More recently, when I began to research Los Angeles as an immigrant-based global city and the effects of a globalizing economy on communities, including food and the culture of food, I realized that John Ogonowski's celebration of immigrant contributions and their love of farming still remained compelling for me. His story contrasted with a post-9/11 climate of hostility and a desire to erect even greater barriers. It spoke of how the immigrant experience could provide new forms of community and why differences should be welcomed rather than feared.

Recreating Community

Los Angeles, a city of immigrants, finds itself once again in conflict over how its newest migrants are redefining and recreating new forms of community within this global city. Community identity based on geographic location has long been a shifting target in Los Angeles, as its residents have continually experienced displacement and dispersal. The Anglo settlers of the mid-nineteenth century waged assaults against the homes and neighborhoods of Mexican residents. Communities in or near downtown Los Angeles—Chinatown, Bunker Hill, and Sonoratown—have been uprooted. During the 1950s, the Chavez Ravine neighborhood made way for the development of Dodger Stadium. This uprooting of neighborhoods in the urban core has often been defined as a form of community betterment that overcomes conditions of blight. The descriptions of the "rubbish and filth" of the congested, heterogeneous immigrant districts inhabited by Mexicans, Chinese, Filipinos, or Russians established a type of social distance between areas based on race and ethnicity—a borderland within the city, as Greg Hise put it. This contrasted with the language of newness and homogeneity associated with the development of the less dense Anglo residential districts, particularly the automobile suburbs, with their version of a purer and cleaner built environment that was connected to a suburban nature.[4]

During the 1950s and 1960s, many of the urban neighborhoods identified as blighted were impacted by the location of the new interurban system of freeways whose routes were often selected to pass through the poorest and politically least powerful neighborhoods. Key freeways were rerouted or never built when white middle- or upper-class areas such as Beverly Hills resisted. In contrast to the neighborhood-impacting urban routes, freeway routes to the automobile suburbs passed through open-space areas and agricultural lands, helping develop and expand the new suburbs as self-contained, homogeneous environments whose link to other parts of Los Angeles began and ended with the drive on the freeway. By contributing to sprawl development and slicing through denser urban core neighborhoods, freeways undermined any sense of place or community identity in the neighborhoods that they passed through. The vast patchwork of freeways, including those that crossed each other in places such as the City of Commerce or Boyle Heights, established new neighborhood boundaries. Along with the Los Angeles River, which also served as a boundary point between east and west, Latino and Anglo, black and white, and other urban divides that fragmented the region, freeways came to represent a geography of division and dislocation.[5]

For the real estate agents and developers who sought to reshape such urban neighborhoods, the notion of blight has also been associated with heterogeneity, often a code word for differences based on race, ethnicity, and class. These code words for difference and immigrant status, rather than civic betterment, have in fact been central to the process of geographic dispersal and settlement in Los Angeles as well as other urban areas around the country. Like other Sunbelt cities such as Atlanta, Houston, and Phoenix, Los Angeles' demographic and class divide has been compounded by the region's sprawling suburban character and presumed lack of intact and stable dense urban neighborhoods.

Today, however, several neighborhoods in Central and East Los Angeles have become more diverse through distinctive and even denser immigrant settlements. That process of dispersal, resettlement, and urbanization has extended to some of the automobile suburbs and working-class suburban-industrial clusters that have also experienced major demographic changes. These changes have in turn reconfigured the character of such areas as the heavily Latino suburbs in the northern and eastern San Fernando Valley, the Asian and Latino suburban bedroom communities in the San Gabriel Valley, and pockets of immigrant neighborhoods throughout suburban Orange County. It also demographically transformed the suburban-industrial corridor in southeast Los Angeles in cities like Huntington Park and South Gate that are heavily Latino immigrant enclaves and among the densest areas in the state. As these changes have intensified in the past couple of decades, Los Angeles could no longer simply be characterized as a racially and ethnically segregated urban core with white middle-class outer-ring suburbs. Instead, the region has come to represent a constellation of diverse, dispersed communities, many of them denser than the typical notion of the spread-out suburb. Moreover, these areas, particularly in the immigrant enclaves, have developed a sense of place that is more urban than suburban and a notion of community that Saskia Sassen has called "a networked politics of place" that is at once local and global.[6]

Mike Davis has called this process of urban reinvention "magical urbanism," referring to the development of new public spaces and connections to places that have been redefined by their newly emerging majority Asian and Latino populations. "Where there is much abstract talk in planning and architectural schools about the need to 're-urbanize' American cities," Davis wrote in 2000, "there is little recognition that Latino and Asian immigrants are already doing so on an epic scale." Immigration, Robert Fishman argued along the same lines, is at the heart of this contemporary reurbanization that Fishman calls a "fifth migration," taking off from Lewis Mumford's concept of the fourth migration from the cities to the suburbs. Community in these reurbanized neighborhoods is "bounded less by local history and city limits than by a regionally and internationally based support network of family, school, and business associations," John Horton pointed out in his 1995 study of the development of the Chinese immigrant neighborhoods in the city of Monterey Park in the San Gabriel Valley. These linked networks in turn have helped remake the built environment in the city—including the commercial corridors, the use of public space and urban landscapes, and the front and outward-facing homes with porches and front yards in Latino neighborhoods. This "enacted environment," as urban planner James Rojas has characterized the Latino immigrant effect on urban neighborhoods, also includes street vendors, plazalike intersections, public art such as murals on the walls adjoining stores, and shrines on streets and in front of homes. "Every change, no matter how small, has meaning and purpose," Rojas argues. "Bringing the sofa out to the front porch, stuccoing over the clapboard, painting the house vivid colors, or placing a statue of the Virgin in the front yard all reflect the struggles, triumphs, and everyday habits of working-class Latinos." This new language of the built environment, Rojas suggests, represents a hybrid city and culture combining Latin American and U.S. urban forms.[7]

Though Los Angeles has come to represent the heart of this immigrant-based transformation of the urban built environment, the immigrant diaspora has extended to small, midsized, and large urban areas throughout the country. James Goodno, in a 2005 article in *Planning*, describes how the Washington,

D.C., suburb of Langley Park evolved from a Jewish suburban community within Prince George County to a largely African American inner-ring suburb. Asian immigrants subsequently used the area as an entry point, followed by Latino immigrants, who now constitute nearly two-thirds of the population of Langley Park. As the area's residents have changed, so have its public spaces and commercial districts, with Goodno describing one such complex near the Wheaton Metro station consisting of "Salvadoran chicken joints, Thai and Filipino groceries, Vietnamese nail parlors, Mexican and Chinese restaurants, and a Korean supermarket."[8]

The 1995 Chicago heat wave is also instructive in how high-density street life establishes mechanisms of community support. Eric Klinenberg, a sociologist who wrote about the heat wave, describes how Latinos, who constituted 25 percent of Chicago's population, accounted for only 2 percent of the heat-wave deaths. "Many Chicagoans attributed the disparate death patterns to the ethnic differences among blacks, Latinos, and whites—and local experts made much of the purported Latino 'family values,'" Klinenberg recalled his research. "But there's a social and spatial context that makes close family ties possible," he argued. "Chicago's Latinos tend to live in neighborhoods with high population density, busy commercial life in the streets, and vibrant public spaces. Most of the African American neighborhoods with high heat-wave death rates had been abandoned—by employers, stores, and residents—in recent decades. The social ecology of abandonment, dispersion, and decay makes systems of social support exceedingly difficult to sustain."[9]

The key to these transformed urban spaces remains the street. "Street life is an integral part of the community fabric because [streets] bring people together [and] serve as plazas by creating a real sense of place in Latino neighborhoods," Goodno quotes Rojas. Street life has long been associated with the immigrant experience, particularly in the late nineteenth century (such as New York's Lower East Side or the nineteenth ward in Chicago, where Hull House was located). The immigrant impact on urban life, Horace Kallen wrote in 1915, changed the very nature of what it meant to be an American, arguing that instead of repressing the immigrant association of nationality, "America has liberated nationality." "What troubles the Anglo-Saxon Americans is *difference*," Kallen said of the nativists who were asserting that America needed to insulate itself from any immigrant influence.[10]

But the immigrant narrative has also long been associated with adaptation to the American vernacular, and street life was transformed into the mall experience or reimagined in Disneyland's fantasy of Main Street, a built environment that has to a certain extent been mimicked by the new urbanist reconstructions of safe and well-kept small-town streetscapes. During the 1950s, Los Angeles was seen as a place that had rid itself of difference by homogenizing the immigrant past (for example, in the celebrations of the fiesta and the creation of commercial districts like Olvera Street) and creating a purely "American" version of the urban and suburban experience, as the 1960 *Los Angeles Times* series of articles suggested. But by the twenty-first century, Los Angeles could not escape its association with the immigrant experience and secured a reputation, along with New York, as the most diverse urban region in the country.[11]

The current immigrant remaking of street life and public spaces in places like Los Angeles underlines how the current immigrant experience has yet to be transformed into an assimilationist narrative. Describing El Nuevo Mundo, a 2001 exhibit of urban landscape portraits by Chilean photographer Camilo José Vergara, anthropologist Susan Phillips commented that to her surprise, "Latino cultures manage to persist, to ward off what I thought would be the overwhelming influence of North American culture. I was amazed to find that symbols and colors from south of the border dominate large sections of Los Angeles County. It was a mystery because it led me to uncover the strengths of what I regarded as the weaker culture. I had assumed that the Latino presence in Southern California would be reasserted for a few years only, after which signs of assimilation would be widespread. Instead we have witnessed Latino cultures going from strength to strength, with no signs of erasure."[12]

Despite this reconstruction of urban space and the related rebirth of urban community, the immigrant experience has also included repression and exploitation, displacement and fear of deportation, and the absence of basic human rights. Urban immigrants face the pressures of the real estate market to gentrify and displace, housing codes that restrict the number of occupants, prohibitions against street vendors or day-laborer activities, expensive housing and medical care, low wages, and exploitative working conditions. New urbanist Andrés Duany warns against the dangers of "the aestheticization of this 'barrio urbanism'" that serves to "romanticize poverty." Duany also criticizes those Latino new urbanists for their failure to recognize that, "if given the choice, poor immigrants would rather have their own car" than be forced to use public transit. Immigrant communities might mimic some of the community-generating goals of the new urbanists, Duany argues, but not out of choice.[13]

But contesting and overcoming the economic and social barriers that immigrants face has also been central to the arguments about the value of the immigrant culture and its contributions to civic life. While the Latino new urbanists have identified the value of what Albert Camarillo has called "barriorization" (referring to the changing ethnic communities that are able to establish "dynamic and expanding social spaces"), they have also (along with other social-justice advocates, such as immigrant-based unions, Catholic Church immigrant congregations, and community-based organizations) focused on the conditions of daily life that have made immigrants, particularly Latino immigrants with their uncertain status, the most vulnerable group in the country. The social-justice demands for affordable housing, a living wage, and the right to quality education and health care have complemented the Latino new urbanist advocacy for more parks, open space, and recreational places, as well as an accessible, low-cost transportation system.[14]

Many of the immigrant and ethnic communities have become new types of battlegrounds regarding the search for and re-creation of community. The Boyle Heights neighborhood, for example, with its rich multicultural historical narrative, illustrates how strong community identities can be found beside the deteriorating social and economic conditions that affect many of Los Angeles's low-income and immigrant-based neighborhoods. In Boyle Heights, pedestrian-oriented and active street life coincides with the low-income status of its residents, its deteriorating housing stock and lack of affordable housing, and the environmental hazards its residents experience due to its intersecting freeways. With its attractive location just to the east of downtown Los Angeles across the Los Angeles River, Boyle Heights has also become a coveted area as a potential new opportunity for a real estate and commercial development-driven push for gentrification. Similarly, the Pico-Union, MacArthur Park, and Koreatown neighborhoods to the west of downtown Los Angeles also represent diverse cultures and immigrant experiences that exist side by side with homelessness, crime, poor housing, and overcrowded schools. Both Boyle Heights and the Pico Union/Koreatown areas, targets for gentrification and displacement, still provide the "networked politics of place" that can help establish the type of civil society and community economic development initiatives, from hometown clubs to microenterprises, that are able to address neighborhood issues while strengthening community identities.[15]

The Latino new urbanist appeal for "*gentification* rather than gentrification," for a new hybrid city and culture that builds on immigrant, ethnic, and class experiences and resources, contrasts with the new urbanist tendency to promote its concept of community as the self-contained reconstructions of the small-town nineteenth-century American experience. While new urbanist planned communities, despite their rhetoric of inclusion and diversity, have become surrogates for gentrification, the Latino new urbanists look to strengthen existing diverse communities *within* the urban core and to link community building with social- and environmental-justice goals. At the heart of the debate over community is the question of urban daily life, including its spatial dimensions. The different urbanist and community-building perspectives also reflect differences over the meaning of place, and, by extension, the question of the right to the city, even as the boundaries of both city and nation are themselves transformed.

Connections to the Land: Food and Immigrants

In 1997, the Urban and Environmental Policy Institute's Center for Food and Justice (then called the Community Food Security Project) facilitated the development of an innovative "farm-to-school" program at McKinley Elementary School in the Santa Monica–Malibu School District (SMMUSD), which my son and daughter had attended. Though one of the wealthier school districts in Southern California, it also included several schools, such as McKinley, with a high percentage of low-income families, as indicated by the number of students who qualified for free or reduced-price lunch meals. In 1997, about 40 percent of McKinley's students were in this group, and as many as forty-five nationalities were represented among its 450 students. Latinos were then (and continue to be) the largest group, consisting of about 50 percent of the student body. In my daughter's kindergarten class in the late 1980s, thirty-eight students spoke a language other than English at home: fifteen spoke Spanish, and the remaining twenty-three children spoke twelve other languages. This "ethnic flowering," I wrote in an opinion piece for the *Los Angeles Times* that year, "does justice to the concept of Greater Los Angeles as a 'world city.' "[16]

While McKinley's principal, the district's food-service director, and the SMMUSD superintendent were all willing to try the farm-to-school pilot program, they were also skeptical that students would be willing to substitute fruits and vegetables for the standard cafeteria meal that often sought to mimic the fast food that students encountered outside school. Farm to school—local and regional farmers selling fresh fruits and vegetables directly to the school district for school lunches—was a new idea at the time. The previous year, the U.S. Department of Agriculture had experimented with a program to help support minority growers, particularly African American farmers in a couple of southern states, by facilitating the purchase of a couple of crop items such as collards recognizable to the students as part of the family diet in schools with large minority populations. The program was designed to increase the revenue stream of the regional farmers, otherwise buffeted by trends in agriculture that included the federal government's disproportionate support of commodity crops like corn, soybeans, and rice, the post-NAFTA and World Trade Organization–encouraged growth of export agriculture, and a fast-food culture that homogenized diets and expanded waist lines. Our farm-to-school concept sought to extend the USDA approach to establish a community-food system framework from farm to table or, in this case, from farm to the school food-service organization that prepared the meals for the schools and their students.[17]

The McKinley Elementary School pilot had the advantage of drawing on a well organized and expansive system of farmers' markets that was run by the City of Santa Monica. The school district could purchase from several different farmers who brought produce to one of the markets, allowing a variety of items to be selected. The city transported the purchases from the market site to school storage facilities, where one-, two-, or three-day-old produce (deliveries were initially twice a week) could be presented in a full salad bar. The program was called "the farmers' market fresh fruit and salad bar" and was offered (with some additional items, such as milk, bread, and a protein option such as tuna salad) as an alternative to the "hot meal." During an experimental run in the summer, the hot meal choice of pizza was offered with the farmers' market salad bar, and, to the astonishment of the school officials, more than 75 percent of the students selected the salad bar.[18]

This auspicious start became a sustained success story in its first year thanks to the work of Lucia Sanchez, a young staff member at UEPI's Center for Food and Justice, who worked closely with the school district's food-service director, Rodney Taylor, a long-time school food-industry figure who became a convert to the farm-to-school approach as he witnessed its popularity among the students and as he developed ties with the farmers who provided the produce. Sanchez was also able to work closely with the McKinley school community, particularly the Latino students and parents, who had established their own networks. When several of the Latino students and their parents questioned how the salad bar was arranged, particularly the absence of lemon or lime slices, Lucia helped shape the salad bar as not only "fresh from the farm" but connected to what some analysts called "the regional diet" or, its extension, an "immigrant diet" based on local and regionally available foods. When the parents and the school food-service staff began to connect with the farmers, they also discovered that several of these farmers were also immigrants. At one of the Santa Monica markets as many as half of the vendors were immigrants, as research by one of my students subsequently indicated.[19]

The farm-to-school experience at McKinley and similar experiences involving other community-food programs around the country suggested an important dimension to the community-food systems approach—the rich connections between immigrants and food. This is perhaps most visible with changing food choices that have been based on immigrant and ethnic cuisines available in both restaurants and markets and that are often presented as a positive aspect of immigrant influences in the United States. Mexican food, like Italian and Chinese food before it, is now ubiquitous, whether in its fast-food or neighborhood-restaurant form or in the items now stocked on grocery shelves around the country. Food and immigrant connections are also associated with the multiplicity of ethnic restaurants in places like Los Angeles, New York, Chicago, and San Francisco that promote such food diversity as part of what makes them attractive *urban* places.

But immigrants, particularly various Latino and Asian groups, have begun to be seen as contributing in an integral way through their connections to food growing and food culture. Up to now, community-food system advocates have highlighted the value of local and regional farming and food processing, the importance of fresh food (particularly for food-insecure communities), and the environmental and social practices and values associated with food locally grown and with strong culture- or place-based associations. Community-food advocates have been sharply critical of a shift towards a "global food system" that grows and manufactures food that travels long distances from seed to table; that transforms diets through a fast-food culture that creates a uniform taste and contributes to the worldwide epidemic of weight gain (even in low-income groups); and that has produced huge food conglomerates engaged in transnational food operations in the growing, manufacturing, and distribution of highly processed (and, more recently, genetically modified) food products. In response to these trends, community-food advocates have also been an important part of the antiglobalization movement and participated in protest demonstrations at globalization gatherings, such as the 1999 protests in Seattle during the meetings of the World Trade Organization.[20]

However, only recently have community-food advocates begun to recognize and emphasize the importance of ethnic and immigrant experiences. In the early 1990s, as the community-food movement began to develop in the United States, the one direct connection along these lines included efforts to identify the value of "culturally appropriate" food, a criterion included in definitions of what constituted community-food security for various groups and communities. By 2001, John Ogonowski's experience in Massachusetts with the Hmong gardeners was one of the first illustrations of how this cultural or ethnic connection could also be translated into "connection to the land."[21]

The United States Department of Agriculture's 2002 Census of Agriculture, in fact, indicated that Latinos and other minorities were the only groups that had experienced an increase in the number of farmers and the amount of land being farmed in the five years since the last census had been taken. This contrasted with an overall loss of farms and farm land in the United States as farms gave way to suburban and exurban developments and long-distance imported food captured an increasing share of the U.S. fresh-food market. In addition, the size of U.S. farms increased, even as the number of farms decreased, an indication of the land concentration and increasing dominance of the commodity groups in U.S. agriculture. In contrast to these national trends, many of the minority, ethnic, or immigrant farmers were small regional farmers who relied increasingly on local markets, such as farmers' markets or ethnic stores. For example, among Latino farmers, more than 70 percent of their farms were smaller than 50 acres, while more than 85 percent had farms smaller than 180 acres. The increase in the number of Latino farmers—more than 50 percent between 1997 and 2002—was especially striking.[22]

As these trends have become increasingly apparent, community-food activists have begun to establish new food and immigration advocacy networks, such as the program that John Ogonowski joined. Several of these groups have come together under the umbrella of the National Immigrant

Farming Initiative (NIFI), a network of immigrant farm groups from around the country that was established in 2003. NIFI-affiliated groups include projects in Maine that work with immigrant farmers from Mexico, Central America, Peru, Somalia, Sudan, Togo, and Cambodia; a group in Maryland that includes both African and Asian farmers who grow such vegetables as amaranth, anchthea, njamma-njamma, and ndole; and the Salinas, California–based Agriculture and Land-Based Training Association (ALBA), which has emerged out of one of the oldest and most prominent of the Latino immigrant farming groups, the Rural Development Center (RDC). ALBA/RDC provides training and technical assistance, including organic farming techniques, for these "new American farmers," as several of the groups characterize their participants. ALBA operates two farms in Salinas and northern Monterey County that also establish a route to farming for former farmworkers who lease land on the ALBA site while learning how to grow organically and engage in other environmental practices, such as soil, water, and habitat conservation. ALBA also has established an innovative marketing approach, notably through a farm-to-college and farm-to-hospital arrangement at the University of California at Santa Cruz and at a nearby hospital. ALBA farmers, as well as many of the immigrant farmers associated with the various NIFI groups, have helped expand the number of participating farmers among urban farmers' markets while simultaneously broadening the base of shoppers attracted to the produce they were now able to find at these markets. In addition, some of the most interesting changes regarding urban farm activity involve innovative efforts by immigrant farmers (many former farmworkers or farm managers) to establish very small plots within the urban core, including on abandoned land and former industrial sites. Research by the Center for Food and Justice in fact identified upward of 40 to 50 percent immigrant vendors at several of the farmers' market sites in Southern California. Recognizing this shift in the demographics of participating farmers at farmers' markets, a key advocacy group, the Project for Public Spaces, has sought to establish an "immigrant farmer" dimension to its support of farmers' markets as valuable public places and diverse sources of fresh food.[23]

For the immigrant farmers, farming has been a part of their history, a vocation as much as a source of work or income. "I can't imagine a better quality of life for my family. I am a farmer. That is what I have done, and I need to stay connected to the land," a Colombian farmer in New York told his NIFI supporters. For those who do not have access to farm land, the community garden in urban areas has emerged as a parallel type of immigrant activity, complementing and extending the notion of the immigrant farmer as caretaker of the land.

One of the more striking examples of this caretaker function involved the initiative of a Dominican Republic immigrant named Daniel Pérez. In 1991, Pérez decided to create a neighborhood food source and "beautify," as he put it, the median strip dividing the traffic flowing on Broadway between 153rd and 154th streets in Manhattan in an immigrant neighborhood situated between Harlem and Washington Heights. Pérez, who grew up in a small farming village in the Dominican Republic, cleared the land that had been filled with old newspapers, garbage, and weeds and planted enough corn seeds to grow 131 corn stalks (some as high as six feet). He also planted some black beans that he had purchased from a Korean grocer. "'I planted with the idea that this is my own little contribution, my own little Cibao,'" Pérez told *New York Times* reporter Steven Lee Meyers, referring to the farming region in the Dominican Republic. As the corn and beans grew, neighbors came to the plot with grocery bags to take home the food and then cooked the food in various Dominican styles. "'They all tasted great to me,'" Pérez happily reported.[24]

Pérez's love of farming and immigrant cuisine contributed to his desire to transform an unlikely area in the urban landscape into fertile land to grow food and establish a new type of community and public space. His action, however unusual, can be considered more representative of a growing trend toward the development of "immigrant gardens" rather than an isolated, idiosyncratic action by an enterprising immigrant. Throughout the 1980s and 1990s and into the new century, a number of community gardens were either initiated or expanded by Latino, Asian, African, and Caribbean immigrants, among others, in urban places around the country. Immigrant day laborers in

Los Angeles and elsewhere, for example, established small garden plots in areas where the hiring took place. These new immigrant gardens helped reinforce and expand the community-garden movement that had first been renewed during the 1970s but that has found itself subject to the shifting whims of private developers, limited support from local government, and the absence of any public policy to support urban agriculture and these types of community spaces. It has also become a part of an urban-revitalization agenda and new politics of place associated with the immigrant-food connection.

THE BATTLE OVER SOUTH CENTRAL FARM

The promise of the immigrant food connection and the realities of urban land use came directly into play in one of the most contentious battles regarding Los Angeles' largest community garden, at a site in South Los Angeles adjacent to the Alameda corridor where railroads carried their freight from the Ports of Los Angeles and Long Beach. In 1986, this fourteen-acre site in Central Los Angeles had been taken over by the city of Los Angeles by eminent domain at a cost of nearly $5 million to construct one of three solid-waste incinerators to relieve capacity problems at its landfill sites. This Los Angeles City Energy Recovery project (known as LANCER), was touted by its backers as the "twenty-first-century solid-waste management solution." A group of UCLA graduate students whom I supervised analyzed the project's potential environmental and health impacts, its land-use issues, and the environmental-justice concerns associated with the project (it was located in a low-income community of color, and opposition to such a facility was assumed to be minimal). The UCLA report was released at a point when community opposition to LANCER had unexpectedly grown considerably, which led Mayor Tom Bradley to withdraw the incinerator project. The struggle against the LANCER project became one of the leading examples of the success of the emerging environmental-justice movement.[25]

For several years, the land remained vacant, and it became filled with trash, tires, appliances, and broken down buildings. Following the 1992 riots and the increasing focus on food issues and the problems of food insecurity, the city and USDA worked out an arrangement with the Los Angeles Regional Food Bank, which was located across the street from the former LANCER site, to establish a large community garden that would allow food-bank recipients and other community residents to grow their own food and be less dependent on emergency food providers like food banks. The city erected fences around the site, put in a water system, and allowed the Food Bank to manage the site temporarily while USDA provided free seeds and soil amendment.

Though the site at first attracted only a limited number of gardeners, after a formal dedication ceremony, a number of farmers, including Latino immigrants who had initially been hesitant to participate, began to establish plots. From the mid-1990s until 2006, the South Central Farm, as it came to be called, evolved into a striking landscape. Eventually as many as 350 immigrant Latino gardeners (or campesinos, as they also asked to be called) participated in growing crops that had also been grown in Mexico and other Latin American countries—greens like *tlapanche* and *papaloon*, herbs like cilantro, staples like corn and beans, and the ubiquitous nopales cactus. "The plants sown only intensified the otherworldliness of the place," reporter Emily Green said of the garden in a 2004 *Los Angeles Times* magazine story that included Don Normark's striking photographs. "There were chayote and passion vines, amaranth, jimson weed and row after row of cactus, or nopales, grown for the cool flesh inside," Green wrote. This farm in the city, characterized as a mini-pueblo by Latino new urbanist James Rojas for its community activities, also included fruit trees such as lemon guavas and papaloquelite, a leafy green herb, that transformed the land, which was surrounded by warehouses, truck traffic, and the rail lines along the Alameda Corridor. The various plants and herbs also produced different types of smells that recreated aspects of the immigrant memory of the environments of their countries of origin.[26]

As an urban place, the South Central Farm also turned into a safe haven—drug-free, gang-free, and graffiti-free. Reconstructing a kind of urban plaza and green space familiar to Latino immigrants, the garden facilitated social networks and provided spaces for families to gather. On any given day, it could attract as many people as might be found in some city parks. The gardeners/campesinos also established traditions like the celebration for Garden Day L.A., where food and plant vendors, musicians, and folklorica dancers gathered for a day-long fair. To photographer Normark, what was striking about the urban farm was "the gritty exoticism of the place, the sheer grace of people living among plants."

Most of the gardeners/campesinos were recent immigrants from Mexico and Latin America. Many had arrived in the country within the decade, according to a survey of the gardeners that UEPI's Center for Food and Justice conducted in 2004. The survey also indicated that over 90 percent of those surveyed identified such values as "feeling connected to the land when gardening," "spending time with family and friends while gardening," "sharing crops with others," "teaching [their] children and grandchildren to garden," and growing foods that were "healthier than many foods that [they could] buy." About half of the participants lived within the Council District, while others took the bus or biked to the site to grow their crops since they had no other place to do so.[27]

Even as the South Central Farm became a showcase for inner-city food security, urban greening, and the creation of a new type of public and community space, a sequence of actions, dating back to the 1990s and including the garden's initial development, had begun and would come to jeopardize the South Central Farm's existence. Immediately following the 1992 civil disorders, a focus on food security emerged as a critical concern in South and Central Los Angeles, where much of the looting and arson had occurred. The Regional Food Bank initiative was a small although limited illustration of that new interest by city policymakers and business groups. With encouragement from the city, a far more ambitious initiative called Rebuild L.A. was launched to harness the power of the private sector in addressing lack of jobs and lack of access to fresh and affordable food. A key component of the Rebuild LA agenda was the development of as many as thirty-two new supermarkets in the inner city to offset the long-term trend of markets to abandon the inner city and move to suburban locations. Despite a series of pledges by four of the major market chains and the enormous publicity generated by the Rebuild L.A. leadership, a 2002 study by UEPI researcher Amanda Shaffer pointed to no net gain of markets in that ten-year stretch. Some new ethnic supermarkets offset the continuing loss, due to consolidation and abandonment, of the big supermarket chains like Vons (later purchased by Safeway) and their shift towards an upscale, high-end customer base. At the same time, other food-security initiatives, such as the city's advisory body's proposal for a citywide community garden policy, which I along with UEPI staff had helped develop, were either ignored or addressed in only an ad hoc or limited manner.

Even as the South Central Farm and its recent immigrant farmers became more visible as an example of how a community garden could serve multiple functions, city interest in this property began to shift due to legal action against the city that had been brought by the property's earlier owner.[28]

In 2001, Ralph Horowitz (the owner of the property in 1986 when it had been first taken over by the city for the incinerator project) decided he wanted the property back. In 1994, the city had already sold the property to the Harbor Department as part of its Alameda Corridor plans, but nothing was directly pursued concerning the property itself. Horowitz wanted to develop a warehouse project that would take advantage of its location along the Corridor and initiated legal action against the city. With the Concerned Citizens organization and the local City Council representative, Jan Perry, both of whom supported the warehouse project, Horowitz also proposed that a small area of the site could be used for soccer fields. In 2003, the City Council, with Perry in the lead, settled the lawsuit with Horowitz and agreed to sell the property back to him for slightly more than city paid in 1986, despite more than seventeen years having elapsed since the original eminent domain action.

A bitter and protracted struggle took place during the next three years, including a lawsuit filed on a pro bono basis by a law firm on behalf of the farmers over the nature of the 2003 sale. The farmers' group, now led by two of the younger farmers-turned-activists, decided that the only way to save the farm was through mobilization and action, including issuing strong attacks against the developer and making the potential elimination of the farm a public issue. Council member Perry and the Concerned Citizens group, working closely with the developer, opposed the farmers, and the tension between the two groups, with its implied racial overtones (African American against Latino immigrants), added to the debate. Although the warehouse project would have significant environmental and health concerns (including for anyone playing soccer, given the proximity of the air-pollution hazards from truck deliveries and warehouse operations), much of the focus of the debate was over the value of an urban farm versus the presumed sanctity of property rights. The *Los Angeles Times* weighed in on the side of the developer regarding the sanctity of property rights and the equation that the farmers were therefore "squatters" who "spouted revolutionary rhetoric."[29]

As the fight grew more intense and the developer sought an eviction notice, efforts were made to purchase the land back from Horowitz. On June 7, a group that included the Trust for Public Land and the Annenberg Foundation offered $16 million to purchase the property, create a land trust to keep the farm going, and create soccer fields. Though the offer tripled the amount paid just three years earlier and met Horowitz's own asking price, the developer, with support from Perry, refused the offer and called on the sheriff to evict.

On June 13, 2006, at 5 a.m., sheriffs' deputies began to arrest people, bulldoze the fences, and tear up the farm. Several of the immigrant farmers who were undocumented fled, while others were arrested. Mayor Antonio Villaraigosa had unsuccessfully sought to facilitate the Trust for Public Land action in the days preceding the evictions. At a press conference later that morning, he called the evictions "unfortunate, disheartening and unnecessary. After years of disagreement over this property," he said wearily, "we had all hoped for a better outcome."[30]

Following the June 2006 evictions, the farmers' lawyers unsuccessfully returned to court, the mayor's aides sought to find land for a smaller community garden, and the land that had been a remarkable part of nature and community in the city became vacant and desolate once again. The question of the right to the city had flared up once again, more dramatically perhaps than previous conflicts, given the immigrant status of the farmers and the assumption of private property rights even when dubious land transactions were involved that could also have significant negative outcomes for health and the environment. The issue of creating public places, the symbolic pueblo in a small space, the magical garden and urban farm, and the hybrid city and culture that extended through neighborhoods and transformed landscapes continued to be unsettling questions, given the absence of a public role for how land could be developed. Whether for truck or warehouse, magical garden or mini-pueblo, the debates over the South Central Farm had provided a glimpse of the battles to come that could determine the future of this global city.

The Modesta Avila Brigades

In 1889, a twenty-two-year-old Mexican woman named Modesta Avila who had been born and raised in San Juan Capistrano, a former site of the mission and the Mexican pueblo, laid two heavy fence posts along the tracks of the Santa Fe railroad, which had recently completed laying those tracks just fifteen feet from Modesta Avila's home. On one of the fences, Avila posted a sign that read: "This land belongs to me. And if the railroad wants to run here, they will have to pay me ten thousand dollars." The posts were removed by San Juan's express agent later that day, but Avila announced to the authorities in the town of Santa Ana, twenty miles to the north, that the railroad had agreed to pay her the money. To celebrate, she held a dance in Santa Ana but was arrested at the dance by the town sheriff, one of the authorities in whom she had confided. Charged with disturbing the peace, Avila told the judge that the dance was in celebration of her victory against the railroad. In response, the judge sentenced Avila to three years in prison, and she was sent to San Quentin, where she died in her twenties. Although she had never received the railroad money and her act of defiance had an unhappy ending, Modesta

Avila would nevertheless later emerge as a symbol of defiance in the struggle against land loss, conquest, and displacement.[31]

In 2005, a number of community groups, found themselves in the path of railroads, freeways, and an ever-expanding global movement of goods. They came together as a new network they called the Modesta Avila Coalition, which included people mobilizing in the communities adjacent to the ports, fighting against the constant traffic involved in transporting goods to and from rail yards and on freeways, and protesting the megawarehouses that dotted the path from port to the stores like Wal-Mart that sold cheap goods coming from abroad. Like the namesake for their coalition, the groups vowed to block the plans of some of the most powerful interests in the region, which were allied with major globalization forces like shipping companies, global producers, transportation interests, and the global retailers that pushed for the continuous expansion of the global flow of goods. Many though not all of the lead organizers in the Modesta Avila Coalition were Latino, representing primarily immigrant communities subject to the increasing community, health, and environmental impacts from this expanding feature of the global economy.

The community groups that constituted the Modesta Avila Coalition were based in places like Wilmington (one of the first of the working-class oil or "black gold" suburbs and heavily affected by the port traffic), the City of Commerce (located near huge rail yards and several freeways), and the town of Glen Avon (near the massive warehouses springing up in Riverside County to handle the cargo containers of imported goods brought from the ports and also the site of one of the earliest and most protracted environmental-justice battles over a hazardous waste site, the Stringfellow Acid Pits). The groups became particularly adept at establishing alliances with key allies. Research scientists identified the health risks faced by those communities and the regional impacts from the air pollution generated by the port and goods movement traffic flow and were influenced by the community groups in their own understanding of a "street science" that helped shape the research agendas of the academic scientists. Mainstream environmental groups sometimes had testy relationships with community groups but became convinced by these same groups that goods movement had become the most critical environmental—as well as environmental-justice—issue in the region. Longshore workers and truck drivers also were concerned with some of the same health and environmental issues. And environmental regulatory agencies like the South Coast Air Quality Management District shifted their approach toward the pollution from the railroads and the rail yards by calling for new regulations and toward the port itself, which it defined as a "stationary" source that could trigger stronger regulatory intervention.[32]

The groups also helped shape the new appointments to the city of Los Angeles Harbor Commission after Antonio Villaraigosa was elected. Commissioners now included environmental advocates who were more sensitive to the concerns of the community groups, a major change given the long association of Harbor Commissioners with global goods-movement interests. The community groups, through their actions and mobilizations, had also changed the discourse and the nature of the debate around goods movement. An expansion of global trade and a goods-movement infrastructure had previously been embraced without qualification by literally all policy makers, the media, and business interests in Los Angeles. Now, thanks to the continuous struggles of the community groups and their multiple alliances, these players were forced to take a more qualified position. This new position acknowledged that community, environmental, and health issues needed to be addressed either in conjunction with (a position that the community groups were skeptical about based on prior failures to do so) or in advance of any further expansion of goods-movement activities. Perhaps most impressively, the community groups slowed the fast-track plan proposed by California Governor Arnold Schwarznegger that toyed with such ideas as double-decker freeways.[33]

These efforts identified a major environmental and social-justice battleground and are particularly compelling given the scarce resources in the communities represented by the Brigadistas, particularly immigrant communities with their limited rights and vulnerabilities. The lead organizers of the groups also provide their own compelling narrative of transformation and empowerment.

In Wilmington, Jesse Marquez emerged as a key figure—a "bulldog" of an activist, as one of his community members called him, who proudly embraced his family's Mexican and Native American roots. Born and raised in Wilmington, one of five children who lived with their parents in a one-bedroom house that included two bunk beds placed inside a walk-in closet, Marquez felt that he had "never really recognized poverty so much or how low-income we were because everyone was the same in the community." After attending a summer youth teen leadership program, Marquez began to see himself as an activist committed to his community and to a broader agenda of social change. He was able to get training as an electrician, which became his day job while he continued to be involved in organizing and mobilizing around such issues as police violence. His involvement in goods-movement issues was first stimulated when the Port of Los Angeles made plans to build a wall twenty feet tall and 1.6 miles long separating the port from Wilmington. At first, he simply appeared at the kind of pro forma meetings that were characterized as community-input sessions but that essentially ignored any community concerns. But Marquez came to realize that the port and the goods-movement issue would become the defining feature of what was in store for Wilmington, the larger port area, and, as he developed ties with other groups along the goods-movement route, with the region itself.[34]

In 2003, Marquez came to speak to my students about his growing concerns and beginning efforts at organizing (a talk that also inspired one of my students to write a profile of Marquez as an example of how environmental-justice figures become leaders in their communities). He created a group called Coalition for a Safe Environment that he ran out of his home with a couple of friends. But his energy and passion were infectious, and the goods-movement concerns became increasingly apparent to community members. In just a couple of years, Marquez pulled together a group with several hundred members, became a constant presence at hearings (where he would fire off ten-, twenty-, even fifty-page responses to proposed plans or rule changes) and conferences (with his native dress, long pony tail, cheerful disposition, and constant energy), built alliances, and alerted vulnerable communities (such community groups in a Mexican port city that was threatened with goods-movement expansion).

While Marquez was expressive and energetic, City of Commerce activist Angelo Logan emerged as one of the most effective, insistent, thoughtful, and systematic figures among the Modesto Avila Coalition leaders. Born and raised in the City of Commerce and proud of his immigrant roots, Logan was a rebellious teenager who flirted with gang activities but eventually drew on his creativity by becoming active in the urban mural art movement. His public-art interest soon translated directly into activism, and he began to work with a prominent environmental-justice organization in Los Angeles, the Communities for a Better Environment (CBE) (CBE was formerly called Citizens for a Better Environment but changed its name as it became more engaged in low-income communities with undocumented immigrants.) In the late 1990s, CBE began to shift some of its organizing activities to the L.A. River and industrial-zone southeast communities of Huntington Park and adjacent cities like Maywood and South Gate. It did not extend its work to the nearby City of Commerce, and Angelo Logan felt the need to reconnect with his own community and his own roots. He established a new organization called East Yard Communities for Environmental Justice (CEJ). Similar to Jesse Marquez's efforts in Wilmington, East Yard CEJ grew from a handful of people to the leading community organization in the City of Commerce, where, among other changes, the group influenced the City Council to create its own Environmental Justice Task Force.[35]

Logan soon emerged as a key environmental-justice leader associated with goods-movement issues, particularly the unregulated activities of the railroads and the enormous community and health impacts from the rail yards that abutted a number of residences in the community. Logan was instrumental in helping establish new alliances, pulling together the Modesta Avila network, and influencing the South Coast AQMD to begin to focus on the lack of regulatory overview of railroad activities (including intense diesel exposures for community residents). Logan was passionate about democratic process and community engagement and empowerment. His advocacy on goods-movement issues, effective use of street science, and alliances with scientists made him a powerful advocate, while his soft-spoken style and intense manner complemented Marquez's exuberance. He also helped shift the nature of the debate about goods-movement impacts from port-related issues to regional issues of transportation and the overall reach of goods-movement impacts.

One compelling link that Logan established was with the environmental-justice activist Penny Newman, the leader of the Inland Empire (or Inland Valley, as she called it) organization, the Center for Community Action and Environmental Justice (CCAEJ). CCAEJ's predecessor, the Glen Avon–based Neighbors in Action, was formed during the late 1970s in response to significant problems associated with a nearby hazardous waste site. Newman, like Marquez and Logan, was deeply attached to her community, had been PTA president, and strongly appreciated the semirural character of the area. The waste site, known as the Stringfellow Acid Pits, was a major repository for a wide range of toxic chemicals from industry, government, and military sources. During the late 1970s, the site experienced a series of spills that spurred the community to action and transformed Newman into a major figure within the emerging environmental-justice movement.[36]

The Stringfellow issue quickly assumed national significance thanks in part to Newman's leadership, which included a confrontation with the Reagan administration's Environmental Protection Agency that helped bring about the eventual conviction for perjury of Rita Lavelle, the official in charge of the Superfund program. Successful in shifting the approach by EPA and state agencies regarding site cleanup, Newman, now a national figure, helped local groups around the country organize to oppose specific hazards in their communities, while still keeping connected to her home base, which had become an increasingly Latino area. As she became aware of the warehouse plans near her community, which included concerns about diesel exposure, she worked closely with research scientists who identified the local area where warehouse plans were taking shape as one of the most polluted areas in the region and began to identify the health and environmental impacts from the pollution.

Teaming up with the activists from the City of Commerce, Wilmington, West Long Beach, and other communities affected by the goods-movement expansion, Newman, who initially created the Modesta Avila Coalition with Angelo Logan, saw this network as a way to link community action with broader regional infrastructure and goods-movement issues. Passionate and savvy, articulate and fierce in her determination, Newman and her growing organization of organizers became a key link in the Modesta Avila Coalition's conception of a locally based but regionally centered network.[37]

By 2006, the Modesta Avila Coalition became the leading edge of a growing though still disparate set of groups that were questioning the principle of the unrestricted, unregulated, and ever-expanding flow of goods through the ports and the regional transportation infrastructure to their destinations across the country. For the first time, workers at the port and truck drivers transporting the goods began to focus on their own health concerns, identifying a potential community-labor alliance that seemed problematic just a few years earlier. On May 1, 2006, the day of action regarding immigrant rights that saw numerous workplaces that relied on immigrant workers shut down as part of the protests, 90 percent of the independent truckers, also Latino and many of them immigrants, refused to transport goods from the ports that day—an action that was even more startling to the goods-movement advocates than the incipient labor-community connections.[38]

What the Modesta Avila Coalition also provided, as part of its insistence that community, environmental, and health issues could no longer be ignored within this context of globalization, was a counternarrative about their community and their region. "We are a movement that seeks to value a sense of place," Penny Newman argued at a ports-related gathering in 2005. At a talk at Occidental College that same year, Newman further elaborated the notion that the environmental-justice focus of the Modesta Avila brigadistas was an environmental entry point for what was essentially a social-justice argument about change from below, whether related to the movement of goods, the availability of affordable housing, the desire for decent paying and sustainable jobs, or the need for parkland. The counternarrative of the brigadistas, as

Angelo Logan and his East Yard organization also stressed, identified the need for a democratic process that engaged all community residents and that allowed the issues of globalization—whether in the form of the movement of goods, the flow of capital, or the crossing of borders—needed to be framed so that the language and outcomes of social and environmental justice could ultimately prevail.

Social Change across Borders

By 2006, social and environmental justice activists who focused on globalization issues continued to point out that the restrictive policies and substandard living conditions of immigrants were occurring simultaneously with the free flow of goods and capital across borders and the increasing loss of national sovereignty over markets and economies. The immigrant rights protests that swept the nation that year sought to counter the punitive approach of the December 2005 immigration restriction legislation passed by the House of Representatives, the inflammatory rhetoric of CNN's Lou Dobbs and the Minutemen border-security vigilantes, and the increasingly virulent anti-immigrant mood that began to appear in such places as Colorado, Georgia, and other states where immigrant populations began to swell. As a consequence of the pro-immigrant demonstrations and the rise of the anti-immigrant forces, a full-scale national debate emerged about immigration, national sovereignty, and assimilation. Advocates framed their arguments around the need to establish minimal rights (including access to health care, food stamps, education, and driver's licenses) that needed to be extended to citizens and noncitizens alike. But cross-border connections—and their value—were often at the margins of any discussion. Several of the immigrant-rights advocates felt the need to frame the discussion in national terms and criticized the protestors who waved the Mexican flag (sometimes along with the American flag) at demonstrations in their declaration of a transnational identity. American denunciations of this assertion of cross-border identities and a Spanish-language version of the U.S. national anthem became part of the renewed debate over the meaning and realities of assimilation and diversity. But the argument about assimilation was focused less on acculturation than on a certain notion of "Americanization," as Stephen Castles pointed out, referring to the idea that "everybody could become part of a new distinct nation through the opportunities offered by expanding free-market economics and through immersion in a democratic civic culture." These debates about assimilation and Americanization were thus limited to the question of immigration and immigrant status while the broader context of globalization undermining traditional notions of national sovereignty appeared as a distinct, separate set of issues.

But were immigration and globalization really separate? The huge increases in border crossings by migrants in the United States, Europe, and throughout the world represent the underside of globalization—its need for cheap labor, lack of basic rights, and disruption of rural, regional, and national economies through globalization-based policies (such as structural adjustment and free-trade policies) that have led to wholesale migrations from countryside to city, from city to border regions, and then across borders into countries where low-wage labor is aggressively sought. One outcome of this transnational migration in the contemporary period—due in part to the enormous scale of the migrations and accessible forms of communication and transportation—has been increased cross-border relationships between families, immigration communities, and communities or villages in the countries of origin and economic, social, and cultural ties that survive the immigrant's journey.

The best known and most substantial of those cross-border ties involve *remittances*—economic, social, and cultural cross-border exchanges. Economic remittances, the most visible of these exchanges, involve funds that are transferred back to the home country. These often support families but also involve a precautionary method of savings in case of deportation. They have also repaid loans and purchased physical assets like goods or a home in the home country. Remittances have taken the form of organizational activity, such as when immigrant "hometown clubs" provide economic resources for a new community economic development in the village or community of origin.[39]

These economic remittances have become substantial as migration patterns have increased significantly throughout the world. For some countries, remittances have constituted a large part of the country's gross domestic product. Robert Lucas estimates that these may range as high as 15 percent of GDP for Moldova, 5 to 10 percent for Morocco, about 7 percent for Sri Lanka, and 9 percent for the Philippines. Even where remittances constitute a smaller percentage of GDP, as in India, they nevertheless, Lucas argues, can be substantial for certain regions, such as Kerala. The World Bank has estimated that as much as $232 billion had been transferred in 2005 by immigrants to their home countries, with an increasing share of those funds coming from female immigrants.[40]

Remittances to Mexico, also characterized as *migrodólares*, have been estimated to be the highest among all receiving countries, growing at an annual rate of almost 13 percent between 1960 and 2003 and becoming the largest source of foreign exchange in the country. By 2004, that amount was estimated at $16.6 billion. Even more striking, remittances by Mexican and Latin American immigrants in the percent United States have not only increased rapidly in recent years (totaling $45 billion in 2006, an increase of more than 50 percent in two years, according to a study sponsored by the Inter-American Development Bank), but they were being sent from literally every region and state within the United States, also a recent and dramatic change in immigrant dispersal and economic flows. The role of remittances as a community economic-development tool, however, has remained uncertain. Whether remittances are used for investment, for savings, or for home ownership, their capacity to constitute a form of economic development capable of reducing poverty and inequality has crucial implications for the country of origin. Those choices, Timothy Hatton argues, may also depend on "where in the hierarchy of class and income the migrants came from, whether they intend to return and what are the opportunities for investment at home."[41]

Many countries, including Mexico, El Salvador, and several other Latin American countries, have strongly encouraged economic remittances as a way to address problems of dislocation due to globalization pressures, such as

NAFTA-inspired changes in the Mexican countryside and the undermining of self-sustaining rural economies. Immigrant-sending countries have, in many cases, encouraged migrations, characterizing their departing immigrants as "heroic citizens" capable of contributing to the home economy through remittances and through "dual citizenship," a concept also reinforced by policies that allow immigrants to vote and maintain citizenship status with their home and receiving countries.[42]

In El Salvador, for example, emigrants have been mythologized as *los hermanos lejanos*—the distant relatives. As Ana Patricia Rodriguez has described it, the cumbia song, "El hermano lejano," pays tribute to the Salvadoran immigrant who recalls his beautiful land, "hoping to overcome the distance in order to work for those he left behind." The Salvadoran newspaper *La Prensa Gráfica* has used the term "Departamento 15" (Department 15), suggesting a fifteenth department of emigrants in addition to the existing fourteen geographic jurisdictions in the country, a reflection of the widespread nature of Salvadoran immigration to the United States as well as other countries in Central America, Mexico, South America, Canada, Australia, Asia, and elsewhere. These Department 15 migrations, which grew considerably during El Salvador's civil war in the 1980s, have become, according to Rodriguez, a "virtual imagined community of sorts, conceptualized by transnational communication networks."[43]

While economic remittances have significantly affected the national economies of the countries of origin, the cross-border relationships have also had social and cultural effects for both receiving and sending countries. Political refugees from countries like El Salvador become political activists in the United States, while migrating workers from Central America and Mexico carry with them the kind of political and organizational knowledge that became part of a new and more assertive immigrant-based union organizing. Immigrant or ethnic cuisine sometimes reflects a hybrid of influences that contribute to biodiversity by keeping local varieties of seeds and plants alive. Geographic places like Chinatowns, situated in urban cores in New York, San Francisco, and Los Angeles, are dense communities that recreate a sense of place that also "transcends geography," representing places that are both independent of the nation of China in the case of Chinatown while maintaining a cultural identity associated with what it means to be Chinese. "In this era of heightened globalization," Peggy Levitt argues, "transnational lifestyles may become not the exception but the rule."[44]

Such cultural expressions, like the corridor ballads, underline a cross-border or transnational longing. Yet when these cultural connections cross borders in both directions, they can also collide with traditional cultural or ideological values prevalent in the home countries. For example, the large Oaxacan community in Los Angeles (estimated at between 150,000 and 300,000 immigrant residents) has sought to continue the tradition of celebrating the guelaguetza, a traditional dance festival held each July in Oaxaca, in Mexico. It is an early summer harvest gathering where the music and dance that are performed have been embedded in the indigenous culture of the region. Over time, however, the guelaguetza became more of a tourist attraction in Mexico, controlled by the government to capture tourist dollars. Participation became limited to the few wealthy Oaxacans as well as tourists who would pay admission fees that the vast majority of the Oaxacans couldn't afford, even as some were enlisted to perform. When the guelaguetza was first celebrated in Los Angeles in 1972, 10,000 Oaxacan immigrants filled the Sports Arena, and the celebration soon became a crucial "Oaxaca in Los Angeles" event. For those who participated, it was a way "to see my culture, the food, the clothing" and "to keep those traditions alive," as Dalia Martinez, a thirty-year-old seamstress, recounted to writer Daniel Hernandez. As it evolved, what made the Los Angeles guelaguetza different from its Oaxacan counterpart was its democratic character, since high fees were not charged and the festival became an articulation of cultural identity rather than a commercial venture and manipulation for a tourism-based economy.[45]

This assertion of a cultural identity has corresponded with the development of the vast array of immigrant-based associations, clubs, centers, and related organizations that have kept alive a cross-border or political, cultural, and economic transnationalism that can be found in Los Angeles and increasingly in other communities in the United States As many as 250 Mexican immigrant hometown clubs (Asociaciones de Pueblos) are located in Los Angeles, representing sixteen different federations. Though the groups vary widely, several of them have expanded their agendas beyond sporting events and cultural celebrations to address social-justice and human-rights issues. For example, the Oaxacan Frente Indigena Oaxaqueno Binacional (FIOB), initially created in 1991 as a coalition of various clubs, has since evolved into a membership-based organization with a broad social-justice agenda. This has included an "access to justice" project (the Procuracion de Justicia y Defensa de los Derechos Humanos de los Indigenas Migrantes) that provides support and empowers its members around labor and human-rights issues. Besides the Mexican and Latin American groups, the Korean Immigrant Workers Alliance (KIWA) organization became a champion not only for Korean immigrant social-justice and labor-rights issues but also for the rights of non-Korean (primarily Latino) workers exploited by Korean employers, particularly among restaurant owners. Similarly, the Thai Community Development Corporation (Thai CDC), based in the East Hollywood neighborhood of Los Angeles (now known as Thai Town), became prominent in the 1995 scandal involving Thai workers brought into this country to work as indentured laborers, a scandal that included Thai labor brokers who arranged for the workers to be brought into the United States for the garment company that employed them and held them hostage. The Thai CDC subsequently helped as many as eighty of those workers reestablish themselves in Los Angeles, including some who started their own Thai restaurants as well as a Thai spa. In return, these former indentured workers used their success to build an elementary school in their village of origin as a type of social remittance.[46]

In 2005, the Thai CDC became engaged with the Migration Policy and Resource Center at the Urban and Environmental Policy Institute on a fair-trade initiative that had been introduced in the United States by the Educational Network for Grassroots and Global Exchange (ENGAGE), a group of former "study abroad" students who had embraced the efforts of an organic rice-growing cooperative in the province of Surin in northeast Thailand near the Cambodian border. These ENGAGE, Thai CDC, and Migration Center discussions focused on a possible cross-border exchange within a "social change across borders" framework. The concept—social change across borders—had previously been elaborated through a Social Change across Borders Institute founded in 1998 by the Latin American and Latino Studies program at the University of California at Santa Cruz. The Institute identified the goal of building bridges for social-justice activists across the Americas as a critical need in an era of increased globalization. In 2004, the Institute moved to Los Angeles and established a new set of partnerships, including with UEPI's Migration Center. Seeking to expand transnational social-justice activism, the L.A. groups shared strategies for increasing the capacity of various community actors, including immigrant groups and their allies who were engaged in transnational social-change work.[47]

While many immigration activists have focused on human- and labor-rights issues in both home countries and the United States, an *economic*-oriented approach to studying cross-border changes has presented challenges, primarily because economic remittances are private in nature or at times influenced and manipulated by the home country. This has been particularly the case when the economic approach to social change across borders involves goods that cross borders from the home country to where immigrants now reside, given the nature of global trade. The ENGAGE initiative to establish a fair-trade Thai jasmine rice cross-border connection extended the social-change-across-borders approach into both these areas—a justice-oriented approach to economic development and an alternative to the dominant global trade system. It was also a strategy that resonated with the Thai CDC, given its experience with the former indentured garment workers.

I first became aware of the ENGAGE initiative when Ellen Roggemann, a student who had spent a semester during her junior year in Surin Province, came back determined to support an organic jasmine rice cooperative there. Her passion around jasmine rice issues led her to undertake a major research project regarding global trade and food-system issues affecting jasmine rice production and markets both in Thailand and the United States. At the time, the United States and Thailand were in negotiations to establish a Thai-U.S. Free-Trade Agreement while a series of global-trade rulings on patents and the place name of products like jasmine rice identified important consequences for the continuing survival of jasmine rice farming. Moreover, the Green Revolution in Thailand in the 1950s favored chemical and mechanized production methods, which raised the costs of production and increased the debts of the vast majority of small farmers involved in jasmine rice production.[48]

Jasmine rice—with its sweet aroma, soft texture, and delicate flavors—is more like a gourmet feature than a side dish, and in Thailand it is both delicacy and staple. Rice in Thailand has deep roots and cultural associations. It has been harvested for more than six thousand years and constitutes more than 50 percent and as much as 80 percent of the total calories consumed by its people. As Roggemann commented in her study (subsequently published as a joint ENGAGE/UEPI report), "to say 'let's eat' in Thai translates literally to 'eat rice.'" "As a source of culture and belief," Roggemann continued, "festivals are tied to the rice season, with Phi Ta Haek paying homage to rice spirits during land preparation and Boon Koon Larn honoring the Mother Spirit of rice during harvest. For all these reasons and because of its exceptional quality, Thai rice, especially specific varieties such as Jasmine rice, remains a source of pride for many Thai people."[49]

Jasmine rice or *Khao Dawk Mali* (white jasmine flower) is also native to the Northeast region of Thailand, where unique growing conditions have allowed jasmine rice to evolve and flourish. Its special qualities have allowed it to contribute as much as 33 percent to the country's total exports. Jasmine rice and other specialty rice products from Asia, such as basmati rice, have also become popular in the United States. In fact, the growth in the U.S. rice market has largely been associated with these specialty products. Moreover, much of the increased market in the United States is due to Asian immigrants, who consume far more rice than the non-Asian population does and the Asian immigrant population's lack of interest in the heavily subsidized

U.S.-grown white rice varieties. Initiatives to break into the U.S. jasmine rice market through the development of new U.S. hybrids that have sought to identify themselves as either jasmine or basmati products (including one brand that called itself "Jasmati") as well as efforts to develop a genetically modified jasmine rice product have been the subject in turn of global trade and food-system controversies (including whether the name "jasmine rice" has a particular place-based association based on its specific cultural and growing features).[50]

Despite the growing popularity of jasmine rice as an export crop, many of the Thai jasmine rice farmers have lacked control of the milling, have had to pay higher costs for fertilizer and other inputs, and have received lower prices for their product as it makes its way through the global trade system. Those types of issues were similar to the experiences of other farmers in developing countries, including for such crops as coffee and bananas that were major export crops for the countries in which they were produced. Green-revolution-inspired changes compounded the problems of the farmers, adding environmental and health concerns to the economic squeeze on small producers that has been a result of this industrial agriculture model. What Ellen Roggemann discovered in Surin, however, was that farmers had developed an alternative approach—an organic farming cooperative and an arrangement with a European group to develop a fair-trade label for their product, which recently has been introduced into several European countries. When Roggemann asked the farmers what she could do to support their cooperative, they replied "that they wanted the Thai jasmine rice that their cooperative produced to become available in the United States and for Americans to learn the importance of jasmine rice in the lives of Thai farmers."[51]

To introduce a fair-trade product in the United States meant working through the various fair-trade networks that had been established in the United States, in Europe, and elsewhere and find potential constituencies and outlets that would support such a product. Although it is a relatively new phenomenon, fair-trade activity has increased significantly during the past ten years, particularly as some of its products, such as fair-trade coffee, have begun to be carried by large outlets such as Starbucks (although it constitutes only a tiny percentage of those outlets' overall product sales). Fair-trade labels have been provided when certain conditions are met, including growing practices (such as growing coffee organically and in the shade), organizational structure (such as cooperatives and farmer-owned farms), and pricing structure (such as insulating growers from global-market fluctuations so that prices do not fall below the actual cost of production, ensuring at least a small profit for farmers). The fair-trade movement has focused on supporting environmental practices and the survival of small farmers in the face of globalization practices that have increased chemical inputs, among other environmental impacts, and undermined the small-farm rural economies in developing countries. The markets for fair-trade products have in turn primarily been environmentally and socially conscious consumers in the United States and Europe who have been willing to pay a premium price for products as an act of solidarity and preference for an organically grown, small-farmer-supporting product.[52]

A fair-trade jasmine rice product, the ENGAGE activists assumed, would be able to appeal to the fair-trade constituency of upper-middle-class consumers. However, the ENGAGE group, as it explored the possibility of a fair-trade campaign in the United States with the Thai Community Development Corporation and UEPI's Migration Policy and Resource Center, also became attracted to the idea of expanding the fair-trade approach to the Asian immigrant community in the United States through the social-change-across-borders philosophy of social-justice-oriented cross-border connections. To pursue such an approach required establishing connections to Thai immigrants, who are the largest American consumers of jasmine rice and who have a direct cultural association with the Surin growers. Three core constituencies for such an approach were identified: Buddhist centers, the Thai restaurant association, and the ethnic grocery stores where the product could be purchased.

In 2003 and again in 2005, the ENGAGE activists sponsored a tour of the Surin farmers, who met with sympathetic audiences (such as university students). The farmers spoke about the development of their fair-trade product and the significance of Thai jasmine rice in their lives and their heritage.

During their 2005 Los Angeles tour of the largest Thai immigrant community in the country, the Thai CDC arranged meetings with groups at the Buddhist Center, and with Thai rice importers, the restaurant association, and retailers. The Thai immigrant participants were interested in an association (one of the importers explained that she was also from Surin province and shared the love of the fragrant rice and the farm communities that grew it). But the biggest barrier involved the size and quantity of the rice that could be delivered since the fair-trade one-pound package developed by the European fair-trade distributor was designed more for small, upscale retail businesses. The more conventionally grown Thai jasmine rice was already being imported into the United States but in far larger wholesale units of fifty pounds and more. While the distribution issues were being sorted out, the Thai CDC moved forward with plans to establish a "virtual market" to promote cooperative-made goods from Thailand and a food-based farmers' market that could highlight both locally grown foods from immigrant farmers and the Thai jasmine rice fair-trade product coming from Surin.[53]

Reorienting towards a large immigrant-based market represented a different type of approach for the fair-trade advocates. It also suggested a broader approach for the food-security advocates—extending their embrace of locally grown food to include food with strong cultural associations through a cross-border connection between farmer and consumer. This approach complements the community-food approach of regional farmers to local consumers or local institutions like schools. At the same time, social change across borders provided a powerful metaphor for how such connections could be defined. And with different communities in the city increasingly assuming the form of transnational communities while the right to the city (and basic human rights in the city) continued to be denied for such communities, the importance of a social change across borders vision and practice had become more compelling than ever.

9

Globalization from Below: A Conclusion

The riches are global, the misery is local.

—*Zygmunt Bauman*[1]

Changing Urban Nature, Changing the Material World

How can Los Angeles be reinvented, given the past 150 years of Anglo domination and displacement, the rise of freeway culture and pouring of concrete into its rivers, the bulldozing of gardens and absence of green space, and the vast divide between rich and poor, nativists and immigrants? Will Los Angeles ever escape from its dystopian reputation as the smog capitol, the land of two-hour commutes and Sig Alerts, a place where no one walks and everyone drives, captured by the film *L.A. Story* when Steve Martin drives his car 100 feet to his neighbor's house? Or can Los Angeles demonstrate that it has become a place of innovation and transformation, where passion flowers break through the concrete and bikers and pedestrians stroll on freeways every Sunday, where a million trees are planted, where blue herons settle on a free-flowing Los Angeles River, and where immigrant landscapes of chayote and passion vines, amaranth and jimson weed, provide the city with the sights and smells of its residents' multiple roots?

Los Angeles, along with its global counterparts, needs an urban nature agenda. On the one hand, places like Los Angeles have long been derided as nature's implacable enemy. Yet Los Angeles is also a place rich in nature's amenities—ocean, mountains, diverse habitats and ecosystems, a Mediterranean climate. It has become a place where reinvention, while difficult, also seems possible—and seems critical to movements for social change. Such an agenda therefore requires an expansive view of nature in the city as constituting part of the material world in its urban form.

While urban nature advocacy—for more parks and green space, community gardens and farmers' markets, and trees and bike paths—is a critical dimension of the struggles for a more livable city, it needs to be seen as part of and not distinct from an agenda of social change. Such a perspective asserts the need for community gardens *and* for affordable housing as well as community gardens that are part of affordable housing developments. It identifies the need for farmers' markets and for job-creating supermarkets and ethnic markets that can stock local foods, culturally appropriate foods, and fresh and affordable foods. And it seeks sustainable parks *and* schools *and* housing *and* viable job-generating businesses that are part of a broader strategy of land use that is based on public goals for livability and for justice—rather than land use that displaces and creates unaffordable places, no green space, overcrowded schools, and unsustainable jobs based on the imperatives of globalization.

An urban nature agenda identifies opportunities that may appear small in scale yet represent a larger shift in how we perceive the urban landscape. A tree in a parking lot becomes a step in the direction of reducing parking's big footprint while beginning the task of establishing a treescape in the city that can have powerful implications for livability. A pocket park or community garden in one neighborhood can become part of a process of reclaiming urban land otherwise abandoned or degraded—whether a median strip on Broadway in New York City where corn and beans are grown, a trash-strewn property along the Alameda Corridor in Los Angeles where a fourteen-acre garden becomes an immigrant farm and ultimately a new type of urban plaza, and alleyways turned into greenspace and gathering places. But an urban nature agenda also needs to seek big changes, suggesting a larger goal of transformation. To re-envision the Los Angeles river, now a common point of reference for both policymakers and activists, is one such transformative goal. A river reenvisioned and reconstructed is tied directly to core issues of landscape and land use and how the connection to greenspace is valued in an urban setting like Los Angeles and its pervasive urban hardscape. And establishing larger park sites, such as the Cornfield, can provide valuable spaces in the city that strengthen neighborhood and regional identities.

Urban nature advocacy lies at the heart of an emerging urban environmentalism that has the capacity to overcome the classic and debilitating divide between nature and human activity that causes environmentalists to assume a defensive posture. It may seem difficult to advocate for "saving" urban nature, since urban environments are often degraded and therefore have to be renewed and reconstructed as a new type of nature. Through this process of renewal, urban nature advocacy is most strengthened through a connection to place. The most compelling arguments of the urban environmentalists often turn out to be about justice and equity linked to issues of place. In this way, an environmental-justice movement can situate its arguments as the environmental dimension of social-justice advocacy, as Penny Newman asserts. To do so helps establish Bruno Latour's notion of a political ecology that incorporates the human and nonhuman and provides a direct route for environmentalism, building on its links to social-justice movements, to engage in the struggles regarding the material world we inhabit and seek to transform.

Communities of Difference, Communities of Struggle

Urban places are places of difference. They house different ethnicities. They include people who have migrated from different countries and places of origin. They constitute different demographics based on class and race. They have different foods and different music. And they assume differences in how everyday lives are lived. Gated communities, as restricted covenants before them, provide a counterpoint to the city as a community of difference. Their search for homogeneity has resulted in exclusion and discrimination on class, racial, and ethnic grounds. The walls created in many gated communities parallel the fences and walls along the United States' southern border and extend to the restrictions on the use of language other than English. These are arguments about keeping out people and defining who belongs. Horace Kallen wrote more than ninety years ago that Anglo-Saxon Americans saw themselves, by their very nature, as *not different* and thus defined what it meant to be an American. Along those same lines, Milton Gordon wrote forty years ago in his classic text *Assimilation in American Life: The Role of Race, Religion and National Origins* that "the white Protestant American is rarely conscious of the fact that he inhabits a group at all. *He* inhabits America. The *others* live in groups."[2]

During the 1960s and throughout much of the remainder of the twentieth century, a cultural pluralist model prevailed in education and other social and cultural policies. This approach contrasted with assimilationist and melting-pot notions that assumed that differences among and between ethnic and racial groups would eventually be submerged, perhaps not so much through strategies as intermarriage as the eventual loss of any ethnic or cultural characteristic other than the dominant mode of Americanization. The cultural pluralists, on the other hand, celebrated diverse cultures, hyphenated identities, and a tolerance for difference. Theirs was a civil rights discourse that assumed that ethnic and racial differences would still provide for equal participation in society. Yet even the cultural pluralists assumed that a larger American identity would allow subcultures to be tolerated and, if not fully submerged, at least sufficiently integrated into that common American identity. Such hyphenated identities were not assumed by most cultural pluralists during the 1960s to be transnational, despite the emerging interest then in African roots among blacks or the concept of Aztlán among Chicanos. Much of cultural pluralism, even as it continued to evolve into its contemporary version of multiculturalism, remained more an ideology of tolerance within the country than transformation across borders.

By the turn of the twenty-first century, the number of migrants crossing borders began to reach the levels of the great surge of Southern and

Eastern European immigration at the turn of the twentieth century (which had led to the nativist backlash that Horace Kallen had sought to address). Beyond the numbers, this recent immigration surge has established a far more extensive cross-border set of relationships in a far more extensive era of globalized economic relationships. Were the new migrants eventually to become a new generation of Mexican Americans or other so-called hyphenated groups? Or were they forever to be deemed illegal or semilegal Mexicans or Central Americans or Asians and cast into a vulnerable no-man's land, subject to forms of exploitation and discrimination? Even Milton Gordon, writing in 1964 at the tail end of the bracero program but before the Immigration and Nationality Act of 1965 unleashed a new wave of undocumented migration from the south, worried that the slow progress in developing a second-generation Mexican American middle class was "an unmistakable sign of retardation in the acculturation process." Moreover, the nativist reaction that was unleashed in the 1990s and the new century was less an argument about biological superiority that the racism of the earlier twentieth century had revealed than a fear of a dilution of nationality in an era of globalization. Embedded in those fears has been the sense of a loss of community, the "foreign" invasion finally undermining the last threads of community connection already shattered by globalization's displacement of stable industries and jobs as well as long freeway commutes or fast-food cultures. The search for community in the United States in the twenty-first century has thus remained compounded by globalization's reach into the everyday lives of its residents.[3]

How then to locate and reconstruct community? The new urbanist appeal for a reconstruction of community based on a nineteenth-century small-town ideal threatens to become another version of the gated community due to its high income requirements. The Latino new urbanist search for community, on the other hand, seeks to identify a new transnational culture of community based on a democratic use of public space and cross-border identities. Los Angeles, a multicultural city in its multiple dimensions, could be a model for this reconstruction of community based on differences. But cultural, economic, and political status in Los Angeles and elsewhere remain subject to income disparities, abuses of civil and human rights, and citizenship status. This underside of globalization constitutes a less visible yet potent barrier to community in this global city and other regions across the country.

The search for communities of difference needs to extend beyond a civil rights discourse of tolerance and acceptance to the need to establish core human rights and overcome social, economic, and political disenfranchisement. New types of communities of struggle need to establish urban farms or community gardens, ensure a living wage, build affordable housing, create new public spaces, and establish basic human rights for education, health, jobs, and sustainable livelihoods. A community agenda similar to the emerging urban nature agenda requires an agenda of change at multiple levels and within multiple institutions. These include neighborhood organizations, regional alliances, hometown clubs, resident groups, media institutions, and schools, including institutions of higher education that have considered themselves apart from rather than of the city. To build community in the global city becomes both a form of struggle and a part of the process of reinvention and transformation of places and institutions that are both small and large, neighborhood and regional, local and global.

Everyone's Right to the City

Communities need to assert a right to the city for everyone. The absence or loss of a right to the city reflects the daily realities of cities divided by segregation, freeways and the reach of the automobile, loss of green space and concrete rivers, deindustrialization, homelessness, and sprawl. These realities include the conditions that have barred access—across ethnicity, race, class, and neighborhood lines—to parts of the city. Barbed-wire fences separate city residents from a channelized river, freeways cut through neighborhoods, wealthy beach communities seek to keep nonresidents away from the ocean, parkland areas like the Santa Monica Mountains are distant from the low-income communities of South, Central, and East Los Angeles where green space is otherwise limited or absent, supermarkets abandon inner-city communities to locate at suburban and exurban sites near freeway exits, and undocumented immigrants are denied access to basic human rights.

In Los Angeles, freeways, perhaps more than any other part of the built environment, have represented a core feature of the urban divide, undermining viable low-income communities such as Boyle Heights while avoiding wealthier communities that have successfully fought to derail freeway projects that would have run too close to homes and local commercial strips. The right to the city, as Henri Lefebvre passionately argued, needs to address "the invasion of the automobile" that has "turned the car into a key object, parking into an obsession, traffic into a priority, harmful to urban and social life." The right to the city thus also asserts the rights of the pedestrian, cyclist, or and bus or transit rider who have been displaced by the automobile, and to begin to take back the vast amount of land that the automobile consumes.[4]

The right-to-the-city argument, as Susan Fainstein has suggested, requires an approach that accounts for justice as well as diversity. But justice is a matter of access as well as rights. Increasing access to the city's physical, social, cultural, and economic places and institutions helps to create a more just city by expanding opportunities for people to participate and engage in public life and public and private spaces in their multiple forms. This requires a new parks policy, a new food policy, new transportation approaches, new housing initiatives, and more innovative and publically supported public schools. It needs policies that can establish jobs that are sustainable and that allow the poor to earn a living wage. In this era of globalization and global cities, it requires an approach to human and civil rights that provides, as Stephen Castles argues, "rights as human beings, not as nationals."[5]

Globalization from Below

> Poor people believe that someone who builds a wall along the border with Mexico is not the kind of person who has faith in the benefits of free trade agreements.
>
> —*Fernando Báez*[6]

In his 1976 book entitled *Keywords: A Vocabulary of Culture and Society*, Raymond Williams did not include the term *globalization*. The concept most proximate in his discussion was *imperialism*, which he sought to situate in its political and economic dimensions and through its historical usage as it emerged in the last decades of the nineteenth century. The area where the contemporary analysis of globalization could intersect with Williams's keywords definitions was in relation to Williams's discussion of imperialism's location as "an economic system of external investment and the penetration and control of markets and sources of raw materials." Williams also referenced, as part of this economic definition, a subtext of imperialism's "indirect or manipulated political and military control." But this 1976 discussion did not address the kinds of global reordering that have made contemporary globalization such a pervasive force:

- Globalization's expanding cultural reach, led by an increasingly global cultural industry that influences words, music, and visual images in literally every continent and nearly every country;

- The transformation of food diets by an increasingly global and integrated food industry (which, aside from other impacts, has led to an extraordinary increase in worldwide obesity in the midst of food insecurity, due to a global fast-food culture);

- An integrated global system of production, distribution, finance, and consumption;

- The rise of the global city, which functions at the intersection of the movement of capital and of goods and which has transplanted the historical concept of the imperialist core and the dominated periphery primarily defined through the nation state;

- The development of a maritime world economy and the global-goods movement that establishes communities of sacrifice, harmed by the health, environmental, and community impacts associated with the transport of those goods; and

- The vast migrations of people across borders, creating a new kind of resident in the global city whose rights are limited and obscure and whose identities also cross borders.[7]

The 1999 protests at the meeting of the World Trade Organization in Seattle, Washington, made visible an emerging set of disparate groups and ideas that at first sought to constitute themselves as antiglobalization movements. The subsequent development of the World Social Forum and its assertion that "another world is possible" suggested the beginnings of a counterargument to globalization that went beyond protest and opposition. But the World Social Forum was itself a hybrid of nongovernmental organizations, developing country governments and heads of state, indigenous groups, oppositional labor and environmental groups, and assorted prominent globalization critics. How could the declarations that "another world is possible" translate into action at the local level, where as Zygmunt Bauman had argued, the misery is most directly experienced and the power of the forces of globalization need to be confronted? And how could that also translate into a strategy for action and transformation—and the need for reinvention—within a global city like Los Angeles?[8]

As this book has suggested, the arguments about urban nature and community and the right to the city are essentially about an approach that has been called *globalization from below.* This includes the need for new human and civic rights as well as new sets of ethical, environmental, labor, and justice-based values within any global process—whether production, distribution of goods, cultural life, use of space, or food choice. The need for new human rights is imperative in the face of the restrictions and exploitations of immigrants. It requires a new conception of citizenship that is transnational and affords rights that cross borders. It requires an approach to work and labor that dignifies rather than demeans it, that utilizes and enhances skills, and that avoids funneling immigrant workers nearly exclusively into the three Ds (dirty, difficult, dangerous—and low-paying—work) through such strategies as guest-worker programs or the highly exploited use of workers with undocumented status. Growing food, for example, illustrates skills that have been submerged in the global city. Immigrant food growers, in a globalization-from-below approach, should thus be celebrated for their contribution to creating more livable—and food secure—places.[9]

Thai jasmine rice provides a small example of what could be considered a value-based approach to any global-oriented production chain. The "value chain" in this context is a sustainable livelihood within a sustainable development framework. In a food cooperative in northeast Thailand, the organic and sustainable food-growing practices of the farmers, the deep cultural connections to the product and the process of growing, the cooperative control over the milling and distribution of the jasmine rice, and an expansion of fair-trade practices establish a direct transnational connection between producer (the country of origin) and consumer (the immigrant community). In this manner, the concept of privileging "local" and "place-based," an essential part of community food security and other social- and environmental-justice goals, assumes a transnational dimension. The global, in this instance, mirrors many of the values of the local.[10]

The struggles that comprise the globalization-from-below approach have invariably been local but have demonstrated global significance. Yet there is no clear coherent counterforce to globalization from above, even as a "global class war" unfolds, precipitated by agreements like NAFTA and other neoliberal economic policies, as Jeff Faux has argued. But despite the triumphalist nature of the arguments of the globalization advocates, a contest is under way regarding globalization's restructuring of communities, nation states, and the global city. Los Angeles represents a key test of whether that contest can result in changes that begin to demonstrate that another world is indeed possible. For a city steeped in conflict throughout its over two-hundred-year history, where the battlegrounds around race, ethnicity, class, and nature and community have been fierce and bitter, those changes will require nothing less than a reinvention of this global city.[11]

Notes

Introduction

1. <http://www.hollywoodforever.com/Cemetery.htm>.

2. See Robert Gottlieb, Andrea Azuma, and Amanda Shaffer, *Re-Envisioning the L.A. River: A Program of Community and Ecological Revitalization*, (Los Angeles: Urban and Environmental Policy Institute, 2001), Final Report to the California Council for the Humanities, 12.

3. The term *Los Angeles*, as it is used in this book, refers to both the city of Los Angeles and the larger region. Though the more appropriate name for the region is Southern California (a region that includes as many as six counties) the term *Los Angeles* is used as a shorthand term for both city and region. Los Angeles has also become a common reference point (though not entirely accurate) for a generic urban form (such as metropolitan sprawl and low-density population dispersal) that was often used to characterize urbanization processes in the twentieth century. But Los Angeles as a multicultural city and region and as a place undergoing rapid political and demographic change has also come to represent the imaginative possibilities of reinvention of the global city and of metropolitan regions in the twenty-first century.

4. William Cronon, *Nature's Metropolis: Chicago and the Great West* (New York: Norton, 1991).

5. William Cronon, foreword to the paperback edition of *Uncommon Ground: Rethinking the Human Place in Nature*, ed. William Cronon (New York: Norton, 1996), 20, 22; see also

William Cronon, "The Trouble with Wilderness; or, Getting Back to the Wrong Nature" (and comments in response), *Environmental History* 1, no. 1, (January 1996): 7–55.

6. Alexander Wilson, *The Culture of Nature: North American Landscape from Disney to the* Exxon Valdez (Cambridge: Blackwell, 1992), 87.

7. The words "where we live, work, and play," were used by the late Dana Alston in her talk at the 1991 People of Color Environmental Leadership Conference and subsequently became a common reference for environmental-justice activists and researchers. A copy of Alston's talk is in the possession of the author and is cited in Robert Gottlieb, *Forcing the Spring: The Transformation of the American Environmental Movement* (Washington, DC: Island Press, 2005), x.

8. David Harvey's comment is cited by Roger Keil and John Graham in their essay "Reasserting Nature," in *Remaking Reality: Nature at the Millenium*, ed. Bruce Braun and Noel Castree (New York: Routledge, 1998), 28.

9. Carey McWilliams, *Southern California: An Island on the Land* (Santa Barbara: Peregrine Smith, 1973), 21; Robert Gottlieb et al., *The Next Los Angeles: The Struggle for a Livable City* (Berkeley: University of California Press, 2006).

10. The global-city literature of the past three decades provides a rich theoretical and descriptive framework for the discussion of the changes described in this book. This includes the writings, among many others, of a "Los Angeles School" that came to prominence in the 1980s and 1990s. These Los Angeles–focused global-city theoreticians included such figures as John Friedmann, Ed Soja, and Allen Scott from the University of California at Los Angeles and Michael Dear from the University of Southern California, among others. In addition, various historians, journalists, policy analysts, and public intellectuals, such as Mike Davis, have added to an increasingly rich and diverse literature about Los Angeles.

11. "Pollution Prevention Education and Research Center: 1991–1995," Pollution Prevention Education and Research Center/Urban and Environmental Policy Institute, Los Angeles, 1997.

12. Urban and Environmental Policy Institute, at <http://departments.oxy.edu/uepi/about/index.htm>; see also Occidental College, "Learning by Doing," at <http://www.oxy.edu/x676.xml>.

13. Federation of American Immigration Reform, "Immigration 101: A Primer on Immigration and the Need for Reform," available at <http://www.fairus.org/site/PageServer?pagename=research_researcha233>; Carl Pope, "The Virus of Hate," *Sierra Magazine*, (May–June 2004), available at <http://www.sierraclub.org/sierra/200405/ways.asp>.

CHAPTER 1

1. Raymond Williams, *Keywords: A Vocabulary of Culture and Society* (London: Fontana, 1976), 2190.

2. Robert Gottlieb and Andrea Misako Azuma, "Re-Envisioning the Los Angeles River: An NGO and Academic Institute Influence the Policy Discourse," *Golden Gate University Law Review*, 35, no. 3, (Spring 2005): 321–342; Paul Stanton Kibel, "Los Angeles' Cornfield: An Old Blueprint for New Greenspace," *Stanford Environmental Law Journal* 23 (2004): 275, 308.

3. Personal communication with Sean Woods and Dianna Martinez, California Department of Parks and Recreation, March 17, 2006. For a discussion of the planning issues that are associated with the development of an urban park and the ways that such parks reflect the tensions in the mission of the parks bureaucracy "between preservation for the future and available public use in the present," see Hal K. Rothman, *The New Urban Park: Golden Gate National Recreation Area and Civic Environmentalism* (Lawrence: University Press of Kansas, 2003), ix.

4. Lauren Bon, <http://www.notacornfield.info>, July 20, 2005.

5. Personal communication with Lauren Bon, March 17, 2006; personal communication with Lewis MacAdams, March 22, 2006.

6. Carey McWilliams, *Southern California: An Island on the Land* (Santa Barbara, CA: Peregrine Smith, 1973), 21.

7. William Wordsworth, *The Prelude: 1799, 1805, 1850*, ed. Jonathan Wordsworth, M. H. Abrams, and Stephen Gill (New York: Norton, 1979), book 6, p. 218; Phil Macnaghten and John Urry, *Contested Natures* (London: Sage, 1998), 200, 183.

8. See the Sierra Club Web site at "Protect Nature," <http://www.sierraclub.org/vision/nature.asp>, and at "A Special Place," <http://www.sierraclub.org/specialplace/yourplaces/>.

9. Bruno Latour, *Politics of Nature: How to Bring the Sciences into Democracy* (Cambridge: Harvard University Press, 2004), 43; Raymond Williams, "Ideas of Nature," in *Ecology, the Shaping Enquiry*, ed. Jonathan Benthall (London: Longman, 1972), 158–59; J. B. Jackson, "The Imitation of Nature," *Landscape* 9, no. 1 (Autumn 1959), reprinted in Ervin H. Zube, ed., *Landscapes: Selected Writings of J. B. Jackson* (Amherst: University of Massachusetts Press, 1970), 79.

10. Bruno Latour, *We Have Never Been Modern* (Cambridge: Harvard University Press, 1993), 12; Maria Kaika, *City of Flows: Modernity, Nature and the City* (New York: Routledge, 2005), 21.

11. Bill McKibben, *The End of Nature* (New York: Random House, 1989); Marla Cone, "Polar Bears Face New Toxic Threat: Flame Retardants," *Los Angeles Times,* January 9, 2006; Brian Payton, "On Thin Ice," *Los Angeles Times*, February 11, 2006.

12. *The Historical Ecology Handbook*, ed. Dave Egan and Evelyn A. Howell, (Washington, DC: Island Press, 2001), 1; Alexander Wilson, *The Culture of Nature: North American Landscape from Disney to the* Exxon Valdez (Cambridge: Blackwell, 1992), 17, 115.

13. William R. Jordan III, "Restoration, Community and Wilderness," in *Restoring Nature: Perspectives from the Social Sciences and Humanities*, ed. Paul Gobster and R. Bruce Hull (Washington DC: Island Press, 2000), 24–25; Eric Katz, "The Problem of Ecological Restoration," *Environmental Ethics* 18, no. 2 (Summer 1996), 223; see also *The Society for Ecological Restoration Primer on Ecological Restoration*, Version 2, October 2004, at <http://www.ser.org/content/ecological_restoration_primer.asp#3>.

14. Cindi Katz, "Whose Nature, Whose Culture? Private Productions of Space and the Preservation of Nature," in *Remaking Reality: Nature at the Millenium*, ed. Bruce Braun and Noel Castree (New York: Routledge, 1998), 57. From an historical perspective, William Cronon also criticized the underlying thesis of the restorationists and their privileging of the concept of "native." "What's problematic about *native* as a word," Cronon told the *New York Times,* commenting on a "return-of-the-natives" program in New York City, "is that it assumes that there is a moment in history we can fix that was stable and benign and anything since then should be reversed. That's much too crude a picture of how history happens." Kirk Johnson, "The Return of the Natives: Playing God in Urban Fields," *New York Times*, November 12, 2000.

15. Aldo Leopold, *Sand County Almanac*, preface, p. viii; Raymond Williams, *Problems in Materialism and Culture*: Selected Essays (New York: Verso Books, 1997 [1980]), 67; see also

Castree and Braun, "The Construction of Nature and the Nature of Construction," in *Remaking Reality*, 6.

16. Lewis Mumford, *The City in History: Its Origins, Its Transformation, and Its Prospects* (New York: Harcourt, Brace, 1961), 459–60, 462; Macnaghten and Urry, *Contested Natures*, 175.

17. Robert Gottlieb, *Forcing the Spring: The Transformation of the American Environmental Movement* (Washington, DC: Island Press, 1993), 16–19, 59–67.

18. Benton MacKaye, "The Townless Highway," *The New Republic* 62, no. 797 (March 12, 1930): 93–95, and *The New Exploration: A Philosophy of Regional Planning* (Urbana: University of Illinois Press, 1962); see also Lewis Mumford, *The Highway and the City* (New York: Harcourt, Brace & World, 1963), 239–40; Kermit C. Parsons and David Schuyler eds., *From Garden City to Green City: The Legacy of Ebenezer Howard* (Baltimore: Johns Hopkins Press, 2002).

19. *Los Angeles Times* publisher Otis Chandler is quoted in "Changing Times," *Time*, April 25, 1960, 85. The *Time* article also characterized Los Angeles as "a city in which explosive change is routine." Deyan Sudjic, *The Hundred-Mile City* (San Diego: Harcourt Brace, 1992).

20. "Taming the Mountains," *Los Angeles Times*, Janury 3, 1933; John McPhee, "Los Angeles against the Mountains," in *Land of Sunshine: An Environmental History of Metropolitan Los Angeles*, ed. William Deverell and Greg Hise (Pittsburgh: University of Pittsburgh Press, 2005), 179–200; Carol L. Forest and Michael V. Harding, "Erosion and Sediment Control: Preventing Additional Disasters after the Southern California Fires," *Journal of Soil and Water Conservation* 49, no. 6 (November 1994): 535.

21. <http://www.beachcalifornia.com/westlake.html>.

22. "First Two Hundred Homes Set for Westlake Village," *Los Angeles Times*, March 13, 1966; "Dam Being Built for Westlake Village Lake," *Los Angeles Times*, August 21, 1966; Frank J. Tysen, "Nature and the Urban Dweller," in *Man and Nature in the City* (Washington, DC: Bureau of Sport Fisheries and Wildlife, 1968), 12.

23. Robert M. Fogelson, *Bourgeois Nightmares: Suburbia, 1870–1930*, (New Haven: Yale University Press, 2005), 180.

24. Greg Hise, *Magnetic Los Angeles: Planning the Twentieth-Century Metropolis* (Baltimore: Johns Hopkins University Press, 1997), 12.

25. Becky M. Nicolaides, *My Blue Heaven: Life and Politics in the Working Class Suburbs of Los Angeles*, 1920–1965 (Chicago: University of Chicago Press, 2002), 26.

26. "National Historic Landmark Designation—Baldwin Hills Village," p. 56, at <http://www.cr.nps.gov/nhl/designations/samples/CA/baldwin.pdf>; Mike Davis, *Ecology of Fear: Los Angeles and the Imagination of Disaster* (New York: Metropolitan Books, 1998), 73.

27. Ellen Stern Harris, "The Hard-Learned Lessons of a Coastal Preserver," *Los Angeles Times*, September 20, 1987.

28. Robert Gottlieb, *Environmentalism Unbound: Exploring New Pathways for Change* (Cambridge, MA: MIT Press, 2001); personal communication with Dorothy Green, June 11, 1991.

29. Gottlieb, *Forcing the Spring*, 161–64.

30. An important example of this emerging—and linked coalition—of advocates who were focused on enhancing a nature in the city agenda while addressing issues of polluted environments was the development of the Working Group for a Just and Sustainable Future. The group's various task forces were established to identify a broad agenda for greening L.A. in the wake of Antonio Villaraigosa's election as mayor in 2005 and his often repeated pledge to "green L.A." Cofacilitated by UEPI's Martha Matsuoka, this process resulted in a document that was designed to serve as a benchmark for advocacy and coalition building as well as to influence future policies. See "A Green Los Angeles: Recommendations to the City of Los Angeles," Working Group for a Just and Sustainable Future Los Angeles, 2006.

31. The American-Lawns organization at <http://www.american-lawns.com/index.html>.

32. Michael Pollan, *Second Nature: A Gardener's Education* (New York: Dell Publishing, 1991), 66; Virginia Scott Jenkins, *The Lawn: A History of an American Obsession* (Washington, DC: Smithsonian Institution Press, 1994); Ted Steinberg, *The Obsessive Quest for the Perfect Lawn* (New York: Norton, 2006).

33. Harold Van Du Zee, "Address at F. W. Taylor Memorial to 'Boxly,'" in *Frederick Winslow Taylor: A Memorial* (New York: Taylor Society, 1920); Charles V. Piper and Russell A. Oakley, *Turf for Golf Courses* (New York: MacMillian, 1929), 215; Jenkins, *The Lawn*; 60; See also "The History of Lawns in America" at <http://www.american-lawns.com/history/history_lawn.html>.

34. By the turn of the twenty-first century, Los Angeles's fastest-growing suburb, the city of Palmdale, had instituted a rule that made it illegal for residents not to maintain their

grass by setting a height limit on weeds and proscribing the use of yards to park cars. Some cities mandated particular types of lawns that were water-intensive and often required significant pesticide use to keep them from turning brown or weed-infested. In Salt Lake City, for example, a zoning ordinance mandated that all front lawns be covered with flat green grass that required constant watering. The ordinance was challenged in 2006. Melissa Sanford, "Salt Lake City Moving toward Less Thirsty Lawns," *New York Times*, August 25, 2006; Martha L. Willman, "'Lawn Police' May Be Coming to Palmdale," *Los Angeles Times*, May 6, 2001; see also Robert Messia, "Lawns as Artifacts: The Evolution of Social and Environmental Implications of Suburban Residential Land Use," in *Suburban Sprawl: Culture, Theory and Politics*, ed. Matthew J. Lindstrom and Hugh Bartling (London: Rowman & Littlefield, 2003), 74–75; Ted Steinberg, "Lawn and Landscape in World Context, 1945–2000," *Magazine of History* 19, no. 6 (November 2005): 62–68; and Ted Steinberg, "Lawn Mores," *Los Angeles Times*, March 18, 2006.

35. Paul Robbins and Trevor Birkenholtz, "Turfgrass Revolution: Measuring the Expansion of the American Lawn," *Land Use Policy* 20, (April 2003): 181–94.

36. *Redesigning the American Lawn: A Search for Environmental Harmony*, ed. F. Herbert Bormann, Diana Balmori, and Gordon T. Geballe (New Haven: Yale University Press, 1993), 68–70, 88–89.

37. Ilan Brat, "Green Thumb's Profits Grow: Scotts Dominates Lawn and Garden Business by Branching out," *Wall Street Journal*, April 17, 2006; Amy Braunschweiger, "Scotts Sees Green as Gardening Booms," *Wall Street Journal*, June 4, 2003; Paul Robbins and Julie T. Sharp, "Producing and Consuming Chemicals: The Moral Economy of the American Lawn," *Economic Geography* 79, no. 4 (October 2003): 425; "Project Evergreen Strives to Protect the Green Industry," *Grounds Maintenance*, 39, no. 10 (October 2004): 9–10.

38. "The Environmental Benefits of Healthy Lawns," Scotts Company, at <http://www.scotts.com/scotts-sites/about/images/pdf/Green_Lawns_broch.pdf>; David Barboza, "Suburban Genetics: Scientists Searching for a Perfect Lawn," *New York Times*, July 9, 2000.

39. Andrew Pollack, "Application on Modified Grass Is Withdrawn," *New York Times*, October 24, 2002; Michael Hawthorne, "Genetically Modified Grass Must Undergo Federal Environmental Study," *San Luis Obispo Tribune*, September 26, 2004.

40. Ron Hall, "Canada Activism Picks on Lawn Care," *Landscape Management* 45, no. 5 (May 2006): 29; Jason Stahl, "Under Attack," *Landscape Management* 43, no. 2 (February 2004): 22–28; Preston Lerner, "Whither the Lawn?," *Los Angeles Times*, May 4, 2003; Joy L. Woodson, "Lawndale's Stand against Natural Yard Withers," *Los Angeles Times*, February 4,

2003; Corey Kilgannon, "For Some, the Grass Is Greener Where There Isn't Any," *New York Times*, November 27, 2005; see also Paul Robbins and Julie Sharp, "The Lawn-Chemical Economy and Its Discontents," *Antipode* 35 (November 2003): 973.

41. Pollan, *Second Nature*, 198.

42. Dolores Hayden, *The Power of Place: Urban Landscapes as Public History* (Cambridge: MIT Press, 1995), 62–63; Dolores Hayden, *Building Suburbia: Green Fields and Urban Growth, 1820–2000* (New York: Pantheon Books, 2003), 73; Dana W. Bartlett, *The Better City: A Sociological Study of a Modern City* (Los Angeles: Neuner Company Press, 1907), 72; William H. Mathews, "The House Courts of Los Angeles," *The Survey* 30 (July 5, 1913): 466, cited in Douglas Monroy, *Rebirth: Mexican Los Angeles from the Great Migration to the Great Depression* (Berkeley: University of California Press, 1999), 37; Douglas C. Sackman, "A Garden of Worldly Delights," in *Land of Sunshine*, ed. Deverell and Hise, 261.

43. Gottieb, *Environmentalism* Unbound, 243–45.

44. "Community Garden Policy for the City of Los Angeles," Los Angeles Food Security and Hunger Partnership, Los Angeles Community Development Department (1997) (in author's possession); see also Robert Gottlieb, "Peas in our Time," *Los Angeles Times*, June 25, 2006.

45. On the value of community gardens and their barriers, see Karl Linn, "Reclaiming the Sacred Commons," *New Village* 1, no.1 (1999): 42–49; Laura J. Lawson, *City Bountiful: A Century of Community Gardening in America* (Berkeley: University of California Press, 2005); Patricia H. Hynes, A *Patch of Eden: America's Inner City Gardeners* (White River Junction, VT: Chelsea Green, 1996).

46. Dan Barry, "Sudden Deal Saves Gardens Set for Auction," *New York Times*, May 13, 1999.

47. Henry W. Lawrence, "The Greening of the Squares of London: Transformation of Urban Landscapes and Ideals," *Annals of the Association of American Geographers* 83, no. 1 (March 1993): 104.

48. Daniel T. Rodgers, *Atlantic Crossings: Social Politics in a Progressive Age* (Cambridge: Belknap Press of Harvard University Press, 1998), 195; John Muir, *The Yosemite* (New York: Century, 1912), 261–62; Peter J. Schmitt *Back to Nature: The Arcadian Myth in Urban America* (New York: Johns Hopkins University Press, 1990), 73–74; Allen F. Davis, *Spearheads for Reform: The Social Settlements and the Progressive Movement, 1890–1914* (New Brunswick, NJ: Rutgers

University Press, 1984), 61–65; Robert M. Ricard, "Shade Trees and Tree Wardens: Revising the History of Urban Forestry," *Journal of Forestry* 103, no. 5 (July–August 2005): 230–33.

49. Henry F. Arnold, *Trees in Urban Design* (New York: Van Nostrand Reinhold, 1993), 147.

50. Andy Lipkis, "Creating an Urban Watershed in Los Angeles," Presentation for the Re-Envisioning the Los Angeles River program, March 8, 2000; Judith Lewis, "Citizen Andy Lipkis Speaks for the Trees," *L.A. Weekly*, April 15–21, 2005; "TreePeople's Andy Lipkis Redefines Environmental Pragmatism," *The Planning Report* (March 1999); "For the Trees, for the People," *Los Angeles Times*, February 8, 2003; Tracy Rysavy, "Tree People," *Yes!* (Winter 2000), available at <http://www.futurenet.org/article.asp?id=318>.

51. The Gary Moll quote is from a commentary by John Alderman, "Do Urban Trees Really Help Reduce Pollution and Clean the Air?," September 2, 2004 (in author's possession). The cost-benefit ratio of urban trees is discussed in Greg McPherson et al., "Municipal Forest Benefits and Costs in Five U.S. Cities," *Journal of Forestry* 103, no. 8 (December 2005): 411–16; see also David J. Nowak and John E. Dwyer, "Understanding the Benefits and Costs of Urban Forest Ecosystems," in *Urban and Community Forestry in the Northeast*, ed. J. Kuser, (New York: Plenum Press, 2000), 11–25.

52. Galen Cranz, *The Politics of Park Design: A History of Urban Parks in America* (Cambridge: MIT Press, 1982), 29; Julie A. Tuason, "*Rus in Urbe*: The Spatial Evolution of Urban Parks in the United States, 1850–1920," *Historical Geography* 25 (1997): 125.

53. A children's right to play was proposed as part of the United Nations Declaration of the Rights of the Child in 1959, cited in Anastasia Loukaitou-Sideris and Orit Stieglitz, "Children in Los Angeles Parks: A Study of Equity, Quality and Children's satisfaction with neighbourhood parks," *TPR* 73, no. 4, (2002): 472; see also Jeff Hayward, "Urban Parks: Research, Planning and Social Change," *in Public Places and Spaces*, ed. Irwin Altman and Ervin Zube (New York: Plenum Press, 1989); Galen Cranz, "Changing Role of Urban Parks: From Pleasure Gardens to Open Space," *Landscape*, 22 (1978): 9–18.

54. Cranz, *The Politics of Park Design*, 123.

55. P. Harnik, *Inside City Parks* (Washington, DC: Urban Land Institute, 2001); Loukaitou-Sideris and Stieglitz, "Children in Los Angeles Parks," 474–475. In Dana Bartlett's 1907 discussion of Los Angeles, his City Beautiful perspective led him to argue that "in proportion to its population, Los Angeles ranks well in the size of its parks," but he warned that its potential for significant population expansion could cause its park-to-population ratio to

change significantly and that the city would need to begin an active program of land acquisition to sustain any kind of "green city" landscape. Barlett, *The Better City*, 45.

56. Personal communication with Penny Newman, August 3, 2004; Gottlieb, *Forcing the Spring*, 10–11.

57. Eric Klinenberg, *Heat Wave: A Social Autopsy of Disaster in Chicago* (Chicago: University of Chicago Press, 2002), 16, 21; see also "Dying Alone: An Interview with Eric Klinenberg," at <http://www.press.uchicago.edu/Misc/Chicago/443213in.html>. The urban heat-island effect entered the vocabulary about nature in the city in the late 1960s, when the first studies pointed to a significant change in temperature due to such factors as paved surfaces. It also contributed to the arguments of some urban environmentalists in Los Angeles about the need to "unpave L.A.," as one environmental group put it. See William P. Lowry, "The Climate of Cities," *Scientific American* 217, no. 2 (August 1967): 15–23.

58. Craig Colten, *An Unnatural Metropolis: Wresting New Orleans from Nature* (Baton Rouge: Louisiana State University Press, 2005), 2; Ian Burton, Robert W. Kates, and Gilbert W. White, *The Environment as Hazard* (New York: Guilford Press, 1993), 31.

59. John McPhee, "Los Angeles against the Mountains," in *The Control of Nature* (New York: Farrar, Straus & Giroux, 1989); see also William Chaloupka and R. McGreggor Cawley, "The Great Wild Hope: Nature, Environmentalism, and the Open Secret," in *In the Nature of Things: Language, Politics, and the Environment*, ed. Jane Bennett and William Chaloupka (Minneapolis: University of Minnesota Press, 1993), 3–4.

60. Raymond Williams, *The Country and the City* (New York: Oxford University Press, 1973); Fraser Harrison, "England, Home and Beauty," in *Second Nature*, ed. Richard Mabey (London: Jonathan Cape, 1984), 169–70.

61. Rachel Kaplan, Stephen Kaplan, and Robert L. Ryan, *With People in Mind: Design and Management of Everyday Nature* (Washington, DC: Island Press, 1998), 99; Stephen Kaplan and Rachel Kaplan, "Health, Supportive Environments, and the Reasonable Person Model," *American Journal of Public Health* 93, no. 9 (September 2003): 1484; see also Richard Louv, *Last Child in the Woods: Saving Our Children from Nature-Deficit Disorder* (Chapel Hill: Algonquin Books of Chapel Hill, 2005).

62. Susan M. Stuart, "Lifting Spirits: Creating Gardens in California Domestic Violence Shelters," in *Urban Place: Reconnecting with the Natural World*, ed. Peggy F. Barlett (Cambridge: MIT Press, 2005), 71–72.

63. "As It Is," *Los Angeles Times*, December 4, 1891, 2.

64. Roger Keil, *Los Angeles: Globalization, Urbanization and Social Struggles* (New York: Wiley, 1998), 35.

65. Latour, *We Have Never Been Modern*, 6; Ader Wilson, *The Culture of Nature*, 33.

66. Latour, *We Have Never Been Modern*, 10; Margaret FitzSimmons and David Goodman, "Incorporating Nature," in *Remaking Reality*, ed. Braun and Castree, 207; Williams, *Keywords*, 220, 221; David Harvey, "The Nature of Environment: Dialectics of Social and Environmental Change," in *Real Problems, False Solutions: Socialist Register 1993*, ed. Ralph Miliband and Leo Panitch (London: Merlin Press, 1993), 41.

67. Peter Evans, "Introduction," in *Livable Cities: Urban Struggles for Livelihood and Sustainability*, ed. Peter Evans (Berkeley, University of California Press, 2002), 1.

Chapter 2

1. Thomas Bender, *Community and Social Change in America* (New Brunswick, NJ: Rutgers University Press, 1978), 6.

2. *The Boyle Heights Oral History Project: A Multiethnic and Collaborative Exploration of a Los Angeles Neighborhood, Interview Summaries and Essays* (Los Angeles: Japanese American National Museum, 2002); see also Boyle Heights: Neighborhood Sites and Insights, an exhibit of the Japanese American National Museum, at <http://www.imls.gov/grants/museum/pdf/nlg01nmc.pdf>; Manal J. Aboelata, *The Built Environment and Health: Eleven Profiles of Neighborhood Transformation* (Oakland, CA: Prevention Institute, July 2004); personal communication with James Rojas, October 16, 2005.

3. Deborah Kong, "Community Action Creates Jogging Path in L.A," *Children's Advocate* (July–August 2005), available at <http://www.4children.org/news/705gse.htm>; see also Hector Becerra, "Setting the Pace of Life," *Los Angeles Times*, March 15, 2005; Lourdes Lopez, "Un sendero para el Este de Los Angeles?," *La Opinion*, April 27, 2003.

4. Peter Dreier et al., "Movement Mayor: Can Antonio Villaraigosa Change Los Angeles?," *Dissent* (Summer 2006), available at <http://www.dissentmagazine.org/article/?article=656>. Robert Gottlieb et al, *The Next Los Angeles: The Struggle for a Livable City* (Berkeley: University of California Press, 2006).

5. Personal communication with Mary Nichols, June 29, 2005.

6. Cara Mia DiMassa, "Differing Views of Race in L.A. Collide in *Crash*," *Los Angeles Times*, March 2, 2006.

7. Bender, *Community and Social Change in America*, 5; Raymond Williams, *Keywords: A Vocabulary of Culture and Society* (London: Fontana, 1976), 76.

8. Ferdinand Tonnies, *Community and Association* (London: Routledge & Paul, 1955); Louis Wirth, "Urbanism as a Way of Life," *American Journal of Sociology* 44, no. 1 (July 1938): 8; Richard Sennett, *The Conscience of the Eye: The Design and Social Life of Cities* (New York: Norton, 1990), 24; Bender, *Community and Social Change in America*, 15; Robert Bellah, *The Good Society* (New York: Vintage Books, 1992).

9. Floyd Norris, "Win-Win? Tell it to the Losers," *New York Times*, January 27, 2006.

10. Iris Marion Young, "The Ideal of Community and the Politics of Difference," in *Feminism/Postmodernism*, ed. Linda Nicholson (New York: Routledge, 1990), 311–12; David Harvey, "Class Relations, Social Justice and the Politics of Difference," in *Place and the Politics of Identity*, ed. Michael Keith and Steve Pile (London: Routledge, 1993), 55.

11. Florence Kelley, *Hull House Maps and Papers* (New York: Crowell, 1895).

12. Robert Gottlieb, *Forcing the Spring: The Transformation of the American Environmental Movement* (Washington, DC: Island Press, 1993), 184–92.

13. "A Call to Action," First National People of Color Environmental Leadership Summit, October 24–27, 1991 (handbill distributed to delegates), *Proceedings of the First National People of Color Environmental Leadership Summit* (New York: Commission for Racial Justice of the United Church of Christ, 1993); Deeohn Ferris, "A Call for Justice and Equal Environmental Protection," and Karl Grossman, "The People of Color Environmental Summit," in *Unequal Protection: Environmental Justice and Communities of Color*, ed. Robert D. Bullard (San Francisco: Sierra Club Books, 1994), 298, 272–79.

14. Sheila Foster, "Race(ial) Matters: The Quest for Environmental Justice," *Ecological Law Quarterly* 20 (1993): 721–53; Robert Gottlieb, *Environmentalism Unbound: Exploring New Pathways for Change* (Cambridge: MIT Press, 2001), 51–72.

15. Peter Marcuse has argued that the types of division that separate social groups create "barriers, limits, boundaries, borders, divisions, [and] separation" and that the goal of a wall-less city requires problem solving at the neighborhood and even the city level and involves a "large-scale confrontation . . . with the fundamental problems of the society" that have contributed to the erection of such walls. Peter Marcuse, "The Goal of the Wall-Less City: New York, Los Angeles and Berlin," Harvey S. Perloff Lecture, Graduate School of Architecture and Urban Planning, UCLA, May 14, 1992, 2–3; see also Setha Low, *Behind the*

Gates: Life, Security, and the Pursuit of Happiness in Fortress America (New York: Routledge, 2003).

16. Raymond Williams, *The Country and the City* (New York: Oxford University Press, 1973), 231.

17. Robert Gottlieb, Mark Vallianatos, Regina Freer, and Peter Dreier, *The Next Los Angeles: The Struggle for a Livable City* (Berkeley: University of California Press, 2005), 11; Mary Ann Callan, "Suburbs in Quest of City," *Los Angeles Times*, June 2, 1960.

18. George J. Sanchez, "'What's Good for Boyle Heights Is Good for the Jews: Creating Multiracialism on the Eastside during the 1950s," *American Quarterly* 56, no. 3 (2004): 663–61; "Boyle Heights: California's Sociological Fishbowl," *Fortnight*, October 20, 1954, 20; Ralph Friedman, "U.N. in Microcosm; Boyle Heights: An Example of Democratic Progress," *Frontier: The Voice of the New West*, March 1955, 12.

19. The Home Owners' Loan Corporation ratings system is discussed in Eric Avila, *Popular Culture in the Age of White Flight: Fear and Fantasy in Suburban Los Angeles* (Berkeley: University of California Press, 2004), 34–35. Latino Urban Forum activist James Rojas described how the Boyle Heights neighborhood where he grew up in the 1960s still had a wealth of local food stores and restaurants but no fast-food outlets. "We had to go to Montebello [a middle-class suburb to the east of Boyle Heights] to get fast food," Rojas recalled, a situation that began to change in the 1980s and 1990s. Boyle Heights today is filled with fast-food outlets that, Rojas contends is indicative of the pressures that undermine the sense of place. Personal communication with James Rojas, October 19, 2005.

20. Carey McWilliams, "Los Angeles," *Overland Monthly and Out West Magazine* 85 (May 1927): 135; Besse Averne McClenahan, *The Changing Urban Neighborhood: A Sociological Study* (Los Angeles: University of Southern California, 1929), 103.

21. ". . . and Now We Plan", an exhibit at the Los Angeles County Museum, October 22, 1941–January 18, 1942; Fred W. Viehe, "Black Gold Suburbs: The Influence of the Extractive Industry on the Suburbanization of Los Angeles, 1890–1930," *Journal of Urban History* 8, no. 1 (November 1981): 3–26.

22. Robert Bruegmann, *Sprawl: A Compact History* (Chicago: University of Chicago Press, 2005); Joel Kotkin, "In Praise of Suburbs," *San Francisco Chronicle*, January 29, 2006.

23. The Illinois suburb Internet ad is discussed by Robert Putnam in *Bowling Alone: The Collapse and Revival of American Community* (New York: Simon & Schuster, 2000), 205.

Information about this suburb can also be found on its Web site at <http://www.concordhomes.com/corporate/about/history.asp?menuOpen=4f11>. Ronald Reagan's Temecula speech can be found at <http://en.wikipedia.org/wiki/Temecula,_California#Ronald_Reagan_and_Temecula>, while information about Temecula's demographics and housing costs can be found at the city's Web site at <http://www.cityoftemecula.org/Temecula/Businesses/Demographics/Housing/>.

24. Robert Putnam, "Bowling Alone: America's Declining Social Capital," *Current*, no. 373 (June 1995): 3–9; Witold Rybczynski, *City Life: Urban Expectations in a New World* (New York: Scribner, 1995), 194–96. On the relationship between sprawl and obesity, see Reid Ewing et al., "Relationship between Urban Sprawl and Physical Activity, Obesity and Morbidity," *American Journal of Health Promotion* 18, no. 1 (September–October 2003). On the jobs-housing mismatch, see Robert Cervero, "Jobs/Housing Balancing and Regional Mobility," *Journal of the American Planning Association* 55, no. 1 (1989): 24–37; Martin Wachs et al., "The Changing Commute: A Case Study of the Jobs-Housing Relationship over Time," *Urban Studies* 30, no. 10 (December 1993): 1711–1729; Richard Arnott, "Economic Theory and the Spatial Mismatch Hypothesis," *Urban Studies* 35, no. 7 (June 1998): 1171–85.

25. In his classic study of the history of suburban development, Kenneth Jackson argues that the suburbs have become perhaps more representative of U.S. culture than "big cars, tall buildings, or professional football. Suburbia symbolizes the fullest, most unadulterated embodiment of contemporary culture; it is a manifestation of such fundamental characteristics of American society as conspicuous consumption, a reliance upon the private automobile, upward mobility, the separation of the family into nuclear units, the widening division between work and leisure, and a tendency toward racial and economic exclusiveness." Kenneth Jackson, *Crabgrass Frontier: The Suburbanization of the United States* (New York: Oxford University Press, 1985), 4. Problems of Central Valley air pollution turned its Sequoia–Kings Canyon wilderness area into the most polluted national park in the country, more polluted than New York City. See Gary Polakovic, "Polluted Paradise," *Los Angeles Times*, September 13, 2005; Robb Gurwitt, "Into the Haze," *Governing* 16, no. 9 (June 2003): 64.

26. William Fulton, *The New Urbanism: Hope or Hype for American Communities?* (Cambridge: Lincoln Institute of Land Policy, 1996).

27. The Congress of New Urbanism mission statement is reprinted in Andrés Duany and Elizabeth Plater-Zyberk, *Suburban Nation: The Rise of Sprawl and the Decline of the American Dream* (New York: North Point Press, 2000), 260–65. The Vincent Scully quote and description of the Celebration town seal come from Douglas Frantz and Catherine Collins, *Celebration USA: Living in Disney's Brave New Town* (New York: Holt, 1999), 23, 8. See also

Robert Steuteville, "An Alternative to Modern, Automobile-Oriented Planning and Development," available at <http://www.newurbannews.com/AboutNewUrbanism.html>.

28. David Mohney, "Interview with Andrés Duany," in *Seaside: Making a Town in America*, ed. David Mohney and Keller Easterling (New York: Princeton Architectural Press, 1991), 62.

29. The "80 percent solid Republican" statement is from Michael Lassell, *Celebration: The Story of a Town* (New York: Disney Editions, 2004), 125; see also Alex Marshall, *How Cities Work: Suburbs, Sprawl, and the Roads Not Taken* (Austin: University of Texas Press, 2000), 201; Andrew Ross, *The Celebration Chronicles*, 227–28. The Celebration Web site is at <http://www.celebrationfl.com/community/community.html>.

30. Steven Brooke, *Seaside* (Gretna, LA: Pelican, 1995), 22; Kurt Anderson, "Is Seaside Too Good to Be True?," in *Seaside: Making a Town in America*, ed. Mahoney and Easterling, 43; see also the Seaside Institute Web site at <http://www.theseasideinstitute.org/net/content/item.aspx?s=8629.0.79.7801>; Fred A. Bernstein, "Seaside at Twenty-five: Troubles in Paradise," *New York Times,* December 9, 2005.

31. Beth Dunlop, "In Florida, a New Emphasis on Design," available at Congress for the New Urbanism Florida chapter Web site at <http://www.cnuflorida.org/nu_florida/dunlop.htm>.

32. Bernstein, "Seaside at Twenty-five," D6.

33. Personal communication with Elizabeth Moule and Stephanos Polyzoides, June 1, 2005; Bradford McKee, "Gulf Planning Roils Residents," *New York Times*, December 8, 2005. Antonio Villaraigosa's "stylish density" comment is cited in Jim Newton, "Tall, Green, Vital: L.A. as Mayor Dreams It," *Los Angeles Times*, February 19, 2006.

34. Hank Ditmar and Gloria Ohland, *The New Transit Town: Best Practices in Transit-Oriented Development* (Washington, DC: Island Press, 2004); Dena Belzer and Gerald Autler, "Countering Sprawl with Transit-oriented Development," *Issues in Science and Technology* 19, no. 1 (Fall 2002): 51–58; Douglas S. Kelbaugh, "Repairing the American Metropolis," *Forum for Applied Research and Public Policy* 16 (Summer 2001): 6–12. The new urbanist approach to inner-city development is discussed in Kristin Larsen, "New Urbanism's Role in Inner City Revitalization," *Housing Studies* 20, no. 5 (September 2005): 795. Information about the Location Efficient Mortgage network is also available at <http://www.locationefficiency.com>. For a critical economic perspective on location-efficient mortgages as an economic tool, see Alan Blackman and Alan Krupnick,

"Location-Efficient Mortgages: Is the Rationale Sound?," *Journal of Policy Analysis and Management* 20, no. 4 (Fall 2001): 633–649.

35. Personal communication with Katherine Perez, March 18, 2005; Anna Holtzman, "Latin Invasion," *Architecture* 93, no. 3 (March 2004): 21.

36. "Urban space tends to be sliced up, degraded and eventually destroyed, [through] the proliferation of fast roads and of places to park and garage cars, and their corollary, a reduction of tree-lined streets, green spaces, and parks, and gardens," Simon Parker has argued, paraphrasing Levebvre's approach. "The contradiction lies, then, in the clash between a consumption of space which produces surplus value and one which produces only enjoyment—and is therefore unproductive." Simon Parker, *Urban Theory and Urban Experience: Encountering the City* (London: Routledge, 2004), 23. On the City Repair movement, see Sarah Kavage, "Governance: Reclaiming the Grid: Portland's City Repair," *The Next American City* 3 (October 2003), available at <http://www.americancity.org/article.php?id_article=64>.

37. David Harvey, "Cities or Urbanization," *City* 1/2 (1996): 38–61; Henri Lefebvre, *Le Droit à la ville* (Paris: Anthropos, 1974 [1968]), *Everyday Life in the Modern World* (New York: Harper & Row, 1971); Henry Lefebvre, *Henri Lefebvre: Key Writings* (London: Continuum, 2003); David Harvey, "The Right to the City," *International Journal of Urban and Regional Research* 27, no. 4 (2003): 939; David Brooks, "A Nation of Villages," *New York Times*, January 19, 2006; see also Mark Purcell, "Excavating Lefebvre: The Right to the City and Its Urban Politics of the Inhabitant," *GeoJournal* 58 (2002): 99–108.

38. The "neither city nor sprawl" comment is from Brooks, "A Nation of Villages."

39. Dolores Hayden, *The Power of Place: Urban Landscapes as Public History* (Cambridge: MIT Press, 1995), 46.

40. Young, "The Ideal of Community and the Politics of Difference," 319. Sophie Watson and Katherine Gibson similarly echo Young's argument, writing that "for cities to become more democratic, space needs to become less privatized, so that individuals can interact in the open, expressing both their differences and their commonalities. These are spaces which are not bounded by walls of exclusion or inclusion—they are spaces without walls." Sophie Watson and Katherine Gibson, "Postmodern Politics and Planning: A Postscript," in *Postmodern Cities and Spaces*, ed. Sophie Watson and Katherine Gibson (Cambridge: Blackwell, 1995), 261.

41. Anastasia Loukaitou-Sideris, "Urban Form and Social Context: Cultural Differentiation in the Uses of Urban Parks," *Journal of Planning Education and Research* 14 (1995): 89–102;

see also Stephanie Pincetl, "Nonprofits and Park Provision in Los Angeles: An Exploration of the Rise of Governance Approaches to the Provision of Local Services," *Social Science Quarterly* 84, no. 4 (December 2003): 979–1001; John M. Baas, Alan Ewert, and Deborah Chavez, "Influence of Ethnicity on Recreation and Natural Environment Use Patterns: Managing Recreation Sites for Ethnic and Racial Diversity," *Environmental Management* 17, no. 4 (1993): 523–29.

42. Jim Hightower, *Hard Tomatoes, Hard Times: A Report of the Agriculture Accountability Project on the Failure of America's Land Grant College Complex* (Cambridge: Shenkman, 1973); Young-bin Lee Yim, *Spatial Trips and Spatial Distribution of Food Stores*, University of California Transportation Center, Working Paper No. 125 (Berkeley: UCTC, 1993); Gottlieb, *Environmentalism Unbound*, 199–203.

43. Harry Brown-Hiegel, "History of Farmers' Markets in Southern California," Minutes, Southland Farmers' Market Association Retreat, Los Angeles, November 17, 1996; "Background to the Gardena Farmers' Market," Memo from Marion Kalb, Director of the Southland Farmers' Market Association, to Robert Gottlieb and Michelle Mascarenhas, June 4, 1996.

44. C. Clare Hinrichs, Gilbert W. Gillispie, and Gail W. Feenstra, "Social Learning and Innovation at Retail Farmers' Markets," *Rural Sociology* 69, no. 1 (March 2004): 31–58; Robin Summer, "Farmers' Markets as Community Events," in *Public Places and Spaces*, ed. Irwin Altman and Ervin Zube (New York: Plenum Press, 1989, 57–82; Thomas Lyson, Gilbert W. Gillespie, and D. Hilchey, "Farmers' Markets and the Local Community: Bridging the Formal and the Informal Economy," *American Journal of Alternative Agriculture* 10 (1995): 108–13.

45. Andrew Fisher, *Hot Peppers and Parking Lot Peaches: Evaluating Farmers' Markets in Low-Income Communities* (Los Angeles: Community Food Security Coalition, 1999).

46. Robert Gottlieb and Irene Wolt, *Thinking Big: The Story of the* Los Angeles Times, *Its Publishers, and their Influence on Southern California* (New York: Putnam, 1977). The letter from Norman Chandler to Norris Poulson is dated December 26, 1952.

47. Benjamin H. Bagdikian, *The New Media Monopoly* (Boston: Beacon Press, 2004).

48. Peter King, "A Dynasty, a City: As Dynasty Evolved, So Did Power in L.A.," *Los Angeles Times*, March 26, 2006.

49. In 2006, debate within the Tribune Company erupted between the Chandler family interests, which still held a significant stake in the company, and the Tribune Company management. Although the issue of local ownership and a Los Angeles identity was not the

major issue raised by the Chandlers, it led to a new focus about the *absence* of that Los Angeles identity, the family's relationship to ownership of the paper, and the continuing pressures to cut costs (including staff positions) to maintain a favorable stock price and return on investment. In September 2006, an open letter signed by twenty leading Los Angeles business, labor, community, and academic figures and reprinted in the *Los Angeles Times* weighed in on these issues, arguing that "the *Los Angeles Times* has a unique ability and responsibility to unify as well as educate what is a very geographically fractured and otherwise extraordinarily diverse community. At various times many of us have met with *Los Angeles Times* representatives to urge more thorough and consistent coverage of the Greater Los Angeles community. We have been assured that the *Times* was most committed in this regard, as well as committed to remaining one of the nation's great newspapers. But we remain concerned." Tribune Company chair Dennis FitzPatrick responded by arguing that "great newspapers must constantly evolve based on changes in the media environment and the communities they serve," while maintaining the need to cut costs, a response that did not allay the concerns. The Open Letter and Dennis Fitzpatrick's response, "Great Newspapers Must Constantly Evolve," are included in "The *Times*, Tribune and the City," *Los Angeles Times*, September 19, 2006.

50. Robert W. McChesney, *The Problem of the Media: U.S. Communication Politics in the Twenty-first Century* (New York: Monthly Review Press, 2004).

51. Lawrence Leamer, *The Paper Revolutionaries: The Rise of the Underground Press* (New York: Simon & Schuster, 1972); Abe Peck, *Uncovering the Sixties: The Life and Times of the Underground Press* (New York: Pantheon, 1985); David Armstrong, *A Trumpet to Arms: Alternative Media in America* (Los Angeles: Tarcher, 1981).

52. Anita Anand, "Alternative Media: Creating a Stir," *Houndmills* 48, no. 2 (June 2005): 92; James Rainey, "Newspaper Finds New Attitude after Katrina: Advocacy Reporting Is Making an Auspicious Return in New Orleans," *Los Angeles Times*, December 29, 2005. The term "weapons of mass deception" was used by media critic, Danny Schechter in his documentary film, *WMD: Weapons of Mass Deception*. Chris Hedges, "Biting the Media's Hand and Demanding Air Time," *Washington Post*, Decemmber 29, 2004. The comment about advocacy journalism by *New Orleans Times-Picayune* reporter Michael Perlstein, "Covering Katrina," appeared in *Reed* (Winter 2006): 13 (*Reed* is a publication of Reed College, and Perlstein, a Reed alumnus, and National Public Radio reporter Robert Smith, also an alum, wrote about their experiences covering Hurricane Katrina and its aftermath).

53. Danny Schecter, "Million-Word March for Media Reform," Alternet, May 19, 2005, <http://www.alternet.org/story/22049>; Robert McChesney, John Nichols, and Ben Scott,

"Congress Tunes In," *The Nation*, May 23, 2005, available at <www.thenation.com/doc/20050523/scott>; Lakshmi Chaudry, "Can Blogs Revolutionize Progressive Politics?," Alternet post, February 8, 2006, <http://www.alternet.org/story/31955/>; David Kline, *Blog! How the Newest Media Revolution Is Changing Politics, Business, and Culture* (New York: CDS Books, 2005).

54. Scott Gold, "Student Protests Echo the '60s, but with a High-Tech Buzz," *Los Angeles Times*, March 31, 2006; Teresa Watanabe and Hector Becerra, "How DJs Put Five Hundred Thousand Marchers in Motion," *Los Angeles Times*, March 28, 2006; Mireya Navarro, "Between Gags, a D.J. Rallies Immigrants," *New York Times*, April 30, 2006.

55. Columbia University's Seth Low, in an 1895 article entitled "A City University," argued that the relationship of Columbia to New York City should not simply be defined as the way the university could "influence the life of New York" but instead how it could "influence the life of New York because it is a part of it." In contrast, Nicholas Murray Butler, Seth Low's successor as head of Columbia University, reversed Low's emphasis on the university's connection to city life and argued that the university needed to liberate "the disciplines to develop in increasing isolation from the city." The Low and Butler cites are in Thomas Bender, *The New York Intellect: A History of Intellectual Life in New York City, from 1750 to the Beginnings of Our Own Time* (New York: Knopf, 1987), 282–83; see also Herman van der Wusten, ed., *The Urban University and Its Identity: Roots, Locations, Roles* (Dordrecht, The Netherlands: Kluwer, 1998).

56. Russell Jacoby, *The Last Intellectuals* (New York: Basic Books, 1987).

57. Roger L. Geiger, *Research and Relevant Knowledge: American Research Universities since World War II* (New York: Oxford University Press, 1993).

58. Judith Rodin, "The Twenty-first Century Urban University: New Roles for Practice and Research," *Journal of the American Planning Association* 71, no. 3 (Summer 2005): 237–49.

59. Timothy K. Stanton, Dwight E. Giles Jr., and Nadinne I. Cruz, *Service Learning: A Movement's Pioneers Reflect on Its Origins, Practice, and Future* (San Francisco: Jossey Bass, 1999); Feinstein Institute for Public Service, *Community Service in Higher Education: A Decade of Development* (Providence, RI: Providence College, 1996); Ira Harkavy, "Service Learning as a Vehicle for Revitalization of Education Institutions and Urban Communities," Paper presented at the American Psychological Association Annual Meeting, Toronto, August 10, 1996.

60. Mark D. Gearan. "Engaging Communities: The Campus Compact Model," *National Civic Review* 94, no. 2 (Summer 2005): 92; Elizabeth L. Hollander, John Saltmarsh, and

Edward Zlotkowski, "Indicators of Engagement," in *Learning to Serve: Promoting Civil Society through Service-Learning*, ed. L. A. Simon et al. (Norwell, MA: Kluwer, 2001), 1–20.

61. Meredith Minkler and Nina Wallerstein, eds., *Community-Based Participatory Research for Health* (San Francisco: Jossey-Bass, 2003). On the Highlander Center and Paolo Freire, see Myles Horton and Paulo Freire, *We Make the Road by Walking: Conversations on Education and Social Change* (Philadelphia: Temple University Press, 1990); Myles Horton and Judith and Herbert Kohl, *The Long Haul: An Autobiography* (New York: Doubleday, 1990); see also Kurt Lewin, "Action Research and Minority Problems," *Journal of Social Issues* 2 (1946): 34–46.

62. Ernest L. Boyer, *Scholarship Reconsidered: Priorities of the Professoriate* (Princeton, NJ: Carnegie Foundation for the Advancement of Teaching, 1990), 81; Ernest L. Boyer, "The Scholarship of Engagement," *Journal of Public Service and Outreach* 1, no. 1 (1996): 11–20. The "between the observer and observed" quote is from L.W. Green and S. L. Mercer, "Can Public Health Researchers and Agencies Reconcile the Push from Funding Bodies and the Pull from Communities?," *American Journal of Public Health* 91, no. 12 (2001): 1926; see also Nina Wallerstein, "Power Dynamics between Evaluator and Community: Research Relationships within New Mexico's Healthier Communities," *Social Science and Medicine* 49, no. 1 (1999): 39–53; Shobha Srinivasan and Gwen W. Collman, "Environmental Justice: Evolving Partnerships in Community," *Environmental Health Perspectives* 113, no. 2 (December 2005): 1814–16.

63. Callie White's study is entitled "Diverging Perspectives: Conflicts between Land Conservancy Efforts and Residents' Sense of Place on Catalina Island, Urban and Environmental Policy Program, Occidental College, Los Angeles, 2004, available at <http://departments.oxy.edu/uepi/uep/studentwork/04comps/white.pdf>.

Chapter 3

1. "Viva Los Angeles," *Los Angeles Times*, September 7, 1905 (on the water bond election).

2. The description of my experiences with the Metropolitan Water District are from personal files based on author's notes, correspondence, and personal communications with MWD board and staff.

3. Robert Gottlieb and Margaret FitzSimmons, *Thirst for Growth: Water Agencies as Hidden Government in California* (Tucson: University of Arizona Press, 1989), 5–7. In 2001, the MWD board was reduced to thirty seven members to try to create a more efficient structure.

4. Christine Reed was later appointed to replace me on the MWD board when I left in December 1987, and she served on the board until her untimely death a little more than

eight years later. While on the MWD board, Reed became a strong voice promoting alternative water-development approaches.

5. Just after I joined the board in 1980, I was approached to serve as a client for a UCLA Urban Planning Department comprehensive project study of the structure and governance of the MWD. This study calculated how my presence lowered the average age of the directors.

6. See my discussion of the water industry in Robert Gottlieb, *A Life of Its Own: The Politics and Power of Water* (San Diego: Harcourt Brace Jovanovich, 1988), xi–xv.

7. Author's notes of MWD chair Earle Blais's comments to the MWD Board of Directors, April 13, 1982; Letter from Earle Blais to Rex Pursell, President, Friant Water Users' Association, September 2, 1981 (in author's possession); Gottlieb and FitzSimmons, *Thirst for Growth*, 19–20.

8. Author's notes, MWD Executive Committee Meeting, March 8, 1982.

9. Author's notes, MWD board of Directors Meeting, April 13, 1982; personal communication with Del Scott, April 13, 1982.

10. The burning of the Cuyahoga River began as early as 1936, when a spark from a blowtorch ignited waste oil floating on the river. Periodic fires on the river took place over the next thirty-five years, including the major episode in 1969 prior to passage of the Clean Water Act in 1972. See Tina Adler, "The Great Lakes: Awash in Policies," *Environmental Health Perspectives* 113, no. 3 (March 2005): A174–77; see also William McGucken, *Lake Erie Rehabilitated: Controlling Cultural Eutrophication, 1960s–1990s* (Akron, OH: University of Akron Press, 2000); "Cities: The Price of Optimism—Cleveland's Polluted Cuyahoga," *Time*, August 1, 1969, 41; Frances X. Cline, "Navigating the Renaissance of an Ohio River That Once Caught Fire," *New York Times*, January 23, 2000.

11. S. M. Driedger and J. Eyles, "Drawing the Battle Lines: Tracing the 'Science War' in the Construction of the Chloroform and Human Health Risks Debate," *Environmental Management* 31, no. 4 (April 2003): 476–88; J. A. Cotruvo, "EPA Policies to Protect the Health of Consumers of Drinking Water in the United States," *Science of the Total Environment* 18 (April 1981): 345–56; Richard E. Jackson, "Recognizing Emerging Environmental Problems: The Case of Chlorinated Solvents in Groundwater," *Technology and Culture* 45, no. 1 (January 2004): 55–79.

12. Brock Evans, "Washington, D.C: Defending the Dam Back Home—Congress and the Politics of Waste," *Sierra Club Bulletin*, no. 62 (1977): 24; Wiley and Gottlieb, *Empires in*

the Sun, 59–62. The iron triangle is elaborated in Arthur Maas, *Muddy Waters: The Army Engineers and the Nation's Rivers* (Cambridge: Harvard University Press, 1951).

13. Gottlieb, *A Life of Its Own*, 37–40; Ann L. Riley, *Restoring Streams in Cities: A Guide for Planners, Policy Makers, and Citizens* (Washington, DC: Island Press, 1998).

14. Maria Kaika, *City of Flows: Modernity, Nature, and the City* (New York: Routledge, 2005), 53; Helen Ingram and Cy R. Oggins, "The Public Trust Doctrine and Community Values in Water," *Natural Resources Journal* 32, no. 3 (Summer 1992): 515–37; Craig Colten, *An Unnatural Metropolis: Wrestling New Orleans from Nature* (Baton Rouge: Louisiana State University Press, 2005).

15. See, for example, Nelson Blake, *Water for the Cities: A History of the Urban Water Supply Problem in the United States* (Syracuse: Syracuse University Press, 1956).

16. David M. Solzman, "Re-Imagining the Chicago River," *Journal of Geography* 100 (2001): 118–23.

17. On the L.A. River's risk to residents, see J. J. Warner, "A Warning," *Los Angeles Times*, July 30, 1882. Warner also spoke of floods moving the flow of the river westward and the lack of understanding about the river's unpredictable nature.

18. Matt Garcia, *A World of Its Own: Race, Labor, and Citrus in the Making of Greater Los Angeles, 1900–1970* (Chapel Hill: University of North Carolina Press, 2001).

19. On Mulholland's comment, see "Los Angeles Resumes the Large Outlook," *Los Angeles Times*, November 14, 1947.

20. "Viva Los Angeles," *Los Angeles Times*, September 7, 1905.

21. Dana W. Bartlett, *The Better City: A Sociological Study of a Modern City* (Los Angeles: Neuner Company Press, 1907), 14, 22; "Vexing Problems before the City," *Los Angeles Times*, May 26, 1906.

22. Vincent Ostrom, *Water and Politics* (Los Angeles: Haynes Foundation, 1953); Norris Hundley Jr., *The Great Thirst: Californians and Water, 1770s–1990s* (Berkeley: University of California Press, 1992), 135; Richard Bigger and James D. Kitchen, *How the Cities Grew: A Century of Municipal Independence and Expansionism in Metropolitan Los Angeles* (Los Angeles: Bureau of Governmental Research, University of California, 1952), 5–27.

23. Tom Sitton and William Deverell, eds., *Metropolis in the Making: Los Angeles in the 1920s* (Berkeley: University of California Press, 2001). The "population mad" quote is from Oliver

Carlson, *A Mirror for Californians* (Indianapolis: Bobbs-Merrill, 1941), 122–23; Harry Chandler, "Imperial Valley's Most Essential Need Is a Flood Control and Storage Dam in the Colorado River," available from the Sherman Foundation, Corona Del Mar, California, n.d.

24. Metropolitan Water District of Southern California, *History and First Annual Report*, compiled and edited by Charles A. Bissell (Los Angeles, 1939), 20.

25. Gottlieb and FitzSimmons, *Thirst for Growth*, 5–11.

26. "Analysis of Annexation Policy," Metropolitan Water District of Southern California, Los Angeles, February 1976; see also Steven Erie and Pascal Joassart-Marcelli, "Unraveling Southern California's Water-Growth Nexus: Metropolitan Water District's Policies and Subsidies for Suburban Development, 1928–1996," *California Western Law Review* 36 (2000): 101–24.

27. "Laguna Declaration Statement of Policy," approved by the board of directors of the Metropolitan Water District of Southern California, December 16, 1952, reprinted in Gottlieb and FitzSimmons, *Thirst for Growth*, 15.

28. William C. Stewart, "Westward the Course . . ." and "Beyond Tomorrow," *Los Angeles Times,* January 2, 1960, and January 3, 1961. These articles were part of the *Times* Midwinter edition, which championed the cycles of growth for the region.

29. The central role of imported water in relation to groundwater is discussed in a law journal article by two key water-industry figures that came to be a framing document regarding the relationship between these two sources. See James H. Krieger and Harvey O. Banks, "Ground Water Basin Management," *California Law Review* 50 (1962): 56–77.

30. Gottlieb and FitzSimmons, *Thirst for Growth*; Albert Lipson, *Efficient Water Use in California: The Evolution of Groundwater Management in California* (Santa Monica, CA: RAND Corporation, November 1978), R-2387/2.

31. Poster titled *Alaska-California Sub-Oceanic Fresh Water Transport System* is in the author's possession.

32. Marc Reisner, *Cadillac Desert: The American West and Its Disappearing Water* (New York: Viking, 1986); "The World's Biggest Ditch," *Forbes* 119, no. 10 (May 15, 1977): 112.

33. Gottlieb, *A Life of Its Own*, 168–87.

34. The deputy director of the California Department of Water Resources at the time of the Peripheral Canal election, Gerald Meral, had previously been a staff member with the

Environmental Defense Fund. His arguments about the water-quality benefits from the canal were not embraced by the environmental groups, however, including EDF, since they were most focused on protecting the environment of the Sacramento Bay Delta. Meral subsequently rejoined the environmental ranks as head of the Planning and Conservation League.

35. Matthew Gandy, *Concrete and Clay: Reworking Nature in New York City* (Cambridge: MIT Press, 2002), 27–28; see also Charles Weidner, *Water for a City: A History of New York City's Problem from the Beginning of the Delaware River System* (Newark, NJ: Rutgers University Press, 1974); Blake, *Water for the Cities*; Charles Jacobsen, Steven Klepper, and Joel A. Tarr, "Water, Electricity, and Cable Television: A Study of Contrasting Historical Patterns of Ownership and Regulation," *Urban Resources* 3 (1985): 9.

36. *Water, Power, and the Growth of Los Angeles: A One-Hundred-Year Perspective* (Los Angeles: Los Angeles Department of Water and Power, 1986); Daniel J. Johnson, "'No-Make Believe Class Struggle': The Socialist Municipal Campaign in Los Angeles, 1911," *Labor History* 41, no. 1 (February 2000): 25–45; Jeff Stansbury, "How Kilowatt Socialism Saved L.A. from the Energy Crisis," *Los Angeles Times*, April 29, 2001.

37. Robert Stavins, *Trading Conservation Investments for Water* (San Francisco: Environmental Defense Fund, 1983); personal communication with Tom Graff, October 3, 1986.

38. At the time of these debates, the MWD staff and board leadership was most concerned that the San Diego approach was designed to secure less expensive Colorado River water for San Diego's benefit while the rest of MWD focused on the more expensive State Water Project sources, including for possible transfers. The issue of San Diego utilizing the MWD delivery system for a separate transfer agreement with IID continued to preoccupy board discussions for another twenty years.

39. "Position Regarding the Possible Transfer of State Water Project Agricultural Water Entitlements," "An Update on Proposed Transfers of State Project Entitlements," Letters from the General Manager to MWD Board of Directors, April 4, 1987, and March 31, 1987.

40. Peter Passell, "A Gush of Profits from Water Sale?," *New York Times*, April 23, 1998; Claudia Deutsch, "Vivendi of France Acquiring U.S. Filter," *New York Times*, March 23, 1999; Michael Hiltzik, "Water Firm Awash in Political Influence," *Los Angeles Times*, February 13, 2006.

41. Jacques Pauw, "The Politics of Underdevelopment: Metered to Death—How a Water Experiment Caused Riots and a Cholera Epidemic," *International Journal of Health Services* 33, no. 4 (2003): 819–30; Kris Christen, "Can Private Sector Funds Resolve U.S. Water

Woes?," *Environmental Science and Technology* 37, no. 1, (January 1, 2003): A10; Erik Swyngedouw, "Dispossessing H2O: The Contested Terrain of Water Privatization," *Capitalism Nature Socialism* 16, no. 1 (March 2005): 81–98; Tony Clarke, "Water Privateers," *Alternatives Journal* 29, no. 2 (Spring 2003): 10–11; Claudia H. Deutsch, "There's Money in Thirst," *New York Times*, August 10, 2006; John Tagliabue, "As Multinationals Run the Taps, Anger Rises over Water for Profit," *New York Times*, August 26, 2002. French-owned global water companies, bolstered by French government intervention that sought to maintain French control, were the most aggressive water-privatization enterprises. They also maintained interests in waste treatment, energy, and transportation-related services. Allesandra Galloni, "National Teams: In Big Takeover Fights, France Retains a Home-Field Advantage," *Wall Street Journal*, March 24, 2006.

42. S. D. Raj, "Bottled Water: How Safe Is It?," *Water Environment Research* 77, no. 7 (November-December 2005): 3013–18; Erik Olson, *Bottled Water: Pure Drink or Pure Hype*? (New York: Natural Resources Defense Council, 1999); Field Institute, "Californians' Views on Water: A Survey of Californians' Opinions about Issues of Water Supply, Development, Quality, and Policy," San Francisco, 1990, 13–28.

43. Jerry Gabriel, "Water for Sale," *Human Ecology* 29, no. 4 (December 2001): 7–9; Andy Opel, "Constructing Purity: Bottled Water and the Commodification of Nature," *Journal of American Culture* 22, no. 4 (Winter 1999): 67–76; C. Potera, "The Price of Bottled Water," *Environmental Health Perspectives* 110, no. 2 (February 2002): A76.

44. Author's notes, Comments by Michael McGuire at the Meeting of the Special Committee on Water Quality, MWD, April 19, 1987.

45. Park City conference session notes in author's possession; Dean Murphy, "DWP Taps into Drinking Habits," *Los Angeles Times*, November 18, 1989; Patrick McGreevy, "DWP Scrutinizes Bottled Water Buying," and "DWP Pays to Drink Sparkletts," *Los Angeles Times*, January 11, 2006, and January 3, 2006.

46. "POWER Board Adopts Goals and Policies," *POWER Newsletter* 1, no. 1 (Fall 1992): 1–2; "Minutes of Water and Growth Conference," Memo from Sabrina Gates to POWER Members, January 17, 1992 (in author's possession); Andrew Cohen, "Water Supply and Land Use Planning: Making the Connection," *Land Use Forum* (Fall 1992): 341–44; Robert Gottlieb, "Our Water Agencies' Historic Growth Agenda Needs a New Mission," *Los Angeles Times*, September 13, 1989.

47. "Unsung Heroes: Dorothy Green," California Community Foundation 2000 Annual Report, available at <http://www.calfund.org/3/unsung_heroes_3.2.4.4.php>; author's

notes, Presentations at the Friends of the Los Angeles River annual event honoring Dorothy Green, September 19, 2003.

48. Lance Dehaven-Smith and John R. Wodraska, "Consensus Building for Integrated Resources Planning," *Public Administration Review* 56 (1996): 367–71; Frederick Muir, "MWD Voices Greater Environmental Concern," *Los Angeles Times*, September 26, 1991; Joy L. Woodson, "Cities Go Green with Native Plants," *Los Angeles Times*, May 24, 2003; personal communication with Adan Ortega, April 13, 2003.

CHAPTER 4

1. Dick Roraback, "Up a Lazy River, Seeking the Source: Your Explorer Follows in Footsteps of Gaspar de Portola," *Los Angeles Times*, October 20, 1985. The series ran intermittently between October 20, 1985, and January 30, 1986.

2. Dick Roraback, "From Base Camp, the Final Assault," *Los Angeles Times*, January 30, 1986; and Dick Roraback, "Bridging the Gap on the L.A. River with a Song in His Heart and a Yolk on His Shoe," *Los Angeles Times*, November 7, 1985.

3. Personal communication with Lewis MacAdams, March 10, 2001; Don Shirley, "Los Angeles River," *Los Angeles Times*, September 27, 1985, D4; Tad Friend, "River of Angels," *The New Yorker*, January 26, 2004, 46; Lewis MacAdams, "Restoring the Los Angeles River: A Forty-Year Art Project," *Whole Earth Review* (Spring 1995): 63; Lewis MacAdams, "From the Friends of the Los Angeles River," in Joe Linton, *Down by the Los Angeles River* (Berkeley: Wilderness Press, 2006), xvi.

4. Lewis MacAdams, "Sharing Memories of the L.A. River," *Los Angeles Times*, November 28, 1985.

5. Roraback, "From Base Camp," Dick Roraback, "The L.A. River Practices Own Trickle-Down Theory," *Los Angeles Times*, October 27, 1985. The "poet laureate of the river" phrase has been used to describe MacAdams on a number of occasions, including the remarks by Joe Edmiston of the Santa Monica Conservancy District in introducing MacAdams on the occasion of the opening of the L.A. River Center in May 2001. Similarly, the "forty-year art project" phrase has been used continuously by MacAdams and those writing about him. MacAdams L.A. River poems are now available in a triology, *The River: Books One, Two, and Three* (Los Angeles: Blue Press Books, 2005).

6. William Deverell, *Whitewashed Adobe: The Rise of Los Angeles and the Remaking of Its Mexican Past* (Berkeley: University of Press, 2004), 99; Jared Orsi, *Hazardous Metropolis: Flooding and Urban Ecology in Los Angeles* (Berkeley: University of California Press, 2004).

7. The Father Crespi quote is from the Portola Expedition diaries of 1769, reprinted in Herbert Eugene Bolton, *Fray Juan Crespi: Missionary Explorer on the Pacific Coast, 1769–1774* (Berkeley: University of California Press, 1971 [1927]), 147–48. Crespi also indicated that the river appeared to generate "great floods in the rainy season, for we saw that it had many trunks of trees on its banks." (148); See also Blake Gumprecht, *The Los Angeles River: Its Life, Death, and Possible Rebirth* (Baltimore: Johns Hopkins University Press, 1999), 38; Patt Morrison, *Rio L.A.: Tales from the Los Angeles River* (Santa Monica, CA: Angel City Press, 2001), 46. W.L. Pollard's ideas were laid out in a series of articles he wrote for the *Los Angeles Times*, including "Parks Linked in Flood Control," October 11, 1931, and "Recreational Plan Outlined," March 8, 1931. The Olmsted-Bartholomew Plan is discussed by Mike Davis in "How Eden Lost Its Garden: A Political History of the Los Angeles Landscape," in *The City: Los Angeles and Urban Theory at the End of the Twentieth Century*, ed. Allen J. Scott and Edward J. Soja (Bekeley: University of California Press, 1996), 160–85; and Davis, *Ecology of Fear: Los Angles and the Imagination of Disaster* (New York: Metropolitan Books, 1998), 64–72. The full text of the plan is reprinted in Greg Hise and William Deverell, *Eden by Design: The 1930 Olmsted-Bartholomew Plan for the Los Angeles Region* (Berkeley: University of California Press, 2000).

8. "As It Is," *Los Angeles Times*, December 4, 1891, 2; "Menaces the City: How the Forestry Question Comes Home to All," *Los Angeles Times*, May 20, 1899.

9. "Floods and the Future," *Los Angeles Times*, February 26, 1914; "Would Make Nature Assist," *Los Angeles Times*, February 26, 1914; Gumprecht, *The Los Angeles River*, 167–71.

10. Hise and Deverell, *Eden by Design*, 93–138. The identification of land-use approaches for flood management were laid out in a 1937 article by Gilbert White, where he argued that "the basic problem in human adjustment to floods is one of land-use planning" and that while "major flood flows result from natural, hydrologic events," "with few exceptions, the damages therefrom are the price of man's encroachment upon natural flood plains." White, a geographer whose work on flood plains would become part of the core arguments of the river restorationists, was ultimately considered to be the "father of flood plain management" and widely recognized for his writings and his advocacy. See Gilbert White, "Notes on Flood Protection and Land-Use Planning," *Planners' Journal* 3 (1937): 61, 57.

11. Judith Coburn, "Whose River Is It Anyway? More Concrete versus More Nature: The Battle over Flood Control on the Los Angeles River Is Really a Fight for Its Soul," *Los Angeles Times*, November 20, 1994; *Hazardous Metropolis*, 102.

12. Gumprecht, *The Los Angeles River*, 127, 246. Soft-bottom areas of the river were not paved over because the groundwater table was sufficiently high that pouring concrete on the bottom of these sections was not deemed viable by the river's engineer managers.

13. Personal communication with Lewis MacAdams, October 15, 1998.

14. Richard Katz, "What's So Silly about a Bargain Freeway?," *Los Angeles Times*, September 8, 1989; *The Los Angeles River*, 273–74; see, for example, Charlie DeDuff, "Los Angeles by Kayak," *Los Angeles Journal*, December 8, 2003.

15. When the engineering agencies first publically revealed their plans to issue their LACDA update, they issued dire warnings about the potential for new flooding in the poor and working-class communities south of downtown Los Angeles. Very few residents up to then—only 9,000 among the 7.7 million people who lived within the L.A. Basin—had purchased flood insurance, which had not previously been required by the flood-control agencies. "We think we have a duty to advise the citizenry that lives along the river that they should take action to protect themselves. About the only action they could take to protect themselves is to buy federal flood insurance," James Easton, the chief deputy director of the L.A. County Public Works Department, told the Los Angeles *Times*. Larry Stammer, "Ominous Look Forward: L.A. River Drainage Deficiencies Create Catastrophic Flood Possibility," July 28, 1985. See also Orsi, *Hazardous Metropolis*, 148–52; Gumprecht, *The Los Angeles River*, 279; and Nancy Wride, "U.S. Flood Insurance Mandate Ends in Southeast L.A. County," *Los Angeles Times*, January 12, 2002.

16. Christopher Kroll, "Changing Views of the River," *California Coast and Ocean* (Summer 1993): 32; Ann Riley, *Restoring Streams in Cities: A Guide for Planners, Policy Makers, and Citizens* (Washington, DC: Island Press, 1998), 34. FoLAR's notion of flood management, influenced in part by Riley, was distinguished from "flood control" and included plantings, spreading grounds, and other strategies to slow down the flow of the river.

17. Gumprecht, *The Los Angeles River*, 283, 298; Orsi, *Hazardous Metropolis*, 156; Duke Helfand, "Controversial L.A. River Project OKd," *Los Angeles Times*, April 7, 1995.

18. Lewis MacAdams, Founder, Friends of the L.A. River, address at Occidental College Environment and Society Class, October 28, 1998.

19. "Pollution Prevention Education and Research Center: 1991–1995" (Los Angeles: Pollution Prevention Education and Research Center/Urban and Environmental Policy Institute, 1997).

20. Urban and Environmental Policy Institute, at <http://departments.oxy.edu/uepi/about/index.htm>; see also Occidental College, "Learning by Doing," at <http://www.oxy.edu/x676.xml>.

21. Larry Gordon, "River, Parks, Shops Proposed," *Los Angeles Times*, March 1, 1998; Personal communication with Lewis MacAdams, March 21, 2006.

22. Carl Ingram, "Senate OK's Bill on Water Conservation," *Los Angeles Times*, May 22, 1998; Personal communication with Tom Hayden, September 13, 1999; Paul Pringle, "The River That L.A. Forgot," *Los Angeles Times*, December 23, 2003; Joe Mozingo, "Urban Oases: A Parks Renaissance Is Regreening the L.A. River; Other Neglected Spots," *Los Angeles Times*, November 7, 2000.

23. Robert Gottlieb, Andrea Azuma, and Amanda Shaffer, *Re-Envisioning the L.A. River: A Program of Community and Ecological Revitalization* (Los Angeles: Urban and Environmental Policy Institute, 2001), available at <http://departments.oxy.edu/uepi/publications/re-envisioning.htm>.

24. Peter Aeschbacher et al., *Cornfield of Dreams: A Resource Guide of Facts, Issues and Principles* (Los Angeles: UCLA Department of Urban Planning, 2000), available at <http://www.sppsr.ucla.edu/dup/research/main.html>; Robert Gottlieb, Elizabeth Braker, and Robin Craggs, "Expanding the Opportunities and Broadening the Constituency for Interdisciplinary Environmental Education," Report to the Andrew W. Mellon Foundation, April 15, 2003; Robert Gottlieb, "Rediscovering the River," *Orion Afield* (Spring 2002): 32.

25. The Marcus and Nichols talks took place at the first forum, "Making It Happen," October 1, 1999; see Gottlieb, Azuma, and Shaffer, *Re-Envisioning the L.A. River*, 4–5.

26. The comments of Blake, Orsi, and Deverell are from a transcript of the History of the River Forum that had the title "A Place We Hardly Knew: The History of the Los Angeles River," October 28, 1999, in the author's possession; see also Gottlieb, Azuma, and Shaffer, *Re-Envisioning the L.A. River*, 11–12.

27. "Water Supply Running Low," *Los Angeles Times*, May 8, 1902; see also "Racing for More Water," *Los Angeles Times*, May 18, 1902, where William Mulholland, arguing in favor of more conservation and water metering, nevertheless suggested that the image of Los Angeles as garden and pleasure ground meant that the city would never "rank among the

cities with small water consumption for the reason that the semi-leisure class is attracted here by the adaptability of both soil and climate for building up beautiful homes by reason of the long intervals of almost absolute cessation of rainfall."

28. Gottlieb, Azuma, and Shaffer, *Re-Envisioning the L.A. River*, 6–7.

29. Riley, *Restoring Streams in Cities*, 228–29.

30. The Tujunga Wash project is described at <http://www.theriverproject.org/tujunga/>.

31. The San Gabriel River Watershed, Los Angeles County Department of Public Works, available at <http://ladpw.org/wmd/watershed/sg/>; RMC legislation SB 216 (Solis) San Gabriel River and Lower Los Angeles River (Statutes of 1999, chapter 789).

32. Joe Mozingo, "San Gabriel River Backers Push for National Ranking," *Los Angeles Times*, July 22, 2001.

33. Kathleen Bullard, "Parameters," in Harvard University Department of Landscape Architecture, *L.A. River Studio Book* (Cambridge, MA: Harvard University Graduate School of Design, 2002), 21; Kathryn Maese, "Harvard Grad Students to Release L.A. River Study," *L.A. Downtown News*, January 31, 2002.

34. Gottlieb, Azuma, and Shaffer, *Re-Envisioning the L.A. River*, 24; David Rozensweig, "Rules of Robles 'Kingdom' Described," *Los Angeles Times*, July 27, 2005; Cara Mia DiMassa, Sam Quinones, and David Pierson, "Plight Brought South Gate Together," *Los Angeles Times*, November 21, 2004.

35. Personal communication with Joe Linton, July 20, 2001; Wendy Thermos, "Cyclists Not-So-Little Secret," *Los Angeles Times*, November 9, 2000.

36. Gottlieb, Azuma, and Shaffer, "Hollywood Looks at the River Panel, April 6, 2000," *Re-Envisioning the L.A. River*, 22. Six of the major Hollywood studios had their headquarters or office complexes at sites that bordered the river and had been targeted by river-renewal advocates for participation in community and environmental renewal, a factor also discussed during the panel by Michael Klausman, the president of CBS Studio Center, which represented one of those six studios.

37. Friends of the L.A. River, The River through Downtown, at <http://www.folar.org/about.html>; Personal communication with Chi Mui, December 11, 2001. Chi Mui subsequently became involved in the Asian community in the San Gabriel Valley and became mayor of the City of San Gabriel shortly before his untimely death in 2006. David Pierson,

"Chi Mui Fifty-three: Was the First Mayor of Asian Descent in San Gabriel's History," *Los Angeles Times*, April 28, 2006.

38. Paul Stanton Kibel, "Los Angeles' Cornfield: An Old Blueprint for New Greenspace," *Stanford Environmental Law Journal*, 23 (2004): 275, 306; Howard Fine, "Can Riordan's Redevelopment Program Work? Mayor Richard Riordan's Genesis L.A. Urban Development Plan," *Los Angeles Business Journal*, March 29, 1999.

39. Lewis MacAdams had first learned about Majestic's plans when he overheard a conversation in City Council member Mike Hernandez's office to the effect that Majestic was moving quickly to obtain a Mitigated Negative Declaration to proceed with its warehouse plans. MacAdams immediately called environmental attorney Jan Chatten-Brown to delay the proceedings and give time for FoLar and other river advocates to mobilize around the issue. The discussion with Nichols and Marcus occurred soon after. Personal communication from Lewis MacAdams, December 11, 2001. The meeting at the Eagle Rock Restaurant took place on October 1, 1999.

40. Aeschbacher et al., *Cornfield of Dreams*, 114–15; Personal communication from Lewis MacAdams, September 14, 2000; Lewis MacAdams and Robert Gottlieb, "Changing a River's Course: A Greenbelt versus Warehouses," *Los Angeles Times*, October 3, 1999.

41. Kibel, "Los Angeles' Cornfield," 325–30; Joe Mozingo, "Firm to Sell Land Near L.A. River for Park," *Los Angeles Times*, September 16, 2001; Jesus Sanchez, "L.A.'s Cornfield Row: How Activists Prevailed," *Los Angeles Times*, April 17, 2001.

42. D. J. Waldie, "Changing the River's Course in Pursuit of Public Spaces," *Los Angeles Times*, October 3, 1999; D. J. Waldie, "As We Gather at the River," *Los Angeles Times*, July 23, 2000; D. J. Waldie, *Holy Land: A Suburban Memoir* (New York: Norton, 2005); see also, David Ferrell, "L.A. River Defies City in Nurturing Wildlife," *Los Angeles Times*, July 26, 2001, where Waldie comments "The problem with the river [has been] its placeness."

43. D. J. Waldie, "Reclaiming a Lost River, Building a Community," *New York Times*, July 10, 2002.

44. Betsy Otto, Kathleen McCormick, and Michael Leccese, *Ecological Riverfront Design: Restoring Rivers, Connecting Communities* (Chicago: American Rivers and American Planning Association, May 2004), 4 (chapter 1 is available at <http://www.americanrivers.org/site/DocServer/chapter_1.pdf?docID=430>).

45. Riley, Restoring Streams in Cities, 202.

46. The Corps activity from the 1940s through 1973 is documented in a report by John Wilkinson, *Report on Channel Modifications to Council on Environmental Quality*, vols. 1–2 (Cambridge: Arthur D. Little, March 1973), cited in Riley, *Restoring Streams in Cities*, 202. The St. Louis study was published in Robert Criss and Everett Shock, "Flood Enhancement through Flood Control," *Geology* 29 (2001): 875–78.

47. The Flats area, where the 1969 Cuyahoga River fire took place, emerged in the 1990s, thanks to the tourism-driven restoration agenda, into a popular entertainment district with clubs, restaurants, and water taxis that ferried nightclub customers across the river from one night spot to the next. Ann Breen and Dick Rigby, *Waterfronts: Cities Reclaim Their Edge* (New York: McGraw-Hill, 1994), 14.

48. Though some of Marco Engineering Company's concepts for a Texas and Mexican colonial-style development had been rejected for their "carnival-like" character, the emphasis on the River Walk as a core tourism attraction prevailed. See Ann Breen and Dick Rigby, "Sons of River Walk: How a Masterpiece of Design Has Inspired Communities Across the Nation, 54, no. 3," *Planning* (March 1988); "River Walk History: 1950–2005," the San Antonio River Walk Web site, available at <http://thesanantonioriverwalk.com/RiverWalkHistory/History4.asp>; presentation by Ann Riley, "City Rivers—The Urban Bankside Restored," Golden Gate University School of Law, San Francisco, November 18, 2005.

49. The Trashed Rivers gathering is discussed by one of its participants, Mike Houck, "In Defense of Trashed Rivers," in *Rivertown: Rethinking Urban Rivers,* ed. Paul Stanton Kibel (Cambridge: MIT Press, 2007); on restoration initiatives, see Harry Austin, "Chattanooga," *Land and People* 7, no. 2 (Fall 1995): 3–7; Craig Savoye, "River Towns Reconnect with Waterfront Potential," *Christian Science Monitor,* November 13, 2001; See also Kathy Dieden, "In the Spotlight: Ann Riley," Water Resource Center Archives, UC Berkeley Library, *WRCA News* 7, no. 1 (February 2000), available at <www.lib.berkeley.edu/WRCA/pdfs/news71.pdf>.

50. Michael Hawthorne, "A Whiff of Success: Million-Dollar Homes along a Long-Polluted Stretch of the Chicago River Fuels New Interest in Cleaning Up Bubbly Creek," *Chicago Tribune,* November 21, 2004; Christopher Theriot and Kelly Tzoumis, "Deep Tunnels and Fried Fish: Tracing the Legacy of Human Interventions on the Chicago River," *Golden Gate University Law Review* 35, no. 3 (Spring 2005): 392; Gottlieb, *Forcing the Spring: The Transformation of the American Environmental* (Washington, DC: Island Press, 2005), 103.

51. In her book on the history of urban park design, Galen Cranz argued that a community-benefits model includes "the stabilization of values around parks in declining

neighborhoods and the revitalization of those neighborhoods [that] clearly provides benefits for the city as a whole by holding, and attracting back to it, people with money to invest and spend." Cranz, *The Politics of Park Design: A History of Urban Parks in America* (Cambridge: MIT Press, 1982), 209; Andrea Azuma et al., *Connecting the Parks to the Community and the Community to the Parks: A Community, Economic, and Environmental Assessment of the Los Angeles State Historical Park (The Cornfield) and Rio de Los Angeles State Park (Taylor Yard)*, A Report to the California Department of Parks and Recreation and the Coastal Conservancy (Los Angeles: Urban and Environmental Policy Institute, July 2006).

52. Personal communication with Lauren Bon, March 17, 2006.

53. "Revitalizing the L.A. River," *Centerscene,* Center for Healthy Communities, The California Endowment (Spring 2006): 6.

54. Azuma, *Connecting the Parks to the Community;* Notes from Andrea Azuma, personal communication with Principal Chuek Yan Choi, Castelar Elementary School, March 24, 2006; personal communication with Chanchanit Martorell, Thai Community Development Center, March 29, 2006; Sustainable Economic Enterprises of Los Angeles, "Cornfield State Park Farmers' Market Feasibility Study," Prepared for the State of California Department of Parks and Recreation," SEE-LA, Los Angeles, August 2006; "Farmlab: Activities Plan Narrative: Phase IV, July 1–December 31, 2006," memo in author's possession.

55. In 2006, in response to the problem of river access at both the Cornfield and Taylor Yard sites, the City of Los Angeles, as part of its Los Angeles River Revitalization Plan, identified potential land purchases at these two sites among the five areas selected to serve as pilot projects. "Twenty years ago we were cutting through chain-link fence to get people to the River," FoLAR's executive director Shelly Backlar told the Los Angeles *Times*. "I may be more optimistic than some, but I think it's only a matter of when restoration is going to happen." Steve Hymon, "City to Name L.A. River Park Sites," *Los Angeles Times,* June 24, 2006.

CHAPTER 5

1. "Raid Siren Test Late: Wrong Button Pushed," *Los Angeles Times,* November 26, 1955; Roy Rivenburg, "SigAlert Creator Dead at Ninety-five," *Los Angeles Times,* June 3, 2004; "What Are Sig-Alerts?," California Department of Transportation, available at <http://www.dot.ca.gov/hq/paffairs/faq/faq18.htm>.

2. Joan Didion, *The White Album* (New York: Pocket Books, 1979), 83.

3. Alfred Sloan, *My Years with General Motors,* ed. John McDonald with Catherine Stevens (Garden City: Doubleday, 1964), 220–21; David Brodsly, *L.A. Freeway: An Appreciative Essay* (Berkeley: University of California Press, 1981), 7 and 4.

4. Andrea Hricko, "Ships, Trucks, and Trains: Effects of Goods Movement on Environmental Health," *Environmental Health Perspectives* 114, no. 4 (April 2006): A204–05.

5. "Table 4, Trends: Annual Delay per Traveler, 1982–2003," in Texas Transportation Institute, *2005 Urban Mobility Study*, available at <http://mobility.tamu.edu/ums/congestion_data/tables/national/table_4.pdf>; Anthony Downs, *Still Stuck in Traffic: Coping with Peak-Hour Traffic Congestion* (Washington DC: Brookings Institution, 2004), 13.

6. George W. Hilton and John F. Due, *The Electric Interurban Railways in America* (Stanford: Stanford University Press, 1960), 406. Huntington's quote from the December 12, 1904, *Los Angeles Examiner* is cited in William B. Friedricks, *Henry E. Huntington and the Creation of Southern California* (Columbus: Ohio State University Press, 1992), 7, also 87–97. See also Bruce Henstell, *Sunshine and Wealth: Los Angeles in the Twenties and Thirties* (San Francisco: Chronicle Books, 1984), 23.

7. Sloan, *My Years with General Motors*, 282; James J. Flick, *The Automobile Age* (Cambridge: MIT Press, 1988), 229–50; "Motor Cars Are Essential," *Los Angeles Times*, April 21, 1920; "The Raw No-Parking Ordinance," *Los Angeles Times*, December 24, 1919; "Seats Coming to Everyone," *Los Angeles Times*, March 7, 1920.

8. "Stops Parking after April 10," *Los Angeles Times*, February 4, 1920; Henstell, *Sunshine and Wealth*, 26. Congestion, however, didn't improve during the next several years, and a downtown parking ban was reinstituted on several occasions during the Christmas period, only to be lifted after the holidays had passed. See Ashley Brilliant, *The Great Car Craze: How Southern California Collided with the Automobile in the 1920s* (Santa Barbara: Woodbridge Press, 1989), 74–75; "Los Angeles and Its Motor Jam," *Literary Digest* 71, no. 4 (April 26, 1924): 68.

9. David W. Jones, *California's Freeway Era in Historical Perspective* (Berkeley: Institute of Transportation Studies, University of California at Berkeley, June 1989), 88; Kelker, De Leuw & Co., *Report and Recommendations on a Comprehensive Rapid Transit Plan for the City and County of Los Angeles* (Chicago: Kelker, De Leuw & Co., 1925). The 1925 election regarding the elevated rail system, which also included competing visions of downtown, is described in Robert Gottlieb and Irene Wolt, *Thinking Big: The Story of the Los Angeles Times, Its publishers, and Their Influence on Southern California* (New York: Putnam, 1977), 152–55; see also "Hit this Thing Hard," *Los Angeles Times*, April 25, 1926.

10. Scott Bottles, "Mass Politics and the Adoption of the Automobile in Los Angeles," in *The Car and the City: The Automobile, the Built Environment, and Daily Urban Life*, ed. Martin Wachs and Margaret Crawford (Ann Arbor: University of Michigan Press, 1991); Charles Seims, *Trolley Days in Pasadena* (San Marino: Golden West Books, 1982); Spencer Crump, *Ride the Big Red Cars: How Trolleys Helped Build Southern California* (Los Angeles: Trans-Anglo Books, 1995); Crump, *Henry Huntington and the Pacific Electric* (Los Angeles: Trans-Anglo Books, 1970); Donald F. Davis, "North American Urban Mass Transit, 1890–1950: What If We Thought about It as a Type of Technology?," *History and Technology* 12 (1995): 309–26.

11. Chapin Hall, "What Goes On," *Los Angeles Times*, March 9, 1940; "Alhambra Loses Rail Service," *Los Angeles Times*, November 30, 1941; "Yellow Car Lines Sold to Chicagoans," *Los Angeles Times*, December 5, 1944; "New Owners Take Over Street Railway Today," *Los Angeles Times*, January 10, 1945.

12. Jonathan Kwitny, "The Great Transportation Conspiracy,", *Harpers* 262, no. 569 (February 1981), 14–21; David J. St. Clair, *The Motorization of American Cities* (Westport, CT: Praeger, 1986); Bradford Snell, *American Ground Transport*, reproduced as an appendix to U.S. Senate Committee on the Judiciary, The Industrial Reorganization Act: Hearings before a Subcommittee of the Senate Committee on the Judiciary on S. 1167, 93rd Cong., 2d sess., 1974, pt. 4A; Sy Adler, "The Transformation of the Pacific Electric Railway: Bradford Snell, Roger Rabbit, and the Politics of Transportation in Los Angeles," *Urban Affairs Quarterly* 27 (September 1991): 58, 61; Scott L. Bottles, *Los Angeles and the Automobile: The Making of the Modern City* (Berkeley: University of California Press, 1987), 229–30; Cliff Slater, "General Motors and the Demise of Streetcars," *Transportation Quarterly* 51 (Summer 1997): 61; Mark S. Foster, *Streetcar to Superhighway: American City Planners and Urban Transportation, 1900–1940* (Philadelphia: Temple University Press, 1981).

13. James Dunn, *Miles to Go: European and American Transportation Policies* (Cambridge: MIT Press, 1981); Flick, *The Automobile Age*, 373. The "mongrel vehicle" quote is from a July 31, 1953, article in the *Seattle Journal of Commerce* and is cited in St. Clair, *The Motorization of American Cities*, 75; see also Zachary M. Schrag, "The Bus is Young and Honest," *Technology and Culture* 41, no. 1 (January 2000): 51–79.

14. "Urban Transit and Municipal Ownership," *Los Angeles Times*, May 14, 1899; Howard C. Kegley, "To Vote on Subway to This City," *Los Angeles Times*, March 30, 1919; "Move to Buy Car Lines," *Los Angeles Times*, September 9, 1922; "City-Owned Street Cars?," *Los Angeles Times*, February 10, 1927; "Becoming Socialistic?," *Los Angeles Times*, November 16, 1926; Tim Brick, "Twaddle, Bunk and Flub-Dub: How Pasadena Got Its Current Form of Government" (in author's possession). Prior to the 1958 sale, the Big Red interurban

streetcar and bus system was purchased (except for its tracks for freight service) by the newly formed Metropolitan Coach Lines (MCL). MCL immediately sought to convert the interurbans to bus service but was denied permission to do so by the state commission that regulated the interurbans. MCL then bailed out with its 1958 sale to the newly established public transit district. The shift in transit system ownership from 4 percent publicly owned in 1959 to 45 percent in 1977 is from the American Public Transit Association's *Transit Fact Book,* cited in Dunn, *Miles to Go*, 76.

15. Hilton and Due, *The Electric Interurban Railways in America*, 409.

16. The Moses statement was made in a talk before the Highway Users Conference in 1964 and is cited by Helen Leavitt in *Superhighway—Superhoax* (New York: Doubleday, 1970), 60; see also Clay McShane, *Down the Asphalt Path: The Automobile and the American City* (New York: Columbia University Press, 1994), 221–23. In a 1870 article in *American Social Science,* Frederick Law Olmsted elaborated on the parkway "pleasure-ride" concept, stating that parkways "should be so planned and constructed as never to be noisy and seldom crowded" and that "the straightforward movement of pleasure-car carriages need never be obstructed, unless at absolutely necessary crossings, by slow-going heavy vehicles used for commercial purposes." "Parks and the Reenlargement of Towns," in *The City Reader*, ed. Richard T. LeGates and Frederick Stout (London: Routledge, 1996), 343.

17. Benton MacKaye, "The Townless Highway," *New Republic* 62, no. 297 (March 12, 1970): 93–95; Lewis Mumford, *The Highway and the City* (New York: Harcourt, Brace & World, 1963), 236.

18. "Experts Offer Traffic Plans," *Los Angeles Times*, July 25, 1924; Grey Hise and William Deverell, *Eden by Design: The 1930 Olmsted: Bartholomen Plan for the Los Angeles Region* (Berkeley: University of California Press, 2000), 30–32; "Two Strides Forward," *Los Angeles Times*, October 9, 1924.

19. Matthew Roth, "Mulholland Highway and the Engineering Culture of Los Angeles in the 1920s," in Tom Sitton and William Deverell, eds., *Metropolis in the Making: Los Angeles in the 1920s* (Berkeley: University of California Press, 2001), 45–76; Hise and Deverell, *Eden by Design*; Terence Young, "Moral Order, Language and the Failure of the 1930 Recreation Plan for Los Angeles County," *Planning Perspectives* 16 (2001): 333–56; W.L. Pollard, "Parks Linked in Flood Control," *Los Angeles Times*, October 11, 1931.

20. The description of parkway design as "bioengineering" was offered by landscape architect Wilbur Simonson, who is quoted in Benjamin Forgey, "Parkway Design: A Lost Art?," *Landscape Architecture* (1989): 2.

21. "Rapid Transit Line Proposed for Freeway," *Los Angeles Times*, May 28, 1946; "Freeway Use as Bus Service Route Urged," *Los Angeles Times*, April 29, 1946; "Los Angeles Resumes the Large Outlook," *Los Angeles Times*, November 14, 1947; Jones, *California's Freeway Era in Historical Perspective*, 235.

22. Martin Webster, "Transportation: A Civic Problem," *Engineering and Science* (December 1949): 14.

23. "State Legislators Told of City's Plan to Add Ten Thousand More Parking Spaces," *Los Angeles Times*, April 3, 1946.

24. Brian Taylor, "When Finance Leads Planning: Urban Planning, Highway Planning and Metropolitan Freeways in California," *Journal of Planning Education and Research*, 20 (2000): 198.

25. Transportation Engineering Board, *Transit Program for the Los Angeles Metropolitan Area* (Los Angeles: City of Los Angeles, December 1939); "Long-Range View Taken for City Traffic Solution," *Los Angeles Times*, December 22, 1939; "Los Angeles Maps Program for Traffic Speedup," *Los Angeles Times*, August 18, 1941; "A Parkway Plan for the City of Los Angeles and the Metropolitan Area," Department of City Planning, Los Angeles, May 1941; Jones, *California's Freeway Era in Historical Perspective*, 47, 171, 187, 235.

26. Flick, *The Automobile Age*, 369; Samuel W. Taylor, "How California Got Fine Roads," in *Freedom of the American Road* (Detroit: Ford Motor Company, 1956), 11, cited in John B. Rae, *The Road and the Car in American Life* (Cambridge: MIT Press, 1971), 184; see also Bruce E. Seely, *Building the American Highway System: Engineers as Policy Makers* (Philadelphia: Temple University Press, 1987). "Since we consider the city primarily as a traffic problem, rather than a place to live, we have engaged the highway engineers to fix it," architecture critic Wolf Van Eckardt wrote in 1967, arguing that the nation's highway builders were "Los Angelizing our cities, a process which automatically keeps increasing pressure for more freeways." Wolf Von Eckardt, *A Place to Live: The Crisis of the Cities* (New York: Delacorte Press, 1967), 336–37.

27. Mark Rose, *Interstate: Express Highway Politics, 1941–1956* (Lawrence: Regents Press of Kansas, 1979), 69. The "Road Gang" is described in Leavitt, *Superhighway—Superhoax*, 52–53. Style changes are discussed in Anthony Young and Mike Mueller, *Chevrolet's Hot One: 1955, 1956, 1957* (Osceolo, WI: Motorbooks, 1955). The Disneyland-freeway-automobile connection is discussed in Eric Avila, *Popular Culture in the Age of White Flight: Fear and Fantasy in Suburban Los Angeles* (Berkeley: University of California Press, 2004), 203; See also Mumford, *The Highway and the City*, 234; Margaret Crawford, "The Fifth Ecology: Fantasy,

the Automobile, and Los Angeles," in *The Car and the City: The Automobile, the Built Environment, and Daily Urban Life*, ed. Martin Wachs and Margaret Crawford (Ann Arbor: University of Michigan Press, 1992).

28. "Record Budget for Roads Announced," *Los Angeles Times*, October 25, 1955; Daniel Moynihan, "Policy vs. Program in the '70s," *Public Interest* no. 20 (Summer 1970): 94.

29. Dwight D. Eisenhower, *Mandate for Change*, 1953–1956 (Garden City, NY: Doubleday, 1963), 548–49; Rose, *Interstate*, 85–94; Jones, *California's Freeway Era in Historical Perspective*, 236–42. Transportation engineering professor Vukan Vuchic argues that, among other impacts on urban transportation, the 90–10 formula of the 1956 legislation also meant that alternative transportation needs (such as pedestrian improvements that were "more conducive to a human-oriented urban environment") were significantly neglected. Vukan R. Vuchic, *Transportation for Livable Cities* (New Brunswick: State University of New Jersey, Center for Urban Policy Research, 1999), 96.

30. Gary T. Schwartz, "Urban Freeways and the Interstate System," *Southern California Law Review* 49 (March 1976): 407–513.

31. "Highway Project Off to Fast Start," *New York Times*, August 17, 1956; Robert Moses, "We Can Lick Congestion," in *Freedom of the American Road* (Detroit: Ford Motor Company, 1956), cited in Schwartz, "Urban Freeways and the Interstate System," 491; Ethan Rarick, *California Rising: The Life and Times of Pat Brown* (Berkeley: University of California Press, 2005), 129–30; Taylor, "When Finance Leads Planning," 210.

32. Scott Newhall's freeway denunciation from the *San Francisco Chronicle* is cited in Jones, *California's Freeway Era in Historical Perspective*, 296. L.A. conflicts are described in Ray Hebert, "Highway Panel Moves to Cancel Two Freeway Plans," *Los Angeles Times*, May 23, 1975; Barbara Baird, "Reaction Mixed on Century Freeway Plan," *Los Angeles Times*, September 4, 1977; Walter P. Coombs, "What's in a Name? With Freeways, Who Knows?," *Los Angeles Times*, April 20, 1979; "State Senate Votes to Rename Nixon Freeway," *Los Angeles Times*, January 29, 1976. The Boyle Heights fight is described in "Boyle Heights: California's Sociological Fishbowl," *Fortnight*, October 20, 1954, 20, and "One Hundred Meet to Protest Proposed Freeway Link," *Los Angeles Times*, October 28, 1953. A subsequent fight nearly fifty years later over diesel trucks driving on surface streets in a Boyle Heights residential neighborhood and adjacent elementary school to the Golden State Freeway entrance is described in Antonio Olivo, "Community Fighting to Ban Diesel Trucks," *Los Angeles Times*, April 23, 2000.

33. Moynihan, "Policy versus Progress in the '70s," 94; Leavitt, *Superhighway—Superhoax*; Kenneth R. Schneider, *Autokind vs. Mankind: An Analysis of Tyranny, a Proposal for Rebellion, a Plan for Reconstruction* (New York: Norton, 1971). However, the car and the freeway system also had its defenders, who argued that freeways reduced congestion (compared to surface streets) and reduced air pollution (as opposed to stop-and-go traffic). Arguments were also made that congestion could signify that "streets are being well-used [and that] the public investment is not wasted." See Scott Greer, "The Functions of Transportation," in *Neighborhood, City, and Metropolis: An Integrated Reader in Urban Sociology*, ed. Robert Gutman and David Popenoe (New York: Random House, 1970), 428; Rae, *The Road and the Car in American Life*. Rae's book was funded by a research grant from the Automobile Manufacturer's Association.

34. Joseph Dimento, Drusilla Van Hengel, and Sherry Ryan, "The Century Freeway: Design by Court Decree," *Access* (University of California Transportation Center) no. 9 (Fall 1996): 7–12; Ray Hebert, "Bradley Urges Wilshire Route for Transit Line," *Los Angeles Times*, August 14, 1975; " 'It Is Time for Decisive Action,' " *Los Angeles Times*, June 10, 1975; "Ward Proposes 268-Mile Rail Network," *Los Angeles Times*, November 19, 1975.

35. William Fulton and John Chandler, "Los Angeles Prime Time: Who's Who," *Planning* 52, no. 2 (February 1986): 4–12; "Los Angeles Transit History," Metropolitan Transportation Authority, available at <http://www.mta.net/about_us/library/transit_history.htm>.

36. John Pastier, "To Live and Drive in L.A.," *Planning* 52, no. 2 (February 1986): 21–25; Ray Hebert, "New Agency Seeking to Get L.A.'s Act Together," *Los Angeles Times*, September 6, 1977.

37. Hank Dittmar, "A Broader Context for Transportation Planning," *Journal of the American Planning Association* 61, no. 1 (Winter 1995): 7–13; Robert Jay Dilger, "ISTEA: A New Direction for Transportation Policy," *Publius* 22, no. 3 (Summer 1992): 67. On the establishment of the STPP coalition, personal communication with Scott Bernstein, July 15, 2004.

38. Elizabeth Clarke, "TEA-21: Something for Everyone," *Civil Engineering* 68, no. 10 (October 1998): 52–55; Susan J. Binder, "The Straight Scoop on SAFETEA-LU," *Public Roads* 69, no. 5 (March–April 2006): 2–8; see also David Burwell, "Ten Years of Progress: Challenges Ahead," *Progress* (a publication of the Surface Transportation Policy Project) 12, no. 1 (March 2002); Angelina Sciolla, "Paving the Way: The Interstate Highway System Began Fifty Years Ago, and AAA Was There," *Westways* 98, no. 3 (May–June 2006): 74–75.

39. Donald C. Shoup, *The High Cost of Free Parking* (Chicago: Planners Press, American Planning Association, 2005), 349.

40. Vuchic, *Transportation for Livable Cities*, 77.

41. Michael Manville and Don Shoup, "People, Parking and Cities," *Access* (University of California Transportation Center) no. 25 (Fall 2004): 7. The designation for Tulsa as a parking-lot city is from Henry F. Arnold, *Trees in Urban Design* (New York: Van Nostrand Reinhold, 1993), 2.

42. In 1920, the Chicago City Council passed legislation that would have banned parking in the Loop, but it was vetoed by Mayor William Hale Thompson, who in his veto message referenced L.A.'s short-lived experiment, arguing that it had "completely killed business [and] made that city look like Walla Walla, Washington." Cited by Paul Barrett in *The Automobile and Urban Transit: The Formation of Public Policy in Chicago, 1900–1930* (Philadelphia: Temple University Press, 1983), 135. The parking meter history is described on the POM Inc. Web site at <http://www.pom.com>.

43. Urban Land Institute, *Office Development Handbook*, 2nd ed. (Washington, DC: Urban Land Institute, 1998), 131; Shoup, *The High Cost of Free Parking*, 5, 31, 139; Stanley I. Hart and Alvin L. Spivak, *The Elephant in the Bedroom: Impacts on the Economy and the Environment* (Pasadena: New Paradigm Books, 1993), 31. Dunphy of the Urban Land Institute argues that "parking truly is the big foot of urban land uses, since the amount of land required for parking and circulation exceeds the building footprint." Robert Dunphy, "Big Foot," *Urban Land* (February 2003): 82.

44. "Not a Walk in the Park: Parking Lots Take Planning, Plus a Fair Chunk of Land," *Akron* (OH) *Beacon Journal*, June 12, 2005, available at <http://www.uli.org/AM/Template.cfm?Section=Search&template=/CM/HTMLDisplay.cfm&ContentID=25559>; Adam Millard-Ball, "Putting on Their Parking Caps," *Planning* (April 2002): 16.

45. Mark A. Delucchi, "Total Cost of Motor Vehicle Use," *Access*, No. 8 (Spring 1996): 7; Shoup, *The High Cost of Free Parking*, 206, 591.

46. The figures for parking at shopping centers in the *Parking Generation* report were developed from a 1999 Urban Land Institute survey that provided only limited data on non-suburban locations. As a result, the *Parking Generation* study noted that the lack of data meant that researchers could not "assess whether parking demand at urban sites was significantly different from that at suburban sites." *Parking Generation*, 3rd ed. (Washington DC: Institute of Transportation Engineers, 2004), 191; See also Amanda Shaffer, *The Persistence of L.A.'s*

Grocery Gap (Los Angeles: Urban and Environmental Policy Institute, 2002), available at <http://www.vepi.oxy.edu>. Robert Gottlieb and Andrew Fisher, "Food Access for the Transit Dependent," *Access* (University of California Transportation Center) no. 9 (Fall 1996): 18–20; Shoup, *The High Cost of Free Parking*, 64–65.

47. Shoup, *The High Cost of Free Parking*, 229.

48. Jeffrey Brown, Daniel Baldwin Hess, and Donald Shoup, "Unlimited Access," *Transportation* 28, no. 3 (August 2001): 233–67; Shoup, *The High Cost of Free Parking*, 252–53.

49. Parking cash-out problems are discussed in Shoup, *The High Cost of Free Parking*, 253; see also Downs, *Still Stuck in Traffic*, 192–93; "City of Boulder's Parking Plus Program Wins Top International Award," *U.S. State News*, June 16, 2006.

50. Tom Prugh, "Car Sharing Grows," *World Watch* 19, no. 2 (March–April 2006): 9; car-share program locations are available at <http://www.carsharing.net/where.html>.

51. Robert Gottlieb and Andrew Fisher, *Homeward Bound: Food-Related Transportation Strategies in Low-Income and Transit-Dependent Communities* (Los Angeles: Community Food Security Coalition, 1996).

52. Ray Bradbury, *The Pedestrian: A Fantasy in One Act* (Hollywood: Samuel French, 1966), 5, 17, 19; Rob Kendt, "Torching the Library: Different Year, Same Temperature," *New York Times*, March 19, 2006.

53. Deyan Sudjic, *The One-Hundred-Mile City* (San Diego: Harcout Brace, 1992), 2008; Jeffrey Tumlin and Adam Millard-Ball, "How to Make Transit-Oriented Development Work," *Planning* (May 2003): 14.

54. National Center for Statistics and Analysis, U.S. Department of Transportation, "Traffic Safety Facts," available at <http://www-nrd.nhtsa.dot.gov/pdf/nrd-30/NCSA/TSF2003/809769.pdf>; and Bureau of Transportation Safety, "Motor Vehicle Injuries," available at <http://www.bts.gov/publications/transportation_statistics_annual_report/2005/html/chapter_02/motor_vehicle_related_injuries.html>; R. O. Petch and R. R. Henson, "Child Road Safety in the Urban Environment," *Journal of Transport Geography* 8 (2000): 197–211; Henri Lefebvre, *The Urban Revolution* (Minneapolis: University of Minnesota Press, 2003), 20; Eric Dumbaugh and J. L. Gattis, "Safe Streets, Livable Streets/Counterpoint," *Journal of the American Planning Association* 71, no. 3 (Summer 2005): 283–85; Caitlin Liu, "These Routes Aren't Made for Walking," *Los Angeles Times*, December 3, 2004.

55. J. R. Crandall, K. S. Bhalla, and N. J. Madeley, "Designing Road Vehicles for Pedestrian Protection," *British Medical Journal* 324, no. 7346 (May 11, 2002): 1145–48; Neel Scott, "The Wrong Foot Forward: The Short Straw—How Bicyclists and Pedestrians Are Deprived of Traffic Safety Funding," *Transportation Alternatives* (June 2000); Gloria Ohland, Trinh Nguyen, and James Corliss, *Dangerous by Design: Pedestrian Safety in California* (Washington, DC: Surface Transportation Policy Project, September 2000); Anastasia Loukaitou-Sideris and Robin Liggett, "Death on the Crosswalk: A Study of Pedestrian-Automobile Collisions in Los Angeles," UCLA Department of Urban Planning, Los Angeles, 2005; "Motor Vehicle Injuries," at <http://www.bts.gov/publications/transportation_statistics_annual_report/2005/html/chapter_02/motor_vehicle_related_injuries.html>.

56. Steve Lopez, "Guilty of 'Crossing While Elderly,'" *Los Angeles Times*, April 15, 2006.

57. Lou V. Chapin, "An Ideal Cycleway," *Los Angeles Times*, November 14, 1897.

58. Sarah Ferguson, "Ghost Riders: Twenty-one Cyclists Killed Last Year in NYC," *Village Voice*, January 9, 2006; Federal Highway Administration, U.S. Department of Transportation, "Bicycle and Pedestrian Provisions of the Federal Aid Program," available at <http://www.fhwa.dot.gov/environment/bikeped/bp-broch.htm>. The New York City spying incident was reported in the *New York Times*, which identified one episode where an undercover officer in biking gear attended a vigil for a bike rider who had been killed by a motorist. The officer wore a button that said "I am a shameless agitator." The *Times* reported that she "carried a camera and videotaped the roughly fifteen people present." Jim Dwyer, "Police Infiltrate Protests, Videotapes Show," *New York Times*, December 22, 2005.

59. Michael Seiler, "L.A. Crawls to Work," "Freeways Snarled," and "The RTD Leaves the Driving (and the Coping) to Them," *Los Angeles Times*, September 15, 16, and 18, 1982; "Buses: More Like It," *Los Angeles Times*, September 20, 1982; "Transportation Workers on Strike in Los Angeles," *The Economist*, September 23, 2000.

60. Sheldon Edner and Bruce D. McDowell, "Surface-Transportation Funding in a New Century: Assessing One Slice of the Federal Marble Cake," *Publius* 32, no. 1 (Winter 2002): 7–24.

61. Anastasia Loukaitou-Sideris, "Hot Spots of Bus Stop Crime: The Importance of Environmental Attributes," *Journal of the American Planning Association* 64, no. 4 (Autumn 1999): 395–411.

62. Joe Grengs, "Community-based Planning as a Source of Political Change: The Transit Equity Movement of Los Angeles' Bus Riders Union," *Journal of the American Planning Asso-*

ciation 68, no. 2 (Spring 2002): 165–78; Jeffrey L. Rabin and Richard Simon, "Court Order Spurs Plan to Buy 278 Buses," *Los Angeles Times*, September 26, 1997.

CHAPTER 6

1. Moore is cited in Wolf Von Eckardt, *A Place to Live: The Crisis of the Cities* (New York: Delacorte Press, 1967), 27.

2. Personal communication with Anastasia Loukaitou-Sideris, June 3, 2004; Robert Gottlieb and Anastasia Loukaitou-Sideris, "The Day That People Filled the Freeway: Re-Envisioning the Arroyo Seco Parkway and the Urban Environment in Los Angeles," *DISP Journal* (Zurich) 159, no. 4 (2004): 13–19.

3. In 2006, much to the dismay of Arroyo activists, a condo development began to be constructed just below the bridge, marring a visual connection to the bridge from the Arroyo Seco. See the Arroyo Seco Foundation Web site at <http://www.arroyoseco.org>. On the Arroyo culture, see Ward Ritchie, *A Southland Bohemia: The Arroyo Seco Colony as the Century Begins* (Pasadena: Vance Gerry, Weather Bird Press, 1996).

4. Lou V. Chapin, "An Ideal Cycleway," *Los Angeles Times*, November 14, 1897; T. D. Denham, "California's Great Cycle-Way," *Goods Roads Magazine* (November 1901), reprinted by U.S. Department of Transportation Federal Highway Administration, available at <http://www.fhwa.dot.gov/infrastructure/the_great_cycle_way_.htm>; Cecilia Rasmussen, "Bikeway Was Ahead of Its Time," *Los Angeles Times*, November 29, 1998.

5. Greg Hise and William Deverell, *Eden by Design: The 1930 Olmsted-Bartholomem Plan for the Los Angeles Region* (Berkeley: University of California Press, 2000) 98; *Arroyo Seco Parkway: A Brief Discussion of the Proposal and Its Relation to a Boulevard from the Mountains to the Sea* (Los Angeles, 1913), 8; Lippincott's comments are cited in H. W. Fraim, "Flood Control and Parkway Project along Arroyo Seco at Los Angeles," *Western Construction News* (June 1938): 233.

6. H. Marshall Goodwin, Jr., "The Arroyo Seco: From Dry Gulch to Freeway," *The Historical Society of Southern California Quarterly* (May 1965): 73–95; S.V. Cortelyou, "Arroyo Seco Six-Lane Freeway," *California Highway and Public Works* 17, no. 6 (June 1939): 10; Historical American Engineering Record (HAER), *Arroyo Seco Parkway*, no. CA-265, 1999; Bruce Henstell, "Happy Birthday, Dear Freeway," *Los Angeles Magazine*, December 1985.

7. The dedication ceremony statement is in S.V. Cortelyou, "Men, Steel and Concrete Work Miracles in the Arroyo Seco," *The Arroyo Seco Parkway Dedication Ceremonies Program* (Los

Angeles, 1940), 9. The 1934 "Master Highway Plan" issued by the Los Angeles City Planning Commission already used the term "freeway traffic artery" in describing the concept of the high-speed roadways being discussed. See Charles Cohan, "Wide Area to be Drawn Nearer Heart of City," *Los Angeles Times*, April 1, 1934.

8. Funding issues are discussed in "New Highway Opens Today," *Los Angeles Times*, July 20, 1940; David W. Jones, *California's Freeway Era in Historical Perspective* (Berkeley: Institute of Transportation Studies, University of California at Berkeley, June 1989), 169.

9. For example, a 1943 document published by the Los Angeles Regional Planning Commission distinguished between freeways and parkways, suggesting that while freeways offered "practical means of removing many of the unsatisfactory characteristics of the traditional highway [which did not provide for uninterrupted traffic flow]," parkways might or might not have offered the same advantages, given their scenic and recreational functions. Earl Esse and Simon Eisner, *Freeways for the Region* (Los Angeles: Los Angeles Regional County Planning Commission, 1943), 4.

10. E. E. East, "Motorist Saves Six Cents On Each Trip Over New Parkway," and R. D. Spencer, "Flood Control Channel Assures Protection of Parkway," in *The Arroyo Seco Parkway Dedication Ceremonies Program*, December 30, 1940. The strongest benefit of the development of the parkway was considered its design as an unobstructed route. "Now I know how a package feels when it gets an unobstructed ride through a chute to the shipping department," an article in the Los Angeles Auto Club publication commented soon after the parkway/freeway opened. "I've just made a run out to Pasadena on the completed Arroyo Seco Parkway. . . . From the relatively narrow Figueroa tunnels you immediately find yourself launched like a speedboat in a calm, spacious, divided channel. Channel is the word, too, for it's in the arroyo, below the level of traffic-tormented streets. No brazen pedestrians nor kids riding bikes with their arms folded! No cross streets with too-bold or too-timid drivers jutting their radiators into your path. And no wonder I made it from Elysian Park to Broadway and Glenarm Street in Pasadena in 10 minutes without edging over a conservative 45 miles an hour." John Cornell, "Riverbed Route, UN-Ltd.," *Westways* (January 1941): 13.

11. S. V. Cortelyou, "First Parkway for Los Angeles," *Engineering News-Record* (July 21, 1938): 79–81; R. E. Pierce, "Study Shows Accidents on Arroyo Are Less Than on Some Los Angeles City Streets," *California Highway and Public Works* 23, no. 7–8 (July–August 1945): 1–3.

12. Michael Leccesse, "Roadways Recovered," *Landscape Architecture* (April 1989): 49–55. The dedication ceremony statement is in S. V. Courtelyou, "Men, Steel and Concrete Work

Miracles in the Arroyo Seco," *The Arroyo Seco Parkway Dedication Ceremonies Program*, Los Angeles, December 30, 1940, 9. Olson's comments are cited in "Six-Million-Dollar Parkway Opened," *Los Angeles Times*, December 31, 1940, and in "Governor Olson Dedicates and Opens Arroyo Seco Freeway," *California Highways and Public Works* (January 1941): 6; see also "Thousands View Dedication of Arroyo Seco Parkway, First Freeway of West," *Los Angeles Times*, December 31, 1940; Jones, *California's Freeway Era*, 174–75.

13. Transportation Engineering Board, in Jones, California Freeway Era, 211. The Caltrans data are available in Anastasia Loukaitou-Sideris and Robert Gottlieb, "A Road as a Route and Place: The Evolution and Transformation of the Arroyo Seco Parkway," *California History* (Summer 2005): 28–40.

14. One Commissioner, in defending the Commission's 1954 name change from the Arroyo Seco Parkway to the Pasadena Freeway, argued that "10 of 100 don't [even] know what Arroyo Seco means." "Freeway Name Change Squawks Laid to Pique," *Los Angeles Times*, December 17, 1954; California Department of Parks and Recreation, *California Parkways: A Plan for a State Parkway System.* (Sacramento, 1967), 21; James Quinn, "Pasadena Freeway Improvement on Drawing Board," *Los Angeles Times*, August 4, 1971; Hall Leiren, "First Freeway Going Strong at Thirty," *Los Angeles Times*, January 1, 1971.

15. See Anastasia Loukaitou-Sideris and Robert Gottlieb, "Putting Pleasure Back in the Drive: Reclaiming Urban Parkways for the Twenty-first Century," *Access* 22 (2003): 2–8, Fig. 5 (Adjusted Number of Total Accidents on Freeways in District 7).

16. Personal communication with Chuck O'Connell, June 9, 2002 and Tim Davis, June 7, 2002.

17. Tim Brick, "Arroyo Planning Timeline," Arroyo Seco Foundation, available at <http://www.arroyoseco.org/History/ArroyoPlanningTimeline.pdf>; Ann Scheid, *Pasadena: Crown of the Valley* (Pasadena: Windsor Publications, 1986), 106; Herbert Eugene Bolton, *Fray Juan Crespi: Missionary Explorer on the Pacific Coast, 1769–1774* (Berkeley: University of California Press, 1971 [1927]), 147.

18. R. W. Stewart, "Controlling an Erratic Stream with Concrete," *The American City* 19, no. 6 (December 1918): 482–84; Stuart W. French, "The Great Arroyo Seco, a Part of Pasadena," *California Southland* (April 1924), available at <http://www.arroyoseco.org/calsouth1924.htm>.

19. "Flood-Curb Aid Sought," *Los Angeles Times*, May 1, 1934; "Thirty Dead in Southland Floods," *Los Angeles Times*, March 3, 1938; Blake Gumprecht, *The Los Angeles River: Its Life,*

Death, and Possible Rebirth (Baltimore: Johns Hopkins University Press, 1999) 215–20; "Flood Control Channel Assures Protection of Parkway," December 30, 1940.

20. Dana W. Bartlett, *The Better City: A Sociological Study of a Modern City* (Los Angeles: Neuner Company Press, 1907), 32–33.

21. Sue Spaid, *Ecovention: Current Art to Transform Ecologies*, Contemporary Arts Center, 2002, available at <http://greenmuseum.org/c/ecovention/sect2.html>; Tim Brick, "The Arroyo Restored," *West Pasadena Residents' Association News* (Fall 2001), available at <http://www.arroyoseco.org/WPRAFall2001.pdf>; personal communication with Tim Brick, April 6, 2006.

22. "Planting Seeds of Concern," *Los Angeles Times*, April 26, 1990; Berkley Hudson, "Quieter Earth Day Expected This Year," *Los Angeles Times*, April 18, 1991.

23. The Hahomongna Park's name was in fact changed (from Devil's Gate Park) on request from a leader of the Gabrieleno Indians who had also served on the planning authority that had developed the plans for the park. Given its reference to "flowing waters," another Planning Authority member argued that the name Hahamongna provided "a whole different connotation than Arroyo Seco or dry riverbed" and that it also suggested "that water resources are valuable and we need to value them." Edmund Newton, "Pasadena: The Making of Hahamongna," *Los Angeles Times*, August 16, 1992; See also, Edmund Newton, "Council Approves 250-Acre Park Plan for Arroyo Seco," *Los Angeles Times*, December 10, 1992.

24. Newton, "Council Approves 250-Acre Park Plan for Arroyo Seco"; Dave Gardetta, "Wild Things, Second Nature," *Los Angeles Times*, August 17, 1997.

25. Joseph Giovannini, "A Lost Canyon Beckons and Pasadena Responds," *New York Times*, December 31, 1998; personal communication with Tim Brick, April 6, 2006.

26. Northeast Trees and Arroyo Seco Foundation, *Arroyo Seco Watershed Restoration Feasibility Study* (Pasadena and Los Angeles: Northeast Trees and Arroyo Seco Foundation, 2002); Joe Mozingo, "Green Dreams Spring from a Dry Gulch," *Los Angeles Times*, August 12, 2001; Gene Maddaus, "Arroyo Seco Plans Approved," *Pasadena Star-News*, October 1, 2003.

27. California Resources Agency, "Governor Davis Praises Creation of Breakthrough Watershed Task Force: Top Ten Watershed Protection Projects in State Recognized," October 5, 2001; Cara Mia DiMassa, "Seeking a Renewal of Arroyo Seco," *Los Angeles Times*, August 24, 2002.

28. Personal communications with Diane Kane, February 28, 2002, and Richard Polanco, July 15, 2002.

29. Memorandum from Ronald Kosinski to Larry P. Loudon, Chief, Traffic Operations Branch, and Lan Jew, Senior Transportation Engineer, California Department of Transportation, November 20, 1992; Memo from Raja Mitwasi to Diane Kane, "Lowering Speeds on Arroyo Seco Parkway," June 13, 1997; Chapter 8–700 ("Traffic Regulations and Ordinances"), *Planning Manual*, California Department of Transportation, 1960; Personal communications with Larry Loudon, June 26, 2002, with Michael Cacciotti, July 8, 2002, with Chuck O'Connell, June 9, 2002, with Dennis Woodbury, June 7, 2002, with Larry Danielson, July 9, 2002, and with Richard Polanco, July 15, 2002.

30. Historical American Engineering Record, *Arroyo Seco Parkway*, no. CA-265, 1999; Personal communications with Nicole Possert, September 12, 2002, and Diane Kane, July 9, 2002; Car Mia DiMassa, "Note to Drivers: Slow Down—Enjoy the Arroyo," *Los Angeles Times*, March 25, 2003.

31. The national parkways conference is described in Larry Gordon, "Should Historic Status Go to a Freeway?," *Los Angeles Times*, March 7, 1998.

32. A comparative analysis of 1990 and 2000 census data of several dozen tracts within the Arroyo corridor, undertaken as part of the three-way "Re-Envisioning the Arroyo Seco Corridor" collaboration between UCLA, Occidental College, and Caltech, identified an increase in the Latino and Asian populations in the Corridor (constituting almost three-fourths of the corridor's population) and a wide disparity in income levels within different neighborhoods (though overall income in the Corridor was 10% lower than the L.A. County average in 2000). Some areas, such as Lincoln Park and sections of Chinatown and Highland Park, also identified substantial transit dependency (more than 20 percent of residents did not have cars). Kari Michele Fowler and Waynbe Wang, "Re-Envisioning the Arroyo Seco Corridor: Socio-Demographic Analysis," UCLA, Occidental College, and California Institute of Technology, Los Angeles, February 2002.

33. Notes, Meeting of the Arroyo Seco Collaborative, June 10, 2001.

34. The three-way course offered by UCLA, Occidental College, and Caltech was called "Re-Envisioning the Arroyo Seco Corridor: Watershed, Transportation, Environmental, and Community-Building Issues." A course description is available at <http://departments.oxy.edu/uepi/uep/courses/uep403arroyo.htm>.

35. Memo from Hector Obeso, Chief, District 7 Office of Permit, California Department of Transportation to Robert Gottlieb, December 18, 2001.

36. Richard Applebaum and Peter Dreier, "Sweat X Closes Up Shop," *The Nation* 29, no. 3 (July 19–26, 2004): 6; "No Sweat," *American Prospect* 13, no. 10 (June 3, 2002): 8.

37. Terry L. Cooper and Pradeep Chandra Kathi, "Neighborhood Councils and City Agencies: A Model of Collaborative Coproduction," *National Civic Review* 94, no. 1 (Spring 2005): 95; Jessica Garrison, "Panels Political Clout Grows," *Los Angeles Times*, October 9, 2004. *The Next Los Angeles* argues that the neighborhood councils set up "a new governmental body (with loosely defined and largely advisory powers) to fill the void created by the disappearance of citywide political coalitions and more responsive elected and appointed officials." Robert Gottlieb, Mark Vallianatos, Regina Freer, and Peter Dreier, *The Next Los Angeles: The Struggle for a Livable City* (Berkeley: University of California Press, 2005), 192. This type of structure also lent itself to internal power struggles over the control of something that appeared to be another tier of government but that lacked the actual powers of government.

38. One of the goals of the January 2002 luncheon was to demonstrate to Caltrans the political support for ArroyoFest. Toward that end, we arranged that the Caltrans district representative be seated between Goldberg and Nichols at the luncheon. ArroyoFest Executive Committee Meeting Minutes, December 6, 2001; Memo from Marcus Renner to Robert Gottlieb, July 3, 2006.

39. Part of the tension with the Caltrans staff was the political nature of the event, which challenged the concept of the domination of the freeway and criticized the Pasadena Freeway specifically as "an inadequate, inefficient, unsafe and inequitable transportation mode." One angry Caltrans official called the ArroyoFest argument "hypocrisy," asking rhetorically, "I take it these people never use the freeways in their daily travel? Maybe these folks should be reminded that the freeways help build the economy and way of life they currently enjoy." Memo from Ray Higa to Diane Kane, Frank Quon, and Sam Esquenazi, July 13, 2001; Renner's comments are from a July 3, 2006 e-mail communication.

40. Memo from David Roth to Robert Gottlieb, September 6, 2002; Harold Hewitt to Robert Gottlieb, August 9, 2002.

41. Notes, ArroyoFest Steering Committee meeting, August 7, 2002.

42. "Work for UEPI and ArroyoFest Steering Committee," Memo from Marcus Renner to Robert Gottlieb, October 31, 2002.

43. E-mail memo, "ArroyoFest/Community Partners," from Sandy Cooper to Ted Mitchell, January 27, 2003.

44. In a later aside, the Highway Patrol staff members told the L.A. Marathon coordinator that they had not focused on the planning for the event prior to the meeting when Caltrans announced that it was issuing its permit. If they had, they told the Marathon staffer, they would have tried to kill it. Personal communication with Marcus Renner, July 3, 2006.

45. See Marcus Renner, "The People's Freeway," *Orion* 23, no. 3 (May–June 2004): 60–63, Robert Gottlieb and Anastasia Loukaitou Sideris, "Take a Freeway Stroll for a New Look at the 110," *Los Angeles Times*, June 7, 2003.

46. Personal communication from James Rojas to the ArroyoFest listserv, June 16, 2003.

47. Marcus Renner, "ArroyoFest Next Steps," Memo to the Urban and Environmental Policy Institute, June 27, 2003.

48. Information about the Alliance for a Livable Los Angeles, Los Angeles Alliance for a New Economy, the Bicycle Kitchen, NELA Bikes!, and CICLE is available at <http://www.livableplaces.org/policy/alliance.html>, <http://www.laane.org>, <http://www.bicyclekitchen.com>, <http://www.nelabikes.com>, and <http://www.cicle.org/cicle_content>.

Chapter 7

1. Richard Sennett, *The Conscience of the Eye: The Design and Social Life of Cities*, New York: Norton, 1990, 123.

2. Henri Lefebvre, *The Urban Revolution* (Minneapolis: University of Minnesota Press, 2003), 96.

3. Josh Kun, "A Good Beat, and You Can Protest to It," *New York Times*, May 14, 2006.

4. Wilson's rhetoric against illegal immigration was effective in his successful reelection campaign, but six months later he suffered the embarrassment of having to acknowledge that he had employed an undocumented immigrant since 1978 and had never paid Social

Security taxes for her. Jennifer Warren and Dan Morain, "Wilson Flap Underscores Shifting Political Standards," *Los Angeles Times*, May 5, 1995.

5. Henry Yu argues that one could think of Los Angeles "as an intersection in a larger grid." "In this world," he continues, "migration is a process without end, comings and goings rather than the singular leaving of one place and arriving at another by which we mythically understand the immigrant's story." Henry Yu, "Los Angeles and American Studies in a Pacific World of Migrations," *American Quarterly* 56, no. 3 (2004): 531. The *Merriam-Webster* definitions are available at <http://www.m-w.com/dictionary/immigrant> and at <http://www.m-w.com/cgi-bin/dictionary?va=in-migrate>. The INS definition of *immigrant* is on its Web site at <http://www.uscis.gov/graphics/glossary2.htm#I>. In fact, scrolling to *immigrant* on the INS's glossary of terms sends you to "permanent resident alien" for the definition and description.

6. Patricia Limerick, *The Legacy of Conquest: The Unbroken Past of the American West* (New York: Norton, 1987), 251; Nicholas De Genova and Ana Y. Ramos-Zayas, *Latino Crossings: Mexicans, Puerto Ricans, and the Politics of Race and Citizenship* (New York: Routledge, 2003), 13.

7. McWilliams still represents the writer that researchers turn to first in exploring the issues of Los Angeles, migrations from the north, and the city's diverse populations. See, especially, Carey McWilliams, *Southern California Country: An Island on the Land* (New York: Duell, Sloan and Pearce, 1946), *Factories in the Field: The Story of Migratory Farm Labor in California* (Santa Barbara: Peregrine, 1971), and *North from Mexico: The Spanish-Speaking People of the United States*, (New York: Greenwood Press, 1968).

8. Theodore Hittell's description is cited in Arnold Hylen, *Los Angeles before the Freeways, 1850–1950* (Los Angeles: Dawson's Book Shop, 1981), 2; the *Merriam-Webster* definition is available at <http://www.m-w.com/dictionary/migration>; McWilliams, *Southern California: An Island on the Land*, 46; John D. Weaver, *El Pueblo Grande: A Nonfiction Book about Los Angeles* (Los Angeles: Ward Ritchie Press, 1973), 26. William Wallace's diary entry, the "belonged to the victor" quote and the phrase "homes of the defeated" by historian Merry Ovnick in *Los Angeles: The End of the Rainbow* (Los Angeles: Balcony Press, 1994), 59, are cited by William Deverell in *Whitewashed Adobe: The Rise of Los Angeles and the Remaking of Its American Past* (Berkeley: University of California Press, 2004), 16, 18.

9. Bret Harte, *The Heathen Chinee: Plain Language from Truthful James* (San Francisco: Book Club of California, 1934); Alexander Saxton, *The Indispensable Enemy: Labor and the Anti-Chinese Movement in California* (Berkeley: University of California Press, 1971), 19; H. D.

Barrows, "Chinese Competition in Farm Productions," *Los Angeles Times*, August 6, 1891. The racial profile of the "heathen chinee" was a staple of the Anglo press. One of the first editions of the *Los Angeles Times*, for example, contains a racially infused description of L.A.'s Chinese community and their "peculiar characteristics" that made the Chinese "strange men" who fail to engage in the "domestic relations so characteristic of higher civilization." "The Heathen Chinee: A Description of Some of Their Characteristics," *Los Angeles Times*, December 8, 1881.

10. "Immigrants in Hordes," *Los Angeles Times*, May 3, 1905; "Alien Rifraff Is Dangerous," *Los Angeles Times*, April 10, 1904; "Shut Out the Scum," *Los Angeles Times*, September 26, 1897

11. The immigration-restriction legislation and subsequent court rulings of the 1920s failed to resolve the issue of a Mexican racial identity. The 1930 Census Bureau reflected this uncertainty in its definition of *Mexicans* as "persons who were born in Mexico and are not definitely white, Negro, Chinese or Japanese." See Joseph A. Hill, "Composition of the American Population by Race and Country of Origin," *Annals of the American Academy of Political and Social Science* 188 (November, 1936), 77; Mae M. Ngai, *Impossible Subjects: Illegal Aliens and the Making of Modern America*, (Princeton, NJ: Princeton University Press, 2004), 23, 52–54; "Check Urged on Mexican Influx," *Los Angeles Times*, January 10, 1926.

12. "Preference over Migrants to Be Urged in Farm Labor," *Los Angeles Times*, March 12, 1940; Edward C. Krauss, "Light on Transit Problem," *Los Angeles Times*, November 12, 1935; "And Still They Come," *Los Angeles Times*, September 8, 1935; "Ebb of Migration," *Los Angeles Times*, December 10, 1931.

13. Peter Booth Wiley and Robert Gottlieb, *Empires in the Sun: The Rise of the New American West* (New York: Putnam, 1982), 249; "The Mexican Laborer," *Los Angeles Times*, November 4, 1930. The "White Americans refuse to do" comment is from a *Los Angeles Times* editorial, January. 23, 1930.

14. Ngai also points out that between 1946 and 1955, as the bracero program continued to be extended, farm wages as a percentage of manufacturing wages in the United States declined from 47.9 to 36.1 percent. Ngai, *Impossible Subjects*, 139. The Santa Ana grower quote is from John Cornell, "Mexican Labor Vital Factor in Southland Agriculture," *Los Angeles Times*, September 27, 1948.

15. Nagi, *Impossible Subjects*, 2. Patricia Limerick argued that even earlier in the century the distinction between newly arrived immigrants from Mexico and long-term Mexican American residents (some of whom also resented the newly arrived, unassimilated migrants) were

not visible to the Anglos. "We are all Mexicans anyway . . . because the *gueros* [blonds or Anglo-Americans] treat all of us alike," according to a 1931 interview cited in Limerick, *The Legacy of Conquest*, 245.

16. Wiley, *Empires in the Sun*, 253; Personal communication with Jorge Bustamante, June 10, 1980; "U.S. Patrol Halts Border 'Invasion,'" *Los Angeles Times*, June 17, 1954; "Roundup of Wetbacks in L.A. Still On," *Los Angeles Times*, June 20, 1954. Although Operation Wetback focused primarily on agricultural labor, L.A. migrants who were swept up in INS raids in 1954 included hotel and restaurant workers, a group that would eventually constitute the majority of workers in these industries and help lead to a renaissance of immigrant-based unionization campaigns fifty years later.

17. George Sanchez, *Becoming Mexican-American: Ethnicity, Culture, and Identity in Chicano Los Angeles, 1900–1945* (New York: Oxford University Press, 1993), 178; Stuart Cosgrove, "The Zoot-Suit and Style Warfare," *History Workshop Journal* 18 (1984): 77–91.

18. Ruben Salazar, "Why Does Standard July Fourth Oratory Bug Most Chicanos?," *Los Angeles Times*, July 10, 1970. On the early 1960s assumptions about the assimilation of Mexican as well as other minority groups, see Milton Gordon, *Assimilation in American Life: The Role of Race, Religion, and National Origins* (New York: Oxford University Press, 1964).

19. Felix Padilla, *Latino Ethnic Consciousness: The Case of Mexican Americans and Puerto Ricans in Chicago* (Notre Dame, IN: University of Notre Dame Press, 1985), 5; see also Nicho de Genova, *Latino Crossings: Mexicans, Puerto Ricans, and the Politics of Race and Citizenship* (Milton Park, UK: Routledge, 2003), 27.

20. Nicholas de Genova, "The Legal Production of Mexican/Migrant Illegality," *Latino Studies* 2, no. 2 (July 2004): 160.

21. Leslie Marmon Silko, "Fences against Freedom," in *Yellow Woman and a Beauty of the Spirit: Essays on Native American Life Today* (New York: Simon & Schuster, 1996), 114, cited in Jose David Saldivar, *Border Matters: Remapping American Cultural Studies* (Berkeley: University of California Press, 1997), xi.

22. The Alexander quote is from Maura Reynolds, "Stirring the Pot, Focusing on the Melting," *Los Angeles Times*, May 21, 2006.

23. Mary Ann Callan, "What Is L.A.? Midwest Heart, Hollywood Face," "Solid Core of Citizens," "Suburbs in Quest of City," *Los Angeles Times*, May 31, June 1, June 2, 1960.

24. The U.S. Census Bureau, *Decennial Census, 1960*, table 13, available at <http://www2.census.gov/prod2/decennial/documents/12533879v1p6ch03.pdf>.

25. United Nations, *World Urbanization Prospects: The 1999 Revision* (New York, 1999), available at <http://www.un.org/esa/population/pubsarchive/urbanization/urbanization.pdf>; Population Reference Bureau, *Human Population: Fundamentals of Growth: Patterns of World Urbanization* (2004), available at <http://www.prb.org/Content/NavigationMenu/PRB/Educators/Human_Population/Urbanization2/Patterns_of_World_Urbanization1.htm>. Michael A. Cohen, then a senior adviser to the vice president for environmental sustainable development at the World Bank, argued in 1996 that as the population shifts to the cities of the South began to occur, a type of urban convergence between the global cities of the North and the South also began to occur, including the growth of unemployment and a low-wage sector as well as "declining infrastructure, deteriorating environment, collapsing social compact, and institutional weakness." "The Hypothesis of Urban Convergence: Are Cities in the North and South Becoming More Alike in an Age of Globalization?," in *Preparing for the Future: Global Pressures and Local Forces*, ed. Michael A. Cohen, Blair A. Ruble, Joseph S. Tulchin, and Allison M. Garland (Washington DC: Woodrow Wilson Center Press, 1996), 25.

26. Henri Lefebvre, *The Urban Revolution* (Minneapolis: University of Minnesota Press, 2003), 147, 169; Saskia Sassen, *The Global City: New York, London, Tokyo*, (Princeton, NJ: Princeton University Press, 2001), 3–4; Neil Smith's comment is in his foreword to Lefebvre, *The Urban Revolution*, xx.

27. Nancy Cleeland, "L.A. Area Going to Extremes as the Middle Class Shrinks," *Los Angeles Times*, July 27, 2006; Howard Fine, "Change of Heart," *Los Angeles Business Journal*, May 15, 2006.

28. Neeta Fogg and Paul Harrington, "Growth and Change in the California and Long Beach/Los Angeles Labor Markets," Center for Labor Market Study, Northeastern University, Boston, May 3, 2001; Robert Gottlieb, Mark Vallianatos, Regina M. Freer, and Peter Dreier, *The Next Los Angeles: The Struggle for a Livable City* (Berkeley: University of California Press, 2005), 83–87.

29. The El Monte virtual slave-labor camp situation is described in William Branigan, "Sweatshop Instead of Paradise," *Washington Post*, September 10, 1995; George White, "Garment 'Slaves' Tell of Hardship," *Los Angeles Times*, August 4, 1995; see also, Edna Bonacich and Richard Applebaum, *Behind the Label: Inequality in the Los Angeles Apparel Industry* (Berkeley: University of California Press, 2000). The restructuring of the janitorial

services industry is discussed in Robert Gottlieb *Environmentalism Unbound: Exploring New Pathways for Change* (Cambridge: MIT Press, 2002), 145–80.

30. *Empires in the Sun*, 36; "Market Spotlight: Los Angeles, Calif.: Glitz, Glamour, and Smog," *National Petroleum News* (June 2004): 16–17.

31. Francisco L. Rivera-Batiz, "Can Border Industries Be a Substitute for Immigration?," *American Economic Review* 76, no. 2 (May 1986): 263–68; Gary Jacobsen, "The Boom on Mexico's Border," *Management Review* 77, no. 7 (July 1988): 20–23; Joseph Grunwald, "Opportunity Missed: Mexico and *Maquiladoras*," *Brookings Review* 9, no. 1 (Winter 1990–1991): 44–48.

32. Joel Millman, "Work in Progress: Prosperity in Home Countries May Not Stem Tide of Migrants to the U.S.," *Wall Street Journal*, May 8, 2006.

33. Kathryn Kopinak, "*Maquiladora* Industrialization of the Baja California Peninsula: The Co-existence of Thick and Thin Globalization with Economic Regionalism," *International Journal of Urban and Regional Research* 27, no. 2(June 2003): 319–36; Tim Koechlin, "NAFTA's Footloose Plants Abandon Workers," *Multinational Monitor* 16, no. 4 (April 1995): 25; David Bacon, *The Children of NAFTA: Labor Wars on the U.S.-Mexican Border* (Berkeley: University of California Press, 2004), 42–49; Jeff Faux, *The Global Class War: How America's Bipartisan Elite Lost Our Future—and What It Will Take to Win It Back* (New York: J Wiley, 2006).

34. Joseph Contreras, "What China Threat?," *Newsweek*, May 15, 2006.

35. Los Angeles Economic Development Corporation, *International Trade Trends and Impacts: The Southern California Region* (Los Angeles: LAEDC, May 2006).

36. Andrea Hricko, "Ships, Trucks, and Trains: Effects of Goods Movement on Environmental Health," *Environmental Health Perspectives* 114, no. 4 (April 2006): 204–05; see also, Andrea M. Hricko, "Problems along the new 'Silk Road,'" *Environmental Health Perspectives* 112, no. 15 (November 2004): A879.

37. Dinesh C. Sharma, "Ports in a Storm," *Environmental Health Perspectives* 114, no. 4 (April 2006): A222–31; David Streets, et al., "Modeling Study of Air Pollution Due to the Manufacture of Export Goods in China's Pearl River Delta," *Environmental Science and Technology* 40, no. 7 (2006): 2099–07; "Linking China's Air Pollution to Exports," *Environmental Science and Technology* (April 1, 2006): 2073.

38. "Port of Long Beach, California—Water Gateway," U.S. Bureau of Transportation Statistics, 2006, available at <http://www.bts.gov/publications/americas_freight

_transportation_gateways/highlights_of_top_25_freight_gateways_by_shipment_value/port_of_long_beach>; "Port of Los Angeles, California—Water Gateway," Bureau of Transportation Statistics, 2006, available at <http://www.bts.gov/publications/americas_freight_transportation_gateways/highlights_of_top_25_freight_gateways_by_shipment_value/port_of_los_angeles>; Wilbur Smith Associates, *Multi-County Goods Movement Action Plan*, Technical Memorandum 3: Existing Conditions and Constraints, Prepared for Los Angeles County Metropolitan Transportation Authority et al. (2006), E-3. The twenty-foot equivalent unit (TEU) refers to the container capacity (of ships and ports). A twenty-foot equivalent unit is a measure of containerized cargo capacity equal to one standard container, which is 20 feet long × 8 feet wide × 8 feet × 6 inches high. In metric units this is 6.10 meters (length) × 2.44 meters (width) × 2.59 meters (height), or approximately 39 cubic meters. These sell for about $2,500 in China, the biggest manufacturer of such containers.

39. Edmund Newton, "This Ship Has Come In," *Los Angeles Times*, April 23, 1995; James A. Fawcett, "A Tale of Two Ports with '2020' Vision," *Oceanus* 32, no. 3 (1989): 79–84. The Bradley quote is from an interview with Steve Erie and is cited in Erie's book, *Globalizing L.A.: Trade, Infrastructure, and Regional Development* (Stanford, CA: Stanford University Press, 2004), 92.

40. Carolyn Cartier, "Cosmopolitics and the Maritime World City," *Geographical Review* 89, no. 2 (April, 1999): 278–89. Cartier argued that global-city and world-city analysts had tended to ignore the importance of port and goods-movement activities across oceans since such globalization-related activity was seen as belonging to "an earlier era of world economic history." On Pier A, see, Gan Mukhopadhyay, "Preparing for Pier A," *Civil Engineering* 68, no. 8, (August 1998): 36–39. On the Alameda Corridor, see Brian Fortner, "The Train Lane," *Civil Engineering* 72, no. 9 (September 2002): 52–60.

41. Presentation by Angelo Logan, Occidental College, October 7, 2004; W. James Gauderman et al., "The Effect of Air Pollution on Lung Development from Ten to Eighteen Years of Age," *New England Journal of Medicine* 351, no. 11 (September 9, 2004): 1–11; Rob McConnell et al., "Traffic, Susceptibility, and Childhood Asthma," *Environmental Health Perspectives (ehponline)*, February 16, 2006, available at <http:dx.doi.org>; P.W. Arnberg et al., "Sleep Disturbances Caused by Vibrations from Heavy Road Traffic," *Journal of the Acoustical Society of America* 88, no. 3 (1990): 1486–93; Deborah Schoch, "Unsightly Evidence of U.S. Trade Gap Piles Up," *Los Angeles Times*, July 9, 2006; Christopher Boone et al., "Creating a Toxic Neighborhood in Los Angeles," *Urban Affairs Review* 35, no. 2 (1999): 163–187.

42. "Darby Leads River Tour to Boost Highway Route," *Los Angeles Times*, August 22, 1947; Ray Hebert, "Long Beach Freeway (California 7) Ascends to National Status as I-710," *Los Angeles Times,* October 24, 1985.

43. Barney Gimbel, "Yule Log Jam," *Fortune* 150, no. 12 (December 13, 2004): 163–64; Meg James, "The Port Settlement," *Los Angeles Times*, November 25, 2002.

44. Deborah Schoch, "MTA Votes to Expand the 710 Freeway," *Los Angeles Times*, January 28, 2005; "Goods Movement in California," California Environmental Protection Agency and Business, Transportation and Housing Agency, January 27, 2006; Richard D. Vogel, "The NAFTA Corridors: Offshoring U.S. Transportation Jobs to Mexico," *Monthly Review* 57, no. 9 (February 2006): 16–29; Ronald D. White, "Mexican Port Gets American Connection," *Los Angeles Times*, June 20, 2006.

45. Janet Wilson, "L.A., Long Beach Ports Produce Plan to Reduce Diesel Emissions," *Los Angeles Times*, June 29, 2006.

46. The notes of the December 11, 1979, meeting of UNO and the Community Committee, taken by two UNO participants, are in the author's possession. Additional information was provided in a personal communication with Lydia Lopez, June 11, 1980.

47. Mario T. Garcia, *Memories of Chicano History: The Life and Narrative of Bert Corona* (Berkeley: University of California Press, 1994); Ngai, *Impossible Subjects*, 139–59.

48. Gottlieb et al., *The Next Los Angeles*, 139–43.

49. Patrick J. McDonnell and Robert J. Lopez, "Some See New Activism in Huge March," *Los Angeles Times*, October 18, 1994; Gottlieb et al., *The Next Los Angeles*, 158–59.

50. Peter Hong, "Bouyed by Respect, Janitors Back at Work," *Los Angeles Times*, April 26, 2000; William Clairborne, "AFL-CIO Changes Tune on Immigrant Workers," *Washington Post*, June 4, 2000; Harold Meyerson, "The Red Sea: How the Janitors Won Their Strike," *L.A. Weekly*, April 28, 2000.

51. Harold Meyerson, "L.A. Story," *American Prospect* 12, no. 12 (July 2001): 2–16; Gottlieb et al., *The Next Los Angeles*, 163–71.

52. Personal communication with Torie Osborn, April 20, 2006.

53. Ashley Powers, "Proposal on Migrant Issues Will Go to Voters," *Los Angeles Times*, May 16, 2006; Daniel B. Wood, "Two Towns, Two Stands on Immigration Reform," *Christian*

Science Monitor, April 5, 2006; Jennifer Delson, "Migrant Fight Taxes O.C. City's Police Chief," *Los Angeles Times*, April 5, 2006.

Chapter 8

1. Cited by Tony Horwitz in "Immigration—and the Curse of the Black Legend," *New York Times*, July 9, 2006.

2. Samuel P. Huntington, "The Hispanic Challenge," *Foreign Policy* (March–April 2004): 30.

3. The background information and comments about Ogonowski are from a communication from Hugh Joseph, a researcher and community-food advocate based at Tufts University and the head of the New Entry project who also worked closely with Ogonowski in setting up the mentor farm. Hugh Joseph, "September 11th Tragedy—Dracut Land Trust," sent on September 15, 2001, to the Comfood listserv, which was managed through the Community Food Security Coalition. See also Hugh Joseph, "Friend to Farmers: John Ogonowski, American Airlines Flight 11 Pilot," Tufts University Friedman School of Nutrition Science and Policy, available at <http://nutrition.tufts.edu/consumer/feature/ ogonowski.html>.

4. Greg Hise, "Border City: Race and Social Distance in Los Angeles," *American Quarterly* 56, no. 3 (2004): 545–58.

5. The east-west divide along the L.A. River and other geographic divides is discussed in William Deverell, *Whitewashed Adobe* (Berkeley: University of California Press, 2004), 91–128; See also Hise, "Border City," 550. "The geography of nowhere" is taken from James Howard Kunstler's title of his book, *The Geography of Nowhere: The Rise and Decline of America's Man-Made Landscape* (New York: Simon & Schuster, 1993). The idea that freeways could strategically uproot low-income (i.e., "blighted" communities) was proposed as early as 1939 by automobile promoter Stanley Hoffman in his essay (with Neil M. Cook) "America Goes to Town," *Saturday Evening Post*, April 29, 1939. Hoffman argued that "in many cases, the blighted areas offer the one opportunity to develop at reasonable cost the type of highway and parking facilities needed to meet the requirements of the motorized communities" (32).

6. Saskia Sassen, foreword to Victor M. Valle and Rodolfo D. Torres, *Latino Metropolis* (Minneapolis: University of Minnesota Press, 2000), xi; Kristen Hill Maher, "Borders and Social Distinction in the Global Suburb," *American Quarterly* 56, no. 3 (September 2004): 781–806;

James P. Allen and Eugene Turner, *The Ethnic Quilt: Population Diversity in Southern California* (Northridge, CA: Center for Geographical Studies, California State University, 1997). The "immigrant enclave" concept is discussed by John R. Logan, Wenquan Zhang, and Richard D. Alba in "Immigrant Enclaves and Ethnic Communities in New York and Los Angeles," *American Sociological Review* 67, no. 2 (April 2002): 299–322.

7. Mike Davis, *Magical Urbanism: Latinos Reinvent the U.S. City* (New York: Verso, 2000), 57; Robert Fishman, "The Fifth Migration," *Journal of the American Planning Association* 71, no. 4 (Autumn 2005): 357–66; John Horton, *The Politics of Diversity: Immigration, Resistance, and Change in Monterey Park, California* (Philadelphia: Temple University Press, 1995), 36; James Rojas is quoted in James B. Goodno, "On Immigrant Street: Newcomers Are Making a Big Impact on the Built Environment," *Planning* 71, no. 10 (November 2005): 4–9; see also James Rojas, "The Enacted Environment: Examining the Streets and Yards of East Los Angeles," in Chris Wilson and Paul Groth, eds., *Everyday America: Cultural Landscape Studies after J. B. Jackson* (Berkeley: University of California Press, 2003).

8. Goodno, "On Immigrant Street," 8.

9. "Dying Alone: An Interview with Eric Klinenberg," at <http://www.press.uchicago.edu/Misc/Chicago/443213in.html>.

10. Horace Kallen, "Democracy versus the Melting-Pot: A Study of American Nationality, Part II," *The Nation* 100, no. 2591 (February 25, 1915): 219.

11. On the fiesta, see, for example, William Deverell, *Whitewashed Adobe: The Rise of Los Angeles and the Remaking of Its Mexican Past* (Berkeley: University of California Press, 2004), 49–90; See also Ed Ainsworth, "An Open Letter to a Mexican," *Los Angeles Times*, September 29, 1960.

12. Susan A. Phillips, "*El Nuevo Mondo:* The Landscape of Latino Los Angeles—Photographs by (Chilean photographer) Camilo José Vergara" (on Vergara's exhibit at the National Building Museum in D.C.), *American Anthropologist* 103, no. 1 (March 2001): 175–182.

13. Duany is cited in Anna Holtzman, "Latin Invasion," *Architecture*, vol. 91, no. 3 (March 2004): 22.

14. Albert Camarillo, *Chicanos in a Changing Society: From Mexican Pueblos and American Barrios in Santa Barbara and Southern California, 1849–1930* (Cambridge, MA: Harvard University Press, 1979).

15. Gloria Ohland, "Renaissance in the Barrio: Will Mixed-Income Housing Revitalize Boyle Heights or Just Chase out the Poor?," *L.A. Weekly*, November 19–25, 2004; Lydia Avila-Hernandez, "The Boyle Heights Landscape: The Pressures for Gentrification and the Need for Grass Roots Action and Accountable Development," Occidental College Urban and Environmental Policy Department, Los Angeles, 2005, available at <http://departments.oxy.edu/uepi/uep/studentwork/05comps/avila-hernandez.pdf>. One illustration of the early signs of gentrification in these neighborhoods has been decreasing enrollments in neighborhood schools, which "offer an early glimpse of demographic trends that won't show up in census data for several years," according to an *Los Angeles Times* article. Families with children that leave those neighborhoods often are replaced by young single professionals, who represent the initial sign of gentrification. Nancy Cleeland, "There Goes the Enrollment," *Los Angeles Times*, June 11, 2006.

16. Robert Gottlieb, "Education a Public Responsibility," *Los Angeles Times*, February 18, 1992.

17. USDA and the Community Food Security Coalition, "The Farm-School Connection: An Informal Collaboration between the Community Food Security Coalition and USDA," Washington D.C., and Los Angeles, 2000.

18. Michelle Mascarenhas and Robert Gottlieb, "Healthy Farms, Healthy Kids: Final Report to the California Wellness Foundation," Center for Food and Justice, Los Angeles, 1999; Robert Gottlieb, "McKinley Students Lead Lunch Revolution," *The Outlook*, September 16, 1997.

19. Marcia Herrin and Joan Gussow, "Designing a Sustainable Regional Diet," *Journal of Nutrition Education* 6 (1989): 270–75; Jennifer Wilkins, "Seasonal and Local Diets: Consumers' Role in Achieving a Sustainable Food System," *Res Rural Social Development*. 6 (1995): 150–52; see also Joan Gussow, *Chicken Little, Tomato Sauce, and Agriculture: Who Will Produce Tomorrow's Food*? (New York: Bootstrap Press, 1991).

20. See, for example, Christina Schiavoni and Peter Mann, "Sharing a Common Struggle: Bridging Borders toward Food Security," *Community Food Security News* (Fall 2006): 1+; Harriet Friedmann, "The Political Economy of Food: A Global Crisis," *New Left Review* 197 (1997): 29–57; Robert Gottlieb, *Environmentalism Unbound: Exploring New Pathways for Change* (Cambridge: MIT Press, 2002), 181–226.

21. See the Community Food Security Coalition's definition, "What Is Community Food Security?," available at <http://www.foodsecurity.org/views_cfs_faq.html>.

22. United States Department of Agriculture, National Agricultural Statistics Service, Table 49 ("Spanish, Hispanic, or Latino Origin Principal Operators: Selected Farm Characteristics 2002 and 1997"); USDA NASS, "Counting Diversity in American Agriculture"; Kent Mullinix, Leonardo Garcia, Alexandra Lewis-Lorentz, and Joan Qazi, "Latino Views of Agriculture, Careers and Education: Dispelling the Myths," Paper presented at the Sustainable Agriculture, Communities and Environments in the Pacific Northwest Symposium, Washington State University, Richland, Washington, May 19, 2006; Sara Schilling and Elena Olmstead, "Once Limited to Field Work, Hispanics Are Now Buying Farms," *Tri-City Herald*, August 28, 2006.

23. Ann S. Kim, "Growing Roots in Maine," *Portland Press Herald*, January 27, 2006; Judith Weinraub, "New Immigrant Farmers: Each Succeeding Wave of Immigrants Brings Its Crops to Our Country's Table," *Washington Post*, October 15, 2003; Agriculture and Land-Based Training Association, "Progress Report, 2001–2005," Salinas, CA, 2005; personal communication with Steve Davies, Project for Public Spaces, October 25, 2005; Howell Tumlin, Southland Farmers' Market Association, October 10, 2006.

24. Steven Lee Meyers, "Farmer Unearthed: He Planted the Corn," *New York Times*, August 15, 1991; also Steven Lee Myers, "Broadway's New Feature: Cornstalks," *New York Times*, August 13, 1991; Barbara Deutsch Lynch, "U.S. Latino Discourses and Mainstream Environmentalism," *Social Problems* 40, no. 1 (February 1993): 108–24.

25. Louis Blumberg et al., "The Dilemma of Municipal Solid Waste Management: An Examination of the Rise of Incineration, Its Health and Air Impacts, the LANCER Project and the Feasibility of Alternatives," Urban Planning Program, University of California at Los Angeles, 1986; Louis Blumberg and Robert Gottlieb, *War on Waste: Can America Win Its Battle with Garbage?* (Washington, DC: Island Press, 1989), 155–88.

26. Emily Green, "Green Dreams," *Los Angeles Times*, October 31, 2004.

27. "Community Garden Survey: Immigrants and the Environment Project," Urban and Environmental Policy Institute, Los Angeles, 2004.

28. Amanda Shaffer, "The Persistence of L.A.'s Grocery Gap: The Need for a New Food Policy and Approach to Market Development," Urban and Environmental Policy Institute, Los Angeles, May 2002, available at <http://departments.oxy.edu/uepi/publications/the_persistence_of.htm>; Amanda Shaffer and Robert Gottlieb, "Promises of Renewal Broken," *Los Angeles Times*, May 10, 2002; Robert Gottlieb, "Community Gardens and Urban Agriculture: Creating Community, Aesthetic, Environmental, Health and Economic Values

in an Urban Setting," Paper prepared for the South Central Farm court case, Urban and Environmental Policy Institute, Los Angeles, 2006; Robert Gottlieb and James Rojas, "Los Angeles Should Cultivate This Rare Urban Seed," *Los Angeles Times*, March 23, 2004; Martin Zimmerman, "Safeway's High-End Concept Lifts Earnings," *Los Angeles Times*, October 13, 2006.

29. "L.A. Gothic," *Los Angeles Times* (editorial), March 11, 2006. The *Times* also commented that since "the land belongs to Horowitz," "he has every right to kick out the people who have been squatting there for more than a decade."

30. "Remarks by Mayor Villaraigosa Regarding the South Central Farm," Office of the Mayor, Los Angeles, June 13, 2006.

31. The Modesta Avila story is told by Lisbeth Haas in *Conquests and Historical Identities in California, 1769–1936* (Berkeley: University of California Press, 1995), 1–2, 89–91.

32. Jason Corburn, *Street Science: Community Knowledge and Environmental Health Justice* (Cambridge, II MA: MIT Press, 2005); Presentation by Andrea Hricko, Coalition for Clean Air Awards Luncheon, Los Angeles, May 31, 2006.

33. Jim Newton, "Once Rivals, Local Ports Clear Air in Partnership," *Los Angeles Times*, July 4, 2006; Deborah Schoch, "Villaraigosa's Port Panel Choices Suggest New Direction," *Los Angeles Times*, July 27, 2005; Paul Rosenberg, "Port-Community Relations: A Year in Review—and Preview," *Random Lengths*, December 22, 2005.

34. Jaime Ruiz, "The Marquez Equation: Knowledge × People = Power," *Random Lengths*, July 23, 2004.

35. A Resolution of the City Council of the City of Commerce California regarding Environmental Justice, Resolution 04-38, City of Commerce, August 2, 2004; personal communication with Angelo Logan, March 7, 2005.

36. Paul Feldman, "Toxic Runoff Made Teacher an Environmental Activist," *Los Angeles Times*, July 31, 1992; Robert Gottlieb, *Forcing the Spring: The Transformation of the American Environmental Movement* (Washington DC: Island Press, 1993) ("On the Move with Penny Newman"), 162–70.

37. Philip Shabecoff, "Rita Lavelle Gets Six-Month Term and Is Fined $10,000 for Perjury," *New York Times*, January 10, 1984.

38. Deborah Schoch, "Labor Lends Its Clout to Port Pollution Battle," *Los Angeles Times*, January 28, 2006; Ronald D. White, "Immigration Rallies Fuel Resolve of Port Truckers," *Los Angeles Times*, May 4, 2006.

39. Catalina Amuedo-Dorantes and Susan Pozo, "On the Use of Differing Money Transmission Methods by Mexican Immigrants," *International Migration Review* 39, no. 3 (Fall 2005): 554–76.

40. United Nations Population Fund, *A Passage to Hope: Women and International Migration* (New York: UNFPA, 2006), available at <http://www.unfpa.org/swp/2006/english/introduction.html>. The report's principal author, Maria Jose Alcala, notes that while female immigrants earn less than their male immigrant counterparts, they also send a larger proportion of their income to their home countries. "What's very interesting about remittances and the whole migration experience," commented Alcala, "is that it transforms traditional gender norms in both public and private life. In more traditional societies, money is the men's field." Cited in Juliana Barbassa, "Female Immigrants Earn Less, Send More," *Santa Monica Daily Press*, September 14, 2006; see also Robert E. B. Lucas, *International Migration and Economic Development: Lessons from Low-Income Countries* (Cheltenham, UK: Edwin Elgar, 2005).

41. Timothy J. Hatton, "International Migration and Economic Development: Lessons from Low-Income Countries," *Economic Record* 82, no. 256 (March 2006): 101–02; Olivia Ruiz, "Migration and Borders: Present and Future Challenges," *Latin American Perspectives* 33, no. 2 (March 2006): 47; Eduardo Porter, "Tighter Border Yields Odd Result: More Illegals Stay," *Wall Street Journal*, October 10, 2003; Eduardo Porter, "Flow of Immigrants' Money to Latin America Surges," *New York Times*, October 19, 2006.

42. Kim Barry, "Home and Away: The Construction of Citizenship in an Emigration Context," *New York University Law Review* 8, no. 1 (April 2006): 11; Sam Quinones, "Mexican Hometown Clubs Vote for L.A. Politics," *Los Angeles Times*, March 5, 2005.

43. Ana Patricia Rodrigues, "Departamento 15: Cultural Narratives of Salvadoran Transnational Migration," *Latino Studies. Houndmills* 3, no. 1 (April 2005): 19.

44. Harriet Friedmann, "Biodiversity and Cultural Diversity in North American Foods," *Food News*, October 17, 2005, available at <http://www.slowfoodforum.org/archive/index.php/t-1018.html>. Friedmann also argues that this immigrant-inspired link between a "polycultural agriculture" and cultural diversity is especially critical given that industrial monocultures increasingly threaten biodiversity. See also Peggy Levitt, *The Transnational Vil-*

lagers (Berkeley: University of California Press, 2001), 4; Jennifer Lee, "In Chinatowns, All Sojourners Can Feel Hua," *New York Times*, January 27, 2006; Anthony M. Orum, "Circles of Influence and Chains of Command: The Social Processes Whereby Ethnic Communities Influence Host Societies," *Social Forces* 84, no. 2 (December 2005): 921–39. Devra Weber, in her essay on Mexican workers who migrated to California, also points to the historical roots of cross-border influences, arguing that the postwar period "witnessed the intensification, not the creation, of transnationalism." Devra Weber, "Historical Perspectives on Transnational Mexican Workers in California," in *Border Crossings: Mexican and Mexican-American Workers*, ed. John Mason Hart (Wilmington, DE: Scholarly Resources, 1998), 212.

45. Daniel Hernandez, "A Sampling of Culture 'Reminiscent of Home,'" *Los Angeles Times*, May 24, 2004; Jose Fuentes-Salinas, "Guelaguetza Primaveral," *La Opinion*, March 19, 2000; Barbara Hansen, "The Oaxaca Connection," *Los Angeles Times*, May 1, 2002; Robert C. Smith, "Migrant Membership as an Instituted Process: Transnationalization, the State and the Extraterritorial Conduct of Mexican Politics," *International Migration Review* 37, no. 2 (Summer 2003): 297.

46. Gaspar Rivera Salgado, "Binational Organizations of Mexican Migrants in the United States," *Social Justice* 26, no. 3 (Fall 1999): 27–38; Alejandro Portes, Luis E. Guarnizo, and Patricia Landolt, "The Study of Transnationalism: Pitfalls and Promise of an Emergent Research Field," *Ethnic and Racial Studies* 22, no. 2 (March 1999): 217–37; personal communication with Chancee Martorell, October 17, 2006.

47. "Jasmine Rice: U.S. Students Plan to Help Thai farmers." *The Nation* (Bangkok), October 26, 2002.

48. Ellen Roggemann, "My Journey," *OneWorld United States*, December 7, 2005, available at <http://us.oneworld.net/article/view/123089/1/>.

49. Ellen Roggemann, *Fair Trade Thai Jasmine Rice: Social Change and Alternative Food Strategies across Borders* (Los Angeles: Urban and Environmental Policy Institute, 2005).

50. Matthew Clement, "Rice Imperialism: The Agribusiness Threat to Third World Rice Production," *Monthly Review* 55, no. 9 (February 2004): 15–22; A. Barnett, "Thai Fury at US 'Piracy' of Rice Gene," *The Observer*, November, 28, 2000; Tom Hargrove, "Jasmine Rice for U.S. Farmers," August 1, 2001, available at <http://www.biothai.org/cgi-bin/content/biopiracy/show.pl?0004>; Vasana Chinvarakorn, "Rice 'n Controversy." *BioThai*, December 2003, available at <http://www.biothai.org/cgi-bin/content/rights/show.pl?0006>.

51. Personal communication with Ellen Roggemann, April 13, 2005.

52. Jennifer Alsever, "Fair Prices for Farmers: Simple Idea, Complex Reality," *New York Times*, March 19, 2006; Marlike Kocken, "Fifty Years of Fair Trade: A Brief History of the Fair Trade Movement," *European Fair Trade Association* (December 2003), available at <http://www.european-fair-trade-association.org/Efta/Doc/History.pdf>.

53. "Fair Trade Goods in the Virtual Marketplace," Thai Community Development Center, Los Angeles, 2006; Personal communication with Chancee Martorell, Thai Community Development Center, March 21, 2006.

Chapter 9

1. Zygmunt Bauman, *Globalization: The Human Consequences* (New York: Columbia University Press), 74.

2. Horace M. Kallen, "Democracy versus the Melting-Pot: A study of American Nationality," *The Nation*, February 25, 1915, 219; Milton M. Gordon, *Assimilation in American Life: The Role of Race, Religion and National Origins* (New York: Oxford University Press, 1964), 5. Writing around the same time as Kallen, John Dewey also argued that "Democracy will be more productive if it has a tendency to encourage differences. Our dream of the United States ought not to be a dream of monotony. We ought not to think of it as a place where all people are alike." John Dewey, "Nationalizing Education," in National Education Association of the United States, *Addresses and Proceedings of the Fifty-fourth Annual Meeting* (1916), 185–86, cited in Gordon, *Assimilation in American Life*, 140.

3. Gordon, *Assimilation in American Life*, 108–09.

4. Henri Lefebvre, *The Urban Revolution* (Minneapolis: University of Minnesota Press, 2003), 18.

5. Susan S. Fainstein, "Cities and Diversity: Should We Want It? Should We Plan for It?," *Urban Affairs Review* 41, no. 1 (September 2005): 3; Stephen Castles, *Ethnicity and Globalization: From Migrant Worker to Transnational Citizen* (London: Sage, 2000), 131.

6. Fernando Báez, "On the Road With Bush and Chávez," *New York Times*, March 11, 2007.

7. Raymond Williams, *Keywords: A Vocabulary of Culture and Society* (New York: Oxford University Press, 1985 [1976]), 159–60.

8. John L. Hammond, "The Possible World and the Actual State," *Latin American Perspectives* 33, no. 3 (May 2006): 122; *World Social Forum: Challenging Empires*, ed. Jai Sen, Anita Anand, Arturo Escobar, and Peter Waterman (New Delhi: Viveka Foundation, 2004).

9. Stephen Castles, "Nation and Empire: Hierarchies of Citizenship in the New Global Order," *International Politics* 42, no. 2 (June 2005): 203–05; Michael Peter Smith and Luis Guarnizo, eds., *Transnationalism from Below: Comparative Urban and Community Research*, vol. 6 (New Brunswick, NJ: Transaction, 1998); Peggy Levitt, *The Transnational Villagers* (Berkeley: University of California Press, 2001), 180–97.

10. The value-chain concept was introduced in the 1990s by Michael Porter as a model for business and economic performance. It was subsequently appropriated by community-food theorists to identify social- and environmental-justice values and community-food-system values within a food-production system. When applied to a production system that crosses borders, a value-chain approach can help enlarge those social and environmental goals associated with the fair-trade approach to a broader *production system* argument about alternatives. Michael Porter, "New Strategies for Inner-City Economic Development," *Economic Development Quarterly* (February 1997): 11–27.

11. Jeff Faux, *The Global Class War: How America's Bipartisan Elite Lost Our Future—and What It Will Take to Win It Back* (New York: Wiley, 2006).

Index

Errata notice: This series listing was mistakenly omitted from the first printing of Robert Gottlieb's *Reinventing Los Angeles* (The MIT Press, 2007).

Urban and Industrial Environments

Series editor: Robert Gottlieb, Henry R. Luce Professor of Urban and Environmental Policy, Occidental College

Maureen Smith, *The U.S. Paper Industry and Sustainable Production: An Argument for Restructuring*

Keith Pezzoli, *Human Settlements and Planning for Ecological Sustainability: The Case of Mexico City*

Sarah Hammond Creighton, *Greening the Ivory Tower: Improving the Environmental Track Record of Universities, Colleges, and Other Institutions*

Jan Mazurek, *Making Microchips: Policy, Globalization, and Economic Restructuring in the Semiconductor Industry*

William A. Shutkin, *The Land That Could Be: Environmentalism and Democracy in the Twenty-First Century*

Richard Hofrichter, ed., *Reclaiming the Environmental Debate: The Politics of Health in a Toxic Culture*

Robert Gottlieb, *Environmentalism Unbound: Exploring New Pathways for Change*

Kenneth Geiser, *Materials Matter: Toward a Sustainable Materials Policy*

Thomas D. Beamish, *Silent Spill: The Organization of an Industrial Crisis*

Matthew Gandy, *Concrete and Clay: Reworking Nature in New York City*

David Naguib Pellow, *Garbage Wars: The Struggle for Environmental Justice in Chicago*

Julian Agyeman, Robert D. Bullard, and Bob Evans, eds., *Just Sustainabilities: Development in an Unequal World*

Barbara L. Allen, *Uneasy Alchemy: Citizens and Experts in Louisiana's Chemical Corridor Disputes*

Dara O'Rourke, *Community-Driven Regulation: Balancing Development and the Environment in Vietnam*

Brian K. Obach, *Labor and the Environmental Movement: The Quest for Common Ground*

Peggy F. Barlett and Geoffrey W. Chase, eds., *Sustainability on Campus: Stories and Strategies for Change*

Steve Lerner, *Diamond: A Struggle for Environmental Justice in Louisiana's Chemical Corridor*

Jason Corburn, *Street Science: Community Knowledge and Environmental Health Justice*

Peggy F. Barlett, ed., *Urban Place: Reconnecting with the Natural World*

David Naguib Pellow and Robert J. Brulle, eds., *Power, Justice, and the Environment: A Critical Appraisal of the Environmental Justice Movement*

Eran Ben-Joseph, *The Code of the City: Standards and the Hidden Language of Place Making.*

Nancy J. Myers and Carolyn Raffensperger, eds., *Precautionary Tools for Reshaping Environmental Policy*

Kelly Sims Gallagher, *China Shifts Gears: Automakers, Oil, Pollution, and Development*

Kerry H. Whiteside, *Precautionary Politics: Principle and Practice in Confronting Environmental Risk*

Ronald Sandler and Phaedra C. Pezzullo, eds., *Environmental Justice and Environmentalism: The Social Justice Challenge to the Environmental Movement*

Julie Sze, *Noxious New York: The Racial Politics of Urban Health and Environmental Justice*

Robert D. Bullard, ed., *Growing Smarter: Achieving Livable Communities, Environmental Justice, and Regional Equity*

Ann Rappaport and Sarah Hammond Creighton, *Degrees That Matter: Climate Change and the University*

Michael Egan, *Barry Commoner and the Science of Survival: The Remaking of American Environmentalism*

David J. Hess, *Alternative Pathways in Science and Industry: Activism, Innovation, and the Environment in an Era of Globalization*

Peter F. Cannavò, *The Working Landscape: Founding, Preservation, and the Politics of Place*

Paul Stanton Kibel, ed., *Rivertown: Rethinking Urban Rivers*

Kevin P. Gallagher and Lyuba Zarsky, *The Enclave Economy: Foreign Investment and Sustainable Development in Mexico's Silicon Valley*

David N. Pellow, *Resisting Global Toxics: Transnational Movements for Environmental Justice*

Robert Gottlieb, *Reinventing Los Angeles: Nature and Community in the Global City*